Lecture Notes in Electrical Engineering

Volume 532

** **Indexing: The books of this series are submitted to ISI Proceedings, EI-Compendex, SCOPUS, MetaPress, Springerlink** **

Lecture Notes in Electrical Engineering (LNEE) is a book series which reports the latest research and developments in Electrical Engineering, namely:

- Communication, Networks, and Information Theory
- Computer Engineering
- Signal, Image, Speech and Information Processing
- Circuits and Systems
- Bioengineering
- Engineering

The audience for the books in LNEE consists of advanced level students, researchers, and industry professionals working at the forefront of their fields. Much like Springer's other Lecture Notes series, LNEE will be distributed through Springer's print and electronic publishing channels.

For general information about this series, comments or suggestions, please use the contact address under "service for this series".

To submit a proposal or request further information, please contact the appropriate Springer Publishing Editors:

Asia:

China, *Jessie Guo, Assistant Editor* (jessie.guo@springer.com) (Engineering)

India, *Swati Meherishi, Senior Editor* (swati.meherishi@springer.com) (Engineering)

Japan, *Takeyuki Yonezawa, Editorial Director* (takeyuki.yonezawa@springer.com) (Physical Sciences & Engineering)

South Korea, *Smith (Ahram) Chae, Associate Editor* (smith.chae@springer.com) (Physical Sciences & Engineering)

Southeast Asia, *Ramesh Premnath, Editor* (ramesh.premnath@springer.com) (Electrical Engineering)

South Asia, *Aninda Bose, Editor* (aninda.bose@springer.com) (Electrical Engineering)

Europe:

Leontina Di Cecco, Editor (Leontina.dicecco@springer.com)
(Applied Sciences and Engineering; Bio-Inspired Robotics, Medical Robotics, Bioengineering; Computational Methods & Models in Science, Medicine and Technology; Soft Computing; Philosophy of Modern Science and Technologies; Mechanical Engineering; Ocean and Naval Engineering; Water Management & Technology)
(christoph.baumann@springer.com)
(Heat and Mass Transfer, Signal Processing and Telecommunications, and Solid and Fluid Mechanics, and Engineering Materials)

North America:

Michael Luby, Editor (michael.luby@springer.com) (Mechanics; Materials)

More information about this series at http://www.springer.com/series/7818

Angelica Reyes-Munoz • Ping Zheng
David Crawford • Victor Callaghan
Editors

EAI International Conference on Technology, Innovation, Entrepreneurship and Education

TIE'2017

Springer

Editors
Angelica Reyes-Munoz
Universitat Politècnica de Catalunya
Castelldefels, Barcelona, Spain

David Crawford
Ravensbourne University
London, UK

Ping Zheng
Business School
Canterbury Christ Church University
Canterbury, UK

Victor Callaghan
University of Essex
Colchester, UK

ISSN 1876-1100 ISSN 1876-1119 (electronic)
Lecture Notes in Electrical Engineering
ISBN 978-3-030-02241-9 ISBN 978-3-030-02242-6 (eBook)
https://doi.org/10.1007/978-3-030-02242-6

Library of Congress Control Number: 2018966799

This Springer imprint is published by the registered company Springer Nature Switzerland AG.
The registered company address is: Gewerbestrasse 11, 6330 Cham, Switzerland

Preface

We are very pleased to introduce the proceedings of the first International Conference on Technology, Innovation, Entrepreneurship and Education (TIE) that was held from the 11th through to the 12th of September 2017 at Canterbury Christchurch Business School, Great Britain.

One of the unique and valuable dimensions of this conference is that it addresses innovation and creativity from a multidisciplinary perspective. Innovation is at the heart of new technologies, successful businesses, better societies and artistic accomplishments. However, while innovation is important to society, and spans many disciplines, it remains somewhat elusive. Thus, the goal of this conference was to bring together various disciplines so that different approaches to innovation could be shared and compared with a view to deducing how creativity and innovation might be better applied and taught to the benefit of all.

In this conference, we discuss innovation from many perspectives ranging from technology through business (entrepreneurship) to education. Most obviously, the effects of innovation can be seen in the world of technology, so that is one of our main focuses. The conference also considers innovation in education, business and society at large (including sociopolitical systems). Of course, technology remains very prominent in any discussion of innovation which is reflected in the title of the workshops. For example, 'Holodeck' examines reality, and related product innovations, from multiple edges: virtual reality, augmented, mixed, diminished and immersive, reaching out to a more distant posthumanism future where wilder product innovation ideas abound, such as storing thoughts and emotions in a cloud. Another workshop, 'Interaction' explores real applications for innovative technologies probing how research and analysis in areas like 'Big Data', the 'Internet of Things' and 'Smart Cities' are linking research projects to economic and technical applications that are already affecting our lifestyles. It also examines virtual reality providing an opportunity for delegates to experience the transportation of their own digital persona into VR. Of course, wild imagination must live within ethical boundaries; hence, innovation also concerns itself with the responsibility that researchers have for the world around them, with some workshop papers addressing this issue.

Apart from the content of papers, we are happy to see TIE's cosmopolitan character, attracting representatives of innovation from many parts of the world: India, Germany, Morocco, Mexico, Portugal, Brazil, Spain, the UK, China, the Netherlands, Finland, Ireland and the USA, among others.

Finally, we wish to acknowledge the numerous people who have made it possible for us to offer this event. Most notably, our many thanks go to the European Alliance for Innovation for their sponsorship. We are also thankful to our conference organisational committee for their many hours of unpaid work. Of course, most of all, we are thankful to you, our authors and readers, as without you, this endeavour would be worthless. We hope you both enjoy and find your engagement with these proceedings valuable.

Barcelona, Spain — Angelica Reyes-Munoz
Canterbury, UK — Ping Zheng
London, UK — David Crawford
Colchester, UK — Victor Callaghan
TIE'17

Contents

Credits

Conference Organization	
General Chairs	
Dr. Ping Zheng	Business and Management, Canterbury Christ Church University, UK
Nicky Leatherbarrow	Canterbury Christchurch University, UK
Local Chair	
Sarah Taylor	Canterbury Christchurch University, UK
Technical Program Committee Chairs	
Prof. David Crawford	Ravensbourne University, UK
Prof. Victor Callaghan	University of Essex, UK
Workshops Chairs	
Dr. Tiina Kymäläinen	VTT Technical Research Centre of Finland
Jim Hensman	University of Coventry, UK
Dr. Nicholas Lambert	Ravensbourne University
Publicity and Social Media Chairs	
Dr. Anasol Pena-Rios	British Telecom Plc
Jen O'Connor	NUI Galway, Ireland
Publications Chair	
Dr. Angelica Reyes-Munoz	Polytechnic University of Catalonia, Spain
Regional Chairs	
Dr. Simon Egerton	La Trobe University, Australia
Prof Minjuan Wang	San-Diego State University and Shanghai International Studies University
Hsuan-Yi Wu	Institute of Business Administration, National Taiwan University, Taiwan
Dr. Shumei Zhang	Shijiazhuang University, China
Dr. Victor Zamudio	Instituto Tecnológico de León, México
Brian David Johnson	School for the Future of Innovation in Society, Arizona State, University
Conference Manager	
Dominika Belisová	European Alliance for Innovation

Part I
Education: Effective Approaches in Learning and Teaching Creativity and Innovation

CSF Dream Academy: Using Fiction, Cardboard and Simple Electrical Circuits as Educational Tools to Lift Disadvantaged Children Out of the Poverty Trap

Víctor Zamudio and Victor Callaghan

1 Introduction

Innovation, creativity and entrepreneurship have become extremely important topics in recent years, driving the success of companies and individuals alike. Thus, it is not surprising to find there is increasing demand for learning such skills. Various tools exist in the market focused on helping children develop their creative skills. However, for children in poverty, the cost of these tools is often beyond their reach, causing them to fall behind their richer contemporaries, thereby making it difficult for them to escape the poverty trap. Thus, supporting the educational needs of such children is a critical element of improving their live prospects

Universities already play an important role in addressing inclusion issues relating to disadvantaged communities, such as providing access to digital technology and networks. Moreover, major efforts have been made where a University adopts a community, offering various supports, such as psycho-pedagogical support for children, delivery of education, and promotion of family and social values [1].

Despite all these efforts, the results are still not acceptable. In the case of Mexico, 55% of 15-year-old Mexican students fail to reach basic proficiency levels in math; 41% of 15-year-old Mexican students fail to reach basic proficiency levels in reading. Only 1% of 15-year-old Mexican students achieve the highest levels of proficiency in mathematics; only 0.5% of 15-year-old Mexican students achieve the highest proficiency levels in reading [2–4].

V. Zamudio (✉)
Instituto Tecnológico de León, Guanajuato, México
e-mail: vic.zamudio@ieee.org

V. Callaghan
University of Essex, Colchester, UK
e-mail: vic@essex.ac.uk

A. Reyes-Munoz et al. (eds.), *EAI International Conference on Technology, Innovation, Entrepreneurship and Education*, Lecture Notes in Electrical Engineering 532, https://doi.org/10.1007/978-3-030-02242-6_1

Science, Technology, Engineering and Mathematics (STEM) are a set of disciplines that are especially important to national economies and a child's future job prospects [5, 6]. Innovation and creative thinking are key factors in enabling individuals and organisation to positively differentiate themselves from their competitors leading to a desire to include Arts (a discipline frequently associated with creativity) within this educational package, transforming STEM into STEAM (Science, Technology, Engineering, Arts and Mathematics [7].

There have been several initiatives that seek to supplement the efforts of educators and government in countering the drift away from sciences which has been observed in upper secondary education [8]. These initiatives have sought to encourage the use of technology within classrooms as well as introducing computational theory through enabling the use of simplified programming languages such as Scratch, Alice and Greenfoot (based on Java). These initiatives have achieved good results, opening up cutting edge technologies, such as robotics, to a young audience [9].

Although some packages such as Lego Mindstorms (a robotics-themed kit) may be relatively expensive, they have been used to good effect with elementary school children using simple languages (e.g. Scratch) developed especially for children and people unfamiliar with technology [10]. These approaches have even been successfully deployed with children with dyslexia [11] or with autism [12].

Beyond the underlying science, certain applications, such as robotics, have played an important role in motivating and attracting children and young people to science and technology. However, in many cases such applications require a strong investment or sponsorship from those with a philanthropic mindset [13]. Academies and organizations that aim to support young people and children in disadvantaged conditions play a fundamental role in helping create a fairer educational landscape based on equality of opportunity plus contributing to the supply of the next generation of scientists in the world [14]. Technology facilitates the transfer of knowledge, even independently of the tutor; once the children's interest is captured by harnessing their creativity and imagination, children can achieve meaningful learning [15].

Innovation and creativity has had a significant impact on both academic and business success. The latter is of great importance since it can help persuade businesses that it is in their interest to support these initiatives. In Mexico, this has had an impact on the Entrepreneurial culture, as well as the creation and certification of many educational entrepreneurship models. In addition, several local governments, private or federal initiatives have been born and have been reinforced during the last years, including *Punto México Conectado* [16], networks of *Technology Parks* [17], and business incubators [16, 18–20]. In Mexico, a number of programmes have been developed to attract young scientific talent at undergraduate and postgraduate level [21–26], in addition to some focused on business [27]. However, few initiatives have focused on creating an aspirational model for children, and even less, those in conditions of marginalization. This article makes a proposal based on the Science

Fiction Prototyping innovation model, originally developed by INTEL, that allows children from marginalized or disadvantaged communities to aspire to develop innovative technology and processes that improve their social mobility.

The federal government of Mexico has also been pushing initiatives aimed at the democratization of science technology and innovation. For example, the Punto México Conectado programme [16] is a network of 32 digital education and training centres, which facilitate training courses, creativity development and support to undertake innovation projects and generation of new companies. Some regions have launched initiatives aimed at promoting scientific and technological vocations, where children and young participants are selected to participate in visits to universities and research centres, development of collaborative projects under the guidance of a research professor.

To create a fairer system that works for the benefit of the society as a whole there needs to be more attention paid to marginalized communities, but, as it stands, there are no models for resource-poor children focused on provision of innovation skills or affordable low-cost prototyping strategies.

2 Science Fiction Prototyping

Science Fiction Prototyping (SFP) was developed by Brian David Johnson, who at that time was a futurist at INTEL, working on the challenges the company faced in having to anticipate the needs of integrated circuits for a market that was some 7–10 years into the future (the typical design and production period for a modern computer processor integrated circuit) [28, 29]. SFP's core methodology is the use of creative arts (creative writing, drawings, film-making, etc.) as a means to introduce innovations into science, engineering, business and sociopolitical systems [30]. It differs to foresight activities as it does not aim to be predictive (forecasting the future), rather it focuses on innovation (creating new concepts, schemes, services and products). The main (but not exclusive) methodology is the use of science-fiction/fantasy stories, grounded in existing practice which are written for the explicit purpose of acting as prototypes for people to explore a wide variety of futures. These "prototypes" (commonly called *Science Fiction Prototypes*[1]), can be created by students, scientists, engineers, business or sociopolitical professionals to stretch their work or, for example, by school children and members of the public to influence the work of professionals. In this way these stories act as a way of involving the widest section of the population to help set the research agenda and thereby empowering everyone to have a hand in shaping the future. Related methodologies include *Design Fiction* [31] (the use of narrative scenarios to explore design related issues) and *Diegetic Innovation Templating* (getting inspiration for innovations from pre-existing fiction written for the purpose of entertainment) [32].

[1] https://en.wikipedia.org/wiki/Science_fiction_prototyping.

These methodologies have been applied successfully not only in engineering but also in English language teaching [33] and the development of new products in the industry [34].

In *Dream Mexico,* we use a variant of Science Fiction prototyping (SFP) called *μSFP (Micro-SFP)* [35, 36]. This uses a genre of writing called Micro-Fiction that aims to write ultra-short stories as small as just six words but more commonly 140 and 160 characters (30 words) which fit into a Twitter and SMS phone message. The general structure of these stories is as follows:

[Person] in [Situation] uses [Innovation] to do [Action] resulting in [Benefit]

An example of a *μSFP* is: *Jack fall asleep in the sun. His* ***smart sun protection sensor*** *woke him up with an alarm & soft vibration. He avoids sun strokes!* [37]

The writing procedure is to (1) start by identifying an innovation (technology, service, etc.), (2) Identify a *user* (use a very short name; e.g. Joe) and finally (3) create an *event* that illustrates the use and *benefit* of the technology, process or service (should include an inflection point). The general rule is to start big, then reduce it to the required size. It is also possible to use drawings with, or in place of, narrative. The role of a *μSFP* is to capture the idea (as a type of innovation shorthand) which can then be expanded into a larger narrative (e.g. a mini-fiction) or a more elaborate drawing, film or paper/cardboard prototype.

3 Dream México

As we have discussed above, it is necessary to incorporate additional strategies that contribute to a student's social development, particularly to children of limited resources. The main objective of the *Dream Mexico* initiative is to provide an option for children and young people in socially disadvantaged environments (economic, cultural, etc.), to empower them with STEAM skill and to capture their interest and, through use of Science Fiction Prototyping, to become motivated to engage with education (especially STEAM) and thereby improve prospects for their life.

Thus, the Dream Mexico initiative focuses on children from poor suburban or rural communities, who generally do not have access to complementary activities (art, culture, innovation, science workshops, etc.). Thus the design of the activities is predicated on the availability of ultra-low budgets, reusing materials, and utilising digital resources in an open way, wherever possible.

3.1 Methodology

The methodology is based on the creation of both narrative and physical prototypes inspired from near-term science fiction stories that capture and spoof product ideas they have for their lives, in a way that enables them to develop various skills such as innovation, teamwork, prototyping, tolerance and respect.

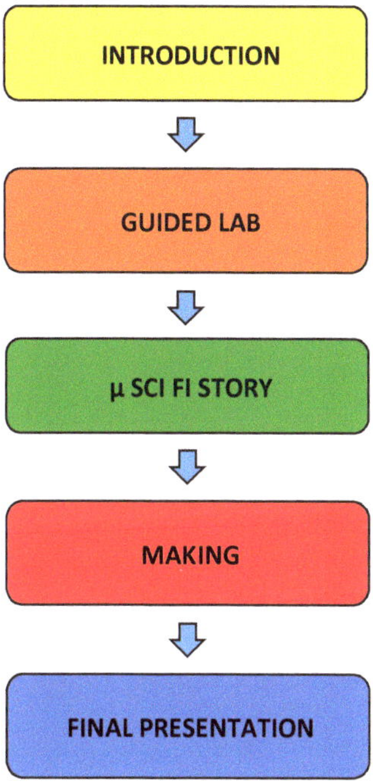

Fig. 1 The methodology used during the workshops consists of five steps. Each step requires the active participation of the students

Each session of the workshops has the following phases (see Fig. 1):

1. *Introduction*: basic elements of science and/or technology are explained briefly and simply.
2. *Guided practice*: Participants are guided by a facilitator to apply what was introduced in phase 1.
3. *µSFP histories*: children imagine products and processes they would like to be available in the future for their communities, describing it by creating a short story and/or drawing (*µSFP* is introduced as an implicit aspect of story-writing).
4. *Making*: Children build a prototype using cheap reusable material (e.g. cardboard), which can be customized (colours, functionality, innovation, etc.) according to their *µSFP*.
5. *Presentation*: The children present their prototype and explain its main characteristics and benefits.

4 Preliminary Results

The Dream Mexico Academy has been operating successfully in León Guanajuato, since 2016. It has attracted children and young people between 8 and 15 years of both sexes, who have participated actively in the workshops. The feedback received

Fig. 2 The participants build their portfolio with a first prototype, in this case a drawing of themselves interacting with their robots

has been extremely satisfactory, and the topic and methodology has proven able to motivate them to develop low-cost prototypes with basic technology (cardboard chassis and simple electrical wiring). So far different themes (called detonators in SFP jargon) have been used for the workshops, including "the house of the future", "robots and my family", "special orchestra", "space travel" and "robotic hands". Figure 2 presents a collage of the student drawings (as part of a preliminary prototype) showing the children's interaction with their robot.

From observations of the children it is clear that, being able to get involved in a workshop that creates a product from their own hands, students have a sense of empowerment. The physical prototypes are very simple, being just paper/cardboard bodies augmented with simple electronics involving elementary skills such as

Fig. 3 Participants of Dream Mexico Academy with their cardboard prototypes and basic technology

Fig. 4 Participants are invited to detail as much as possible their drawings, in this case with the theme The House of the Future

trimming, drawing, colouring, and screwing. The children are generally very proud of what they have achieved and the low costs involved, allows them keep it their home to show their family and friends (see Fig. 3).

Figure 4 is a drawing where participants have shown how they visualize their home in the future (when they become adults). In this case, the use of alternative energies was emphasized, encouraging savings and reducing pollution.

Figures 5 and 6 show participants working on the inclusion of LEDs in their cardboard robots. While the science is relatively simple it generates a lot of enthusiasm in the participants, as well as introducing them to the fundamentals of electrical science (circuit theory).

Figure 7 shows the children constructing a robotic hand from low cost materials

One of the characteristics of this methodology is that it fosters collaborative work among the participants, favouring the development of tolerance, respect, listening skills, etc. In Fig. 8 the participants develop a science fiction story with the theme "The House of the Future":

Fig. 5 Participating children and youngsters enriching their cardboard prototypes with low cost basic technology, in this case a simple circuit with an LED and AA batteries

Fig. 6 The inclusion of LEDs in their robots generated a lot of enthusiasm in the participants

It will be a 3 story orange home, with four bedrooms; clapping 2 times the music and lights are turned on; the chimney is turned on; the house has six toilette, and when you go to the toilette it flushes automatically; the windows will be connected in such a way they can show the images you want; with a swimming pool and a Jacuzzi; everything is bullet-proof; the door bell recognizes your footprint and let you in; it has an automatic crystal lift.

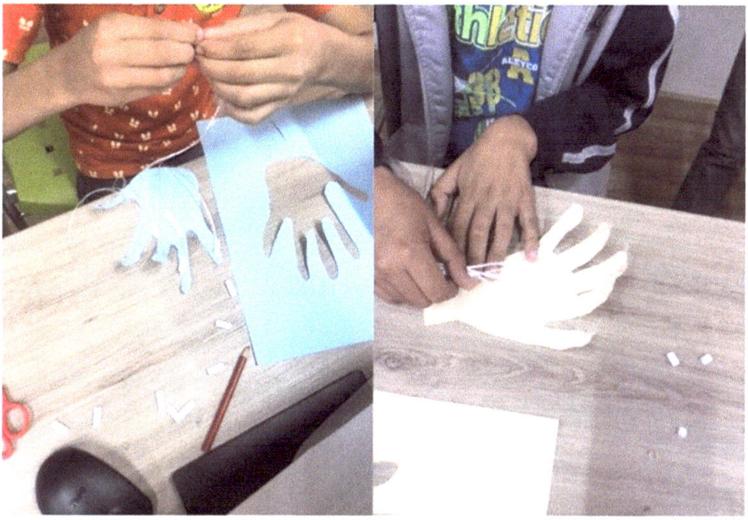

Fig. 7 One of the prototypes developed was a robotic hand, using paper, straws, tape and rope

4.1 Feedback and Perception of the Participants

During the term January–May 2017, 13 sessions were scheduled. Six kids were enrolled during this period (although some of them enrolled late). Two students participated 100% of the sessions. The details are shown on Table 1.

During the first four sessions only two students attended the workshop; however, from session 5 onwards the group stabilized around five participants. The details are shown on Fig. 9

In order to obtain information and feedback from the participants, short interviews were performed focusing on four axes: (1) perception of the work (before and after the course), (2) learning, (3) enjoyment and (4) motivation and engagement.

Perception of the Workshop

Question: Before the beginning of the workshops, what did you think about the advertised activities?

- "I thought they were going to be fun"
- "I was really scared, and I didn't want to come. I thought I was going to be told off by the teacher; now I am very happy as I am learning many things about the space, robots, but sometimes building things is not easy for me."
- "I was sure the sessions were going to be fun; I love robots and I like the workshops."
- "I was very excited, as I was sure the workshops were going to be Ok"

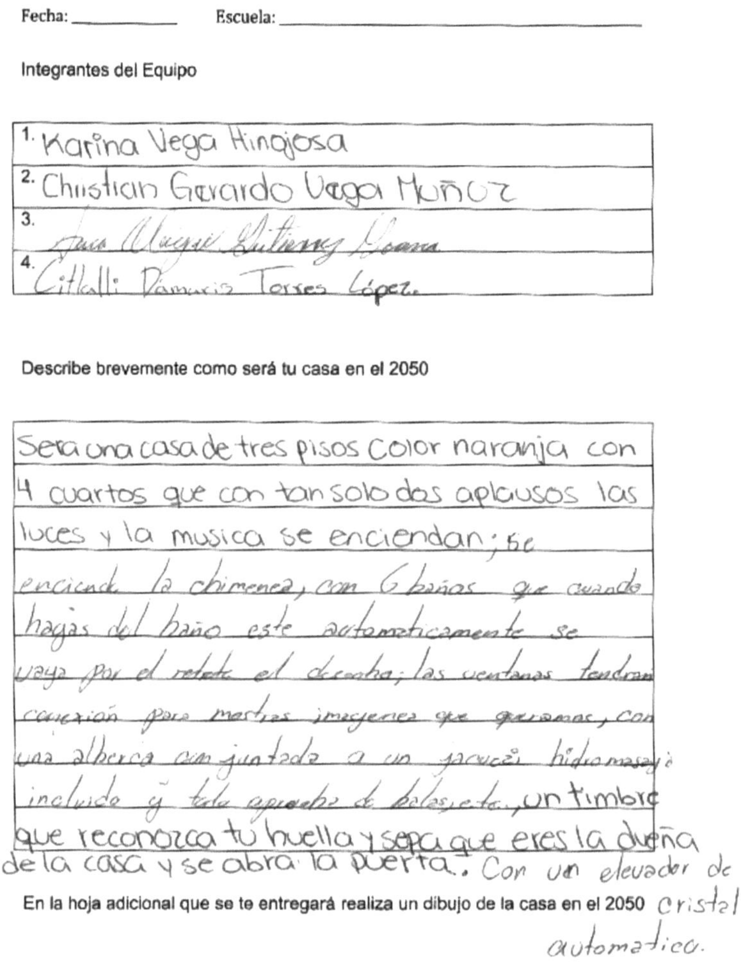
Fecha: __________ Escuela: __________

Integrantes del Equipo

1.	Karina Vega Hinojosa
2.	Christian Gerardo Vega Muñoz
3.	[illegible]
4.	Citlalli Damaris Torres López

Describe brevemente como será tu casa en el 2050

Sera una casa de tres pisos color naranja con 4 cuartos que con tan solo dos aplausos las luces y la musica se enciendan; se encienda la chimenea, con 6 baños que cuando hagas del baño este automaticamente se vaya por el retrete el desecho; las ventanas tendran conexión para mostrar imagenes que queramos, con una alberca conjuntada a un jacuzzi hidromasaje incluido y todo aparato de bicicleta, un timbre que reconozca tu huella y sepa que eres la dueña de la casa y se abra la puerta. Con un elevador de cristal automatico.

En la hoja adicional que se te entregará realiza un dibujo de la casa en el 2050

Fig. 8 Example of a μSFP where the participants visualize the house where they will live in the future

Table 1 Assistance Report during the term January–May 2017

Student no.	Percentage of assistance
x_1	100
x_2	100
x_3	30.77
x_4	69.23
x_5	46.15
x_6	53.84

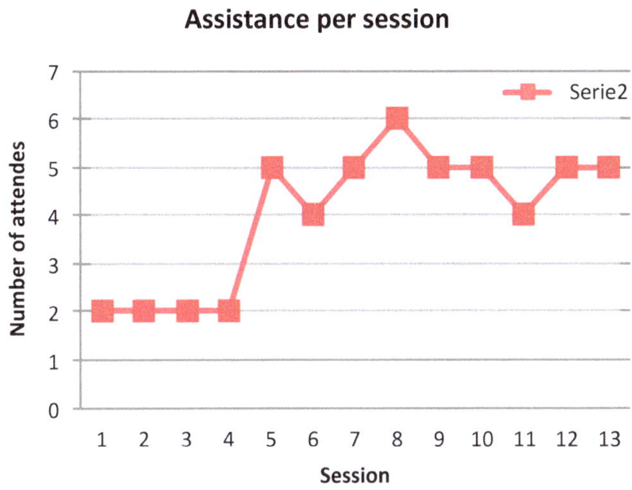

Fig. 9 From January to May 2017, thirteen workshop sessions (1/week) were scheduled. During the first session two kids attended the sessions; at the beginning of the second month three more students joined the workshops, reaching a maximum of six students on the 8th session

Teacher observation: During the first session the participants were very quiet, and a little bit nervous. Some of the kids were escorted to the classroom by their mothers.

Question: Having attended the workshops *What do you think about them?*

- "I really enjoy the workshops, especially because I like to build prototypes."
- "I like the challenges (building prototypes)."
- "I like to build things using recycled materials."

Teacher observation: Children taking part of the workshops enjoy active participation, specially building things, using different materials and sharing tools (scissors, rules, etc.). Long lectures can be boring for them (which is very understandable).

Learning

Question: What have you learned from the workshops?

- "I have learned about robots and basic electronics."
- "I have learned about robots, LEDs, which are small bulbs. "
- "I have learned about the space, robots, electronic circuits (LEDs and batteries)"

Teacher observation: Children taking part of the workshops learned different topics including space, electric energy, robots and circuits. They show more interest when the topic is linked to a task or problem they could solve using the knowledge acquired.

Enjoyment

Question: Have you enjoyed the workshops?

- "Yes, for many reasons. I like building things."
- "Yes, because I like robots."
- "Yes, because I like to build robots, and put them LEDs."
- "Yes, I like the classes very much."

Teacher observation: The participants enjoyed the workshops very much; they seem surprised when the sessions were finishing as they were in "flow state". Some of the kids that during the first sessions were shy or nervous gained confidence in a couple of weeks.

Motivation and Engagement

Question: Would you like to continue learning things about science?

- "Yes, because I would like to become a football player and doctor."
- "Yes, because I want to be a mechanical engineer and build and design cars."
- "Yes, as I want to become a builder as my father, and build and design a house for my family."

Teacher observation: The participants are willing to continue with this workshop (they even want to have sessions during bank holidays). Kids have different building skills; however, they ask for help and support when they struggle with specific tasks. They visualize themselves in the future using science and technology as part of their activities, and supporting their family using the skills gained.

4.2 Discussion of Results

With only six students we have a very small sample. Statistics would not be meaningful with such a small group so, instead, we chose to give you the opinions of the students in their own words. From these words it can be seen that the students and the teachers focus is slightly different. The children are focused on, and excited by the activities (e.g. building robots), whereas the teachers have a much broader view, taking into account the acquisition of social skills and increasing the students aspiration for science. Clearly, this work is only in its early stages and more data will be required to come to more meaningful conclusions but we hope that this initial trial, while only having six students and one teacher will, nevertheless, act as encouragement sustaining and growing this project so that more evidence can be acquired regarding its performance and the methodology itself may be refined and improved. Perhaps readers of this paper might be motivated to deploy these ideas in other regions of the world, enriching our data while helping the numerous educationally disadvantage children of this world, and thereby helping their families and countries; if you are reading this paper and would like to deploy our methods, please contact the corresponding author, Victor Zamudio.

5 Conclusions and Future Work

The paper describes a low-cost methodology aimed at connecting with children's imagination in order to motivate their interest in science, technology and innovation. It has a special focus on educationally marginalized children and youth, emphasising the importance of low-cost material and basic technology. This project and methodology been developed in partnership with the Creative Science Foundation as an application of its Science Fiction Prototyping tool.

Using this approach Dream México Academy has offered a series of workshops for socially disadvantage children in León Guanajuato, Mexico, since 2016. In its first trial deployment, it has attracted six children between 8 and 15 years of both sexes, who have participated enthusiastically in the workshops. The data collected, while limited, is encouraging since it shows that the children have not only built and reflected on science and technology, but enjoyed the process, fostering an atmosphere of camaraderie and a safe place where to making mistakes is part of learning. We are inviting kids from a primary school in the area to increase the number of participants in the workshops. The student positive feedback seems to support our belief that SFP and low-cost cardboard prototyping has the ability to motivate children's interests in STEM topics but, clearly, this is only a small preliminary trial and so our next step is to expand this so as to acquire more convincing data and further hone our methods. To this ends, and as part of our plans for the future, during the summer of 2017, it is planned to use this methodology in an itinerant way. Two small communities in the northwest of Mexico have been chosen: *Teacapán* in the State of Sinaloa and *Nombre de Dios* in the State of Durango. Teacapán[2] is a small fishing village in the municipality of Escuinapa, with around 4000 inhabitants, with a significant amount of indigenous population living in the town. Nombre de Dios[3] is in the state of Durango, and has a population of a little more than 5000 inhabitants. We intend to apply our methodology in both locations, taking advantage of the fact that the children are on vacation during the months of July and August. One of the challenges is the lack of infrastructure (as we do not count with community centres); however, we will intent to contact local churches in order to run the workshops in their premises. The objective will be to continue validating the methodology proposed previously in marginalized kids and youngsters in a less controlled environment compared to what has been currently implemented.

[2]https://goo.gl/maps/F3qidbHDd2t.

[3]https://goo.gl/maps/An51V3mtZPs.

We hope in the future to be able to have access to funds and to make this project sustainable, and to allow us to share the material used with the academic community (teachers, institutions, students doing social service, community centres, etc.) who might be interested in using this methodology for the benefit of children and young people.

This initiative is at an early stage, and we look forward to continuing to report our progress in future publications.

Acknowledgements The authors would like to thank the IMAGINA Foundation for providing their facilities to carry out these workshops, as well as the Creative Science Foundation for the support received.

References

1. C.E. Canovas-Marmo, C. Guerrero, V. Zamudio, in *UNIVA: A University Open to the Community, Ambient Intelligence and Smart Environments*, ed. by J.C. Augusto, T. Zhang. Workshop Proceedings of the 10th International Conference on Intelligent Environments, vol. 18 (IOS Press, 2014), pp. 120–126
2. Teaching and Learning International Study (TALIS) (OECD) (2013)
3. Programme for International Student Assessment (PISA) (OECD) (2012)
4. Comisión Económica para America Latina (CEPAL) (2011)
5. What Is STEM Education? http://www.livescience.com/43296-what-is-stem-education.html
6. C. McLaughlin, STEM: It's elementary, too! Int. Technol. Edu. Assoc. **14**(1), 2–18 (2009)
7. The STEM to STEAM initiative, http://stemtosteam.org
8. P. Zokowski, K. Geramita, J. Ashdown, B. Brooks, A. Thompkins, in *Connecting Kids to STEM Through Entrepreneurship and Innovation*. 2016 IEEE Integrated STEM Education Conference (ISEC) (Princeton, NJ, 2016), pp. 71–74. https://doi.org/10.1109/ISECon.2016.7457557
9. A. Rao, in *The Application of LeJOS, Lego Mindstorms Robotics, in an LMS Environment to Teach Children Java Programming and Technology at an Early Age*. Integrated STEM Education Conference (ISEC) 2015 (IEEE, Princeton, NJ, 2015), pp. 121–122. https://doi.org/10.1109/ISECon.2015.7119902
10. D. Barrios-Aranibar, et al., in *RoboEduc: A Software for Teaching Robotics to Technological Excluded Children Using LEGO Prototypes*. 2006 IEEE 3rd Latin American Robotics Symposium (Santiago, 2006), pp. 193–199. https://doi.org/10.1109/LARS.2006.334332
11. W.L. Martens, in *Designing the Spatial Math Tutor: A Non-Symbolic, On-Line Environment for Teaching Mathematical Skills to Dyslexic Children*. Proceedings Second International Conference on Cognitive Technology Humanizing the Information Age (Aizu-Wakamatsu City, 1997), pp. 262–270. https://doi.org/10.1109/CT.1997.617706
12. S.R. Dehkordi, R.M. Rias, in *Using Mobile Game Application to Teach Children With Autism Spectrum Disorder (ASD) Multiple Cues Responding: A Pilot Study, User Science and Engineering (i-USEr)*. 2014 3rd International Conference on Shah Alam, 2014, pp. 216–220
13. Official webpage of the program FIRST (For Inspiration and Recognition of Science and Technology), http://www.firstinspires.org
14. Bringing STEM Education to Underserved Communities, http://www.usnews.com/news/stem-solutions/articles/2014/05/29/bringing-stem-education-to-underserved-communities
15. Building a School in the Cloud, http://www.ted.com/talks/sugata_mitra_build_aschool_in_the_cloud
16. Punto México Conectado, http://www.pmc.gob.mx
17. Ecosistema de Emprendimiento e Innovación de Guanajuato, http://novaera.com.mx

18. Instituto Nacional del Emprendedor, https://www.inadem.gob.mx
19. Tec de Monterrey, https://tec.mx/index.php/es/diferencia-tec/emprendimiento
20. Start Up México, http://www.startupmexico.com
21. Club de Ciencia, http://www.concyteg.gob.mx/clubciencia
22. Programa Interinstitucional para el Fortalecimiento de la Investigación y el Posgrado del Pacífico, http://www.programadelfin.com.mx
23. scademia Mexicana de Ciencias, Verano de la Investigación Científica, http://amc.edu.mx/amc/index.php?option=com_content&view=article&id=139&Itemid=80
24. Verano de la Ciencia Región Centro, http://www.veranoregional.mx/
25. Verano de la Investigación Científica de la Península de Yucatán Jaguar, http://www.verano-jaguar.dgda.uady.mx/
26. Veranos Universidad de Guanajuato, http://www.veranos.ugto.mx/
27. Veranos por la Innovación en la Empresa, http://veranos.fese.org.mx/
28. B.D. Johnson, in *Science Fiction Prototyping: Designing the Future With Science Fiction* (Morgan and Claypool, San Rafael, CA, 2011), ISBN 978-1608456550
29. https://www.computer.org/portal/web/computingnow/science-fiction-prototyping. Retrieved June 27, 2017
30. V. Callaghan, Creative Science: Injecting innovation into the IT industry. ITNOW **57**(2), 52–55
31. Design Fiction, https://en.wikipedia.org/wiki/Design_fiction
32. P. Zheng, V. Callaghan, in *Diegetic Innovation Templating*. CS'16, London, 12–13 September 2016
33. S. Zhang, V. Callaghan, H. Wang, B. Hedong, in *Learning Computer English Using a Creative Science Approach* (Tsinghua University Press, China, 2016), ISBN: 9787302422730
34. H.Y. Wu, V. Callaghan, in *From Imagination to Innovation: A Creative Development Process*. Presented at CS'16, London, 12–13 September 2016
35. V. Callaghan, in *Micro-Futures*. Creative-Science 2014 (Shanghai Jiaotong University, China 2014)
36. Micro-SFP, https://en.wikipedia.org/wiki/Micro-SFP
37. S. Zhang, V. Callaghan, H. Wang, in *Improving English as a Foreign Language Education in China With Creative-Science,* SOFIEE'16, London, 12–13 September 2016

Using a Creative Science Approach for Teaching English as a Foreign Language to Postgraduate Students

Víctor Zamudio, María del Pilar Pérez Mata, Victor Callaghan, Shumei Zhang, and Carlos Lino

1 Introduction

Currently innovation has become a "hot topic", not only in the technological world but also in education. Creativity is the engine of Innovation and a skill which makes our lives more pleasurable and companies more profitable. Thus, academia searching for effective ways of including innovation studies within most disciplines is hardly surprising.

The capacity to innovate and to create is inherent to every human being and has enabled humanity to make the various cutting-edge technological advances which have made possible the fairly sophisticated world we now inhabit.

Thus, to ensure a bright future, it is fundamental to nurture upcoming generations with the ability to think creatively, ensuring that the world can continue to overcome difficult challenges, develop new technologies, create prosperous businesses and fashion pleasurable living environments.

Of course, we all live in an increasingly global world, where communication plays a vital role in driving learning and business. In this world English has become the de facto international medium and, as a consequence, the acquisition of English as a second language has become a vital skill for graduates. In this world students

V. Zamudio (✉) · M. del Pilar Pérez Mata · C. Lino
Instituto Tecnológico de León, Guanajuato, México
e-mail: vic.zamudio@ieee.org; carloslino@itleon.edu.mx

V. Callaghan
University of Essex, Colchester, UK
e-mail: vic@essex.ac.uk

S. Zhang
Leeds Beckett University, Leeds, UK
e-mail: S.Zhang@leedsbeckett.ac.uk

A. Reyes-Munoz et al. (eds.), *EAI International Conference on Technology, Innovation, Entrepreneurship and Education*, Lecture Notes in Electrical Engineering 532, https://doi.org/10.1007/978-3-030-02242-6_2

will be the agents transforming companies, creating new technologies and doing business in a very complex and globalized world.

The study that we report here was based at The León Technological Institute which is part of the National Technological Institute of Mexico, a multi-campus system with 266 campuses around the country, and with more than half a million students, mainly engineering students [1] (over half of Mexico's engineering graduates). The Technological Institute of León offers the courses of Engineering in Computer Systems, Engineering in Information Systems, Industrial Engineering, Mechatronics Engineering, Electronic Engineering, Logistics Engineering, and Business Management Engineering, as well as postgraduate studies with an MSc in Computing and a doctorate in Computer Science. The master's degree in Computer Science at Instituto Tecnológico de León has two research areas: Intelligent Environments and Intelligent Systems. The programme is accredited by Conacyt as a programme of quality and excellence, so that admitted students receive a scholarship to pursue their postgraduate studies [2].

2 Literature Review

The methodology of creative science has been successfully tested in different scenarios, including business innovation, training, teaching and driving business start-ups [3]. It is rooted in the concept of creating high-fidelity stories that exercise the use of an imagined product in such a realistic way as to be able to test it; such a story becomes a kind of prototype. The methodology was proposed by Brian David Johnson who, at the time, was Intel's futurist. He called the methodology Science Fiction Prototyping (SFP), setting out the principles in a 2011 book [4]. Science Function Prototypes vary in size from Twitter-sized stories, Micro-SFP [5] to scientific paper-sized stories, Mini-SFP [4]. We will adopt both of these formats at different stages of learning.

Creative Science has been successfully applied in teaching the English language to computer students in China, substantially improving the traditional sentence-by-sentence translation strategy. In the Chinese work Zhang used Science Fiction Prototyping with classes of almost 200 students to develop there oral and written skills in "Computer English", together with improving their computer science knowledge and innovation skills [6–8].

Wu has developed a methodology based on Creative Science called Imagination Workshops, with the aim of teaching business innovation to entrepreneurs in the Asian region [9]. Our context in Leon is quite similar, since the subjects involved are professionals (many with industry experience, such as IT consultants, or software developers).

In project-based learning, students develop skills using a real problem (or one inspired by reality) as an anchor so that by using tools from different disciplines it is possible to find a solution. This approach has been successfully tested in different initiatives [10], where students from several disciplines have been able to apply their knowledge in the design of a car for the elderly. Such project-based learning presents the students with real-world problems and challenges, thereby enabling them to acquire new knowledge with real significance and deep understanding in a realistic context where, most of the times, they would need to build or create a device or prototype [11]. Being a science and engineering faculty most of our work would be termed as being part of STEM (Science, Technology, Engineering and Maths) but by adding in ARTs, in which creativity plays a more obvious role, it might better be described as STEAM (Science, Technology, Engineering, Art and Mathematics), a combination which has been shown to work well in the classroom (https://www.edutopia.org/blog/pbl-and-steam-natural-fit-andrew-miller).

Problem based learning is another generic skill that has been successfully applied at university level in various universities [12] (https://www.maastrichtuniversity.nl/education/why-um/problem-based-learning), developing multidisciplinary skills, focused on solving problems in contexts as close to reality as possible. For this reason, the participation of experts from different areas is desirable, including specialists from the industry. It has also been applied in young children, with excellent results. Naturally, creative thinking is also useful for solving problems in innovative and clever ways.

3 Methodology Proposed

The methodology planned for this initiative is founded in the principles of Creative Thinking, Problem-Solving and Creative Science.

Given how crammed modern curriculums are, the ideal methodology would combine in the teaching-learning creative thing along with English as a second language while advancing the area that the student is majoring in. There are a number of possible mechanisms we might incorporate which we now briefly review.

First, there is Communicative Language Teaching (CLT) which focuses on how language is used [13]. An approach in which functional language is relevant and the plentiful exposition to language in use and plenty of opportunities to use it are vitally important for a student´s development of knowledge and skill [14]. Activities in CLT typically involve students in real or realistic communication, where the successful achievement of the communicative task they are performing is at least as important as the accuracy of their language use [14].

Second, there is Task Based Learning (TBL) which is pertinent to the implementation of our new programme. TBL makes the performance of meaningful tasks central to the learning process and it holds the belief that if students are focused on the completion of tasks they are just likely to learn language as if they are focusing on language forms [15].

Consequently, as students deal with the completion of the task the use of certain grammar structures, vocabulary and proper pronunciation will become evident thus the teaching-learning process of it will emerge naturally, providing the students then with meaningful skills.

Although there has been some criticism about the use of tasks as the base of all pedagogical methodology in reason of the specific needs of students and the fact that other functions of the language would be left out [16], it is important to outline the profile of our students to fully understand how relevant the application of this methodology will be. The subjects of the present methodology are adult learner students of MSc in Computing Sciences in the postgraduate department. Their area of expertise is related with varied engineering bachelor's degrees and therefore their aptitudes, interests and learning styles are alike. Having in mind the previous considerations the implementation of the proposed methodology will make students competent in the area of interest they all have in common, Computing Science, helping them tackle their communicative needs as MSc students and later on as professionals of their areas. The fact that all students solve common tasks under the principles of Creative Science has the capability to provide them with the functional language they need along with creative skills and additional discipline knowledge.

Thus it is planned to pilot this method following the approaches mentioned before, considering the four skills in teaching English (reading, listening, writing and speaking) and the language systems: grammar and vocabulary (see Fig. 1).

Since the approaches of TBL and CLT are being followed, the methodology will contain a specific task or challenge to be accomplished, involving research, brainstorming, design and implementation. Some input of vocabulary will be given to the students when stating the problem/task to be solved. This is to be accomplished through the receptive skills listening and reading. Selected texts and/or audio files are to be presented to students with the purpose of informing about the task and clarifying terms related to it (Pre-task).

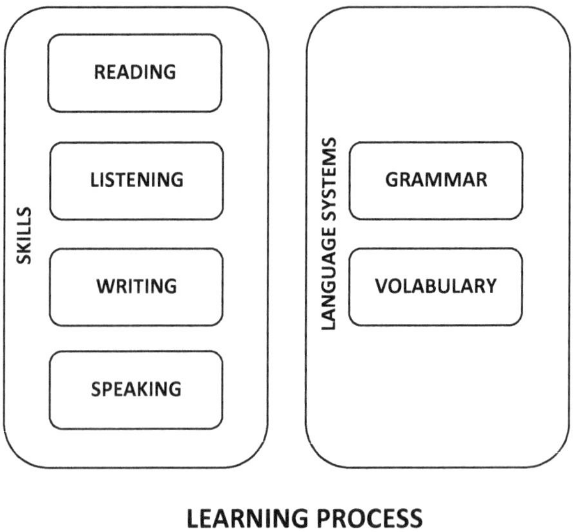

Fig. 1 The English learning process is based on four skills (reading, listening, writing and speaking) and language systems (grammar and vocabulary)

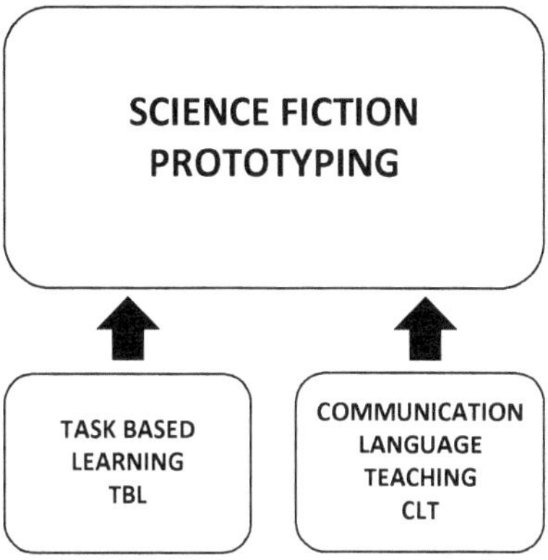

Fig. 2 The methodology proposed is based on the Science Fiction approach, together with Task Based Learning (TBL) and Communication Language Teaching (CLT)

In a second phase, students are to be focused in the solution of the problem/task, planning and reporting about their findings. In this phase it is useful to utilise Creative Science principles because the need to research, analyze and use the information in a creative way, will be fundamental to solve the issue. During the process of researching, analyzing and presenting results, students are required to use both receptive and productive skills. Teacher's function is to be monitoring the process (Task Cycle).

During the last stage, teacher will be required to help students analyze specific grammar structures, vocabulary, syntax elements, etc. which they encountered during the earlier problem-solving stage (language focus).

It should be noted that, since the Creative Science approach is fundamental to our methodology, the problems or tasks which are to be given to student have a direct relationship with a technological and scientific area, and the development of an innovative device/model (see Fig. 2). Because of this, we are considering using a team teaching approach which would include both an English Language and technology facilitator/expert.

Our intention is that this methodology should help students achieve international certification as competent English users, which is essential since reaching the B1 level is the requirement for students to graduate with an MSc in Computer Science.

Being engineering students (i.e. studying systems, computer, electronic, mechatronics or related), the students are required translate their Science Fiction Prototypes into hardware or software engineering prototypes using tools such as:

1. Apps development: Students in computer-related areas have the skills to program on Android or IOS; however, it could also be used tools for the accelerated development of prototypes, such as App Inventor [17], Proto IO [18].
2. 3D Printers: Additive manufacturing has revolutionized the development of agile projects using open source design tools such as Blender [19] or CURA [20].
3. Internet of Things: with the large number of embedded computational tools, it is possible to quickly create prototypes using devices such as Raspberry Pi [21], Arduinos [22] and other kits such as Little Bits [23].

4 Discussion

As was explained earlier in this paper, we are at the early stages of this work and the project is a work in progress. In this paper we report on our plans, pedagogical models and the background information. We plan to implement and deploy this methodology during the academic term August–December 2017 with the 1st year students of the MSc in Computer Science at Instituto Tecnológico de León. Clearly there will be a considerable amount of work involved but we are fortunate to be building upon earlier work in China undertaken by Dr Shumei Zhang learning from her experiences at Shijiazhuang University [7]. We plan to use the same evaluation methodology (including questionnaires) with our Spanish students as were used with the Chinese student, enabling us to make some direct cross-cultural comparisons which we hope will provide useful information for deploying the Creative Science methodology more widely across world languages all of which we plan to report in further papers.

5 Conclusions and Future Work

The teaching of English is a strategic subject for competitiveness, especially in the area of engineering. Graduate students require English language proficiency, which will allow them to improve their employability conditions when they obtain their MSc in Computer Science. Likewise creative thinking and innovation skills are important skills for graduates in order for the companies they work for to achieve the necessary levels of competiveness in modern global economies. In this paper we have argued for this by using a Creative Science methodology it is possible to combine learning innovation, technology and language within an efficient pedagogical framework that uses engaging strategies that are directly connected to the student specialism.

This methodology will be implemented during the next term (August–December 2017) with the 1st year students of the MSc in Computer Science. And we hope to report our results in future conferences.

References

1. Tecnológico Nacional de México, http://www.tecnm.mx/informacion/sistema-nacional-de-educacion-superior-tecnologica
2. Posgrados de Calidad Conacyt, http://www.conacyt.gob.mx
3. V. Callaghan, Creative science: Injecting innovation into the IT industry. ITNOW **57**(2), 52–55 (2015)
4. B.D. Johnson, *Science Fiction Prototyping: Designing the Future With Science Fiction* (Morgan and Claypool, New York, 2011), ISBN: 9781608456550

5. V. Callaghan, *Micro-Futures (Presentation), Creative-Science* (Shanghai Jiaotong University, China, 2014)
6. S. Zhang, V. Callaghan, H. Wang, *Improving English as a Foreign Language Education in China With Creative-Science,* SOFIEE'16, London, 12–13 September 2016
7. S. Zhang, V. Callaghan, *Using Science-Fiction Prototyping as a Means to Motivate Learning of STEM Topics and Foreign Languages.* IEEE International Conference on Intelligent Environments (IE, 2014), pp. 353–356
8. S. Zhang, V. Callaghan, H. Wang, B. Hedong, *Learning Computer English Using a Creative Science Approach* (Tsinghua University Press, China, 2016), ISBN: 9787302422730
9. *From Imagination to Innovation: A Creative Development Process.* Presented at CS'16, Hsuan-Yi Wu and Victor Callaghan, London, 12–13 September 2016
10. J.M. Lopez, A. Sánchez, V. Zamudio, V. Callaghan, in *AVANTI: An Intelligent Vehicle Project as a Catalyst for Education and Business Collaboration and Regional Economic Regeneration*, eds. by V. Callaghan, A. Kameas, S. Egerton, I. Satoh, M. Weber. The 6th International Conference on Intelligent Environments IE10, 19–21 July 2010 (Monash University (Sunway Campus), Kuala Lumpur, Malaysia). pp. 360–363, ISBN: 978-0-7695-4149-5. Print ISBN: 978-1-4244-7836-1
11. Project Based Learning, https://www.edutopia.org/project-based-learning
12. F. Kjersdam, S. Enemark, *The Aalborg Experiment, Project Innovation in University Education* (Faculty of Technology and Science, Aalborg University Press, Aalborg, 1994) ISBN 87-7307-480-2
13. Communicative Language Teaching, https://en.wikipedia.org/wiki/Communicative_ language_teaching
14. D. A. Wilkins, *Notional Syllabuses: A Taxonomy and Its Relevance to Foreign Language Curriculum Development* (Oxford University Press, Oxford 1977), ISBN-13: 978-0194370714, ISBN-10: 0194370712
15. J. Harmer, *The Practice of English Language Teaching*, 4th edn. (Pearson, London, 2007)
16. P. Seedhouse, Task-based interaction. ELT J. **53**(3), 149–156 (1999)
17. App Inventor, http://appinventor.mit.edu/explore/
18. Proto.IO, https://proto.io
19. Blender, https://www.blender.org
20. Cura, https://ultimaker.com/en/products/cura-software
21. Raspberry PI, https://www.raspberrypi.org
22. Arduino, https://www.arduino.cc
23. Little Bits, http://littlebits.cc

Creating Content for Educational Testing Using a Workflow That Supports Automatic Item Generation

Mark J. Gierl, Donna Matovinovic, and Hollis Lai

1 Introduction

The principles and practices that guide the design and development of educational tests are undergoing dramatic changes. One of the major catalysts for these changes stem from the application of technology to assessment best exemplified with the rapid development and application of computer-based testing (CBT) [1–4]. A computer-based test is an exam containing digitally formatted items that are administered with a computer using the Internet. CBT is growing in popularity because it is widely recognized that paper-based testing is an exceedingly time- and resource-intensive process. The printing, scoring, and reporting of paper-based tests require tremendous efforts, expenses, and human interventions. Moreover, as the demand for testing continues to increase, the cost of developing, administering, and scoring paper-based tests will also continue to increase. CBT not only curtails these costs but also provides many added benefits. By administering tests on computers over the Internet, instructors are liberated from performing the costly and time-consuming administration processes associated with disseminating, scanning, and scoring paper-based tests. Computers permit testing on-demand thereby allowing students to take the exam using a flexible test administration schedule. This benefit means that students can write exams at different locations due to the flexibility

M. J. Gierl (✉)
Faculty of Education, University of Alberta, Edmonton, AB, Canada
e-mail: gierl.mark@ualberta.ca

D. Matovinovic
ACT Inc., Iowa City, IA, USA

H. Lai
School of Dentistry, University of Alberta, Edmonton, AB, Canada

A. Reyes-Munoz et al. (eds.), *EAI International Conference on Technology, Innovation, Entrepreneurship and Education*, Lecture Notes in Electrical Engineering 532, https://doi.org/10.1007/978-3-030-02242-6_3

of Internet access. Computer-based tests can be accessed with different types of devices ranging from mobile technologies such as tablets to standalone desktop computers typically found in computer labs or testing centres. Items on computer-based tests are scored immediately by the computer. If permitted by the instructor, students can even be provided with feedback on their performance immediately upon the completion of the exam because all exams are scored by the computer. Computers also support the development of innovative item types—such as drag-and-drop, reordering, and multiple select—that are only made possible with a computer administration because they require technology-based features such as interactivity or multimedia.

Because of these important benefits, many well-known educational tests that were once given in a paper format are now administered by computer using the Internet including the Graduate Management Achievement Test, the Graduate Record Exam, the Test of English as a Foreign Language, the American Institute of Certified Public Accountants Uniform CPA examination, ACT Aspire, the Medical Council of Canada Qualifying Exam Part I, the National Council Licensure Examination for Registered Nurses and the Canadian Practical Nurse Registration Examination. In short, CBT is dramatically changing educational assessment because the use of expanded item types combined with the growing popularity of digital media and the explosion in Internet use is creating the foundation for a new type of technology-based testing system. CBT has become the hallmark of twenty first century assessment and most testing companies are either administering their tests using a computer or migrating to CBT [4].

But the transition to CBT has also raised formidable challenges, particularly in the area of test item development [5–7]. Thousands of new items are needed to create the banks necessary for CBT because items are continuously administered and, therefore, exposure to students. With a limited bank, item exposure poses a serious problem to test security. Hence, banks must be frequently replenished with new items to ensure that students receive a continuous supply of unique content while, at the same time, limiting item exposure within the testing environment to maintain security. Unfortunately, the large, content-specific, banks required for CBT are not readily available. Moreover, the means by which large numbers of new items can quickly be developed to satisfy these banking requirements is unclear. The current approach to item development relies on a method where a subject-matter expert (SME) creates each item *individually*. Item development is currently viewed as a professional practice where SMEs use their experiences and expertise to produce new items on a one-item-at-a-time basis. Then, after the items are written, they are edited, reviewed, and revised by another group of SMEs until the items meet the required standards of quality [5].

The traditional SME approach to item development has two noteworthy limitations. First, item writing is time consuming and expensive because it relies on the item as the unit of analysis [2]. That is, each item in the development process is unique and therefore each item must be individually written, edited, reviewed, and revised. This process is time consuming and labor intensive because many different components of item quality can be identified. For example, item quality can be determined by the content alignment of the item to a curriculum, by the difficulty level of the item, by the accuracy of the content in each item, by the use of a clear stem in each item, by the use of an appropriate key, and by the plausibility of the distractors of each item. Because each element in an item is unique, each component of item quality must be reviewed and, if necessary, each item must be revised by the SME. This approach to writing and reviewing is very expensive because it is conducted by highly trained SMEs. Rudner [8] estimated that the cost of developing operational items for a large-scale test using the traditional SME approach ranged from $1500 to $2500 USD *per item*. Breithaupt et al. [9] claimed that a high-stakes 40-item CBT with two administrations per year would require a bank containing, at minimum, 2000 items. If we combine the Rudner's cost per item estimate with the Breithaupt et al. bank size estimate, then we can project that it would cost between $3,000,000 and $5,000,000 USD to develop the item bank for a single computer-based test in an assessment program.

Second, the traditional SME approach to item development is challenging to scale in an efficient and economical way [10]. The scalability of the traditional approach is again linked to the item as the unit of analysis. When one item is required, one item is written and reviewed by the SME. When 10,000 items are required, 10,000 items must be written and reviewed by the SMEs. Hence, a large numbers of SMEs who can write and review items are needed to scale the development process. Using a traditional SME approach can result in an increase in item production when large numbers of SMEs are available. But, as noted earlier, item writing and reviewing is a time-consuming process due to the human effort needed to create, review, edit, and revise large numbers of new items. As a result, building banks using this approach is extremely slow and costly for any testing organization.

One method that may be used to address these challenges is with *automatic item generation* (AIG) [6, 11–13]. AIG is a rapidly evolving research area where cognitive theories, computer technologies, and psychometric practices are used to create models that produce test items with the aid of computer technology. It requires two general steps. First, SMEs create item models that highlight the variables or elements in the assessment task that can be manipulated. An item model is similar to a template that specifies the elements in the task that must be manipulated to produce new items. Second, the elements in the item model are varied using computer-based algorithms to generate new items. The purpose of this study is to describe and illustrate a workflow between a university research team who contribute their expertise in item generation technology with a large testing company who contribute their expertise in content development to promote the production of large numbers of high-quality, content-specific test items is an efficient, timely, and cost-effective manner.

2 The Emerging Testing Technology of Item Generation

Drs. Gierl and Lai are recognized leaders in the area of AIG[1]. Their research program at the University of Alberta in Edmonton, AB, Canada is focused on developing AIG methods as well as applying these methods to address practical item development problems. They have used the outcomes from their AIG research to generate educational test items in many content areas including mathematics, science, dentistry, non-verbal reasoning, business management, medicine, and nursing [11]. They have created algorithms for generating both the correct and incorrect options as well as the rationales requires to solve the generated items [14, 15]. They have generated items in multiple languages including French, Spanish, and Chinese [16]. Drs. Gierl and Lai work closely with government agencies, testing companies, and publishers to implement AIG principles into their testing practices including ACT Inc. (USA), Alberta Education (Canada), the American Dental Association (USA), the Australian Council for Educational Research (Australia), Cengage Learning (USA), the College Board (USA), CTB-McGraw/Hill (USA), the Educational Records Bureau (USA), the Medical Council of Canada (Canada), and Reed-Elsevier (The Netherlands).

Item modeling provides the foundation for AIG [17, 18]. An item model is comparable to a template or mold that highlights the elements in an assessment task that must be manipulated to produce new items. Elements can be found in the stem and the options. The stem is the part of an item model that contains the context, content, and/or the question the student is required to answer. Options are the alternative answers that include one correct and one or more incorrect option. To illustrate these concepts, an item model from a mathematics achievement test is shown in Fig. 1. This example is presented for the purpose of illustrating the basic concepts that underlie AIG throughout our paper. While mathematics is used in the current study, it should also be noted that AIG has been implement across a range of content areas including science, dentistry, non-verbal reasoning, business management, medicine, and nursing. Hence, the results presented in this study can be can be used in many different content areas by testing companies to generate test items. The mathematics model in Fig. 1 is designed to generate items that measure students' ability to solve problems involving multiplication and division. The stem in this example contains seven elements (NAME, PRODUCT NAME, PRODUCT MATERIAL 1, PRODUCT MATERIAL 2, NUMBER 1, NUMBER 2, NUMBER 3). The options, labelled A–D, are generated using formulas that include the values from the elements NUMBER 1, NUMBER 2, and NUMBER 3. The elements in the stem and options are then combined using computer algorithms to create items.

[1]To date, Drs. Gierl and Lai have received over $5.5 million in external grants for their AIG research program. They have published 1 book, 12 book chapters, and 19 refereed articles as well as delivered 29 invited presentations and 53 conference presentations on the topic of AIG. Drs. Gierl and Lai received the National Council on Measurement in Education Annual Award in recognition of their AIG research program in 2016.

Stem

<NAME> has a total of (<NUMBER 1> * <NUMBER 2> + <NUMBER 3>) <PRODUCT MATERIAL 1> to make <PRODUCT NAME>. Each <PRODUCT NAME> will have <NUMBER 2> <PRODUCT MATERIAL 2>. What is the maximum number of <PRODUCT NAME><NAME> can make?

Elements

NAME: Alex
NUMBER 1: 5 to 9, increments of 1 (Number of <PRODUCT NAME>)
NUMBER 2: 10 to 30, increments of 5 (Material per <PRODUCT NAME>)
NUMBER 3: 1 to 9, increments of 1 (Remaining material)
PRODUCT MATERIAL 1: crayons
PRODUCT NAME: paper mural
PRODUCT MATERIAL 2: crayon colors

Options

A. <NUMBER 1>*
B. <NUMBER 1>+1
C. <NUMBER 2>
D. <NUMBER 3>

*Correct Option

Fig. 1 A seven-element mathematics item model

Gierl et al. [19] developed a JAVA-based computer program called IGOR (which stands for **I**tem **G**enerat**OR**) that automatically creates items using the element combinations specified in the item model. For instance, one element combination produces the stem: *Alex has a total of 51 crayons to make a paper mural. Each paper mural will have 10 crayon colors. What is the maximum number of paper murals Alex can make?* The simple model in Fig. 1 produced 225 different items.

3 How Can Item Generation Promote Organizational Innovation

In their seminal chapter on "Test Development" in the fourth edition of the handbook *Educational Measurement*, Cynthia Schmeiser and Catherine Welch begin with this provocative question—Test Development: art or science? [20]. AIG could be interpreted as a shift away from the "art" of item development where assessment tasks are created solely from content expertise, experience, and judgment toward a new "science" of item development where these tasks are created algorithmically by systematically combining elements in the item model. Unfortunately, this characterization is too simplistic. SMEs play a critical role in AIG. The SME is responsible for the creative task of identifying, organizing, and evaluating the content needed to create test items. That is, the SME is essential in AIG for identifying the knowledge and skills required to think about and solve problems, designing meaningful item models, and specific the content for these models. Computer technology also plays a key role in AIG. Computer technology is required for the generative task of systematically combining information in each item model. This algorithmic task is essential in AIG because large amounts of content specified by the SME must be combined to generate the items. AIG should therefore be considered a merger between the art and science of item development. The outcomes from the content-based creative task can be combined with the technology-based generative task to rapidly expanding the item development process. This new technology-based approach to item development is characterized by speed, efficiency, quality, and cost effectiveness.

ACT is a mission-driven, nonprofit organization dedicated to helping people achieve education and workplace success. Headquartered in Iowa City, Iowa, ACT is trusted as the nation's leader in college and career readiness, providing high-quality achievement assessments grounded in more than 50 years of research and experience. ACT offers a uniquely integrated set of solutions that help people succeed from elementary school through career, providing insights that unlock potential. The partnership will draw on Drs. Gierl and Lai's experience with AIG and the ongoing development of the AIG engine IGOR with ACT Inc.'s assessment and content expertise to create a new level of research and service in the area of AIG. The alliance will also help transform item and passage development at ACT Inc. from what is currently a manual, labor intensive, non-scalable process to a specification driven, automated, highly scalable process.

With AIG, well-defined responsibilities that adhere to specialized skills according to the appropriate division of labor contribute to the production of large numbers of high-quality, content-specific test items. The strategic partnership between Drs. Gierl and Lai at the University of Alberta with content specialists at the testing company ACT Inc. demonstrate how these well-defined responsibilities can be specified, coordinated, and leveraged to create content for educational testing using automatic item generation technology. A five-stage workflow is presented next that describes and illustrates how these responsibilities are managed between the two partners.

3.1 Stage 1: Create Stems, Elements, and Options

The workflow begins with content expertise. The SMEs at ACT Inc. create the stems, elements, and options. Often as a starting point, a parent item is identified. Parent items, which are typically drawn from existing operational tests, highlight the underlying structure of the item model thereby providing a point-of-reference for creating alternative items. The SME's task is to identify the elements and then manipulate the content in the elements to create new items. The stems, elements, and options are created systematically using rules and rationales. The correct option requires implementing rules and rationales that yield the right answer. The incorrect options are based on rules and rationales that yield wrong, but plausible, answers. For example, a plausible but incorrect option could be generated using the rationale that a student may mistakenly confuse the correct option of NUMBER 1 (correct option A in Fig. 1) with one of the numbers used in the calculation which could be NUMBER 2 (incorrect option C in Fig. 1).

Once the content for the stems, elements, and options are specified, each element is placed in an Access database. The elements contain the content and, in some cases, additional instructions, for how to manage and use the information in the stems, elements, and options for item generation. The database also serves as a tracking system so the workflow can be monitored. When Stage 1 is complete, the relevant information from the database is extracted for use in Stage 2.

3.2 *Stage 2: Build Item Model in IGOR*

IGOR is a JAVA-based program designed to assemble the content specified in an item model, subject to elements and constraints described by the SMEs [19]. Iterations are conducted in IGOR to assemble all possible combinations of elements and options, subject to the constraints. Constraints serve as restrictions that must be applied during the assembly task so that meaningful items are generated. In the mathematics item in Fig. 1, for example, a constraint is that the range of values for the element NUMBER 1 is limited to 5 numbers (i.e., 5–9, increments of 1). Stage 2 begins when information from the Stage 1 database extraction is read into IGOR. Then, the item model, along with all associate content in the model, is created and formatted in IGOR. The content for the item model is organized using the same structure shown in Fig. 1 (i.e., stem, elements, options). An Item Model Editor window in IGOR permits the developer to structure each item model by manipulating content and adding the required constraints in the stems, elements, and options. Then, to generate items from a model, a Test Item Generator dialogue box in IGOR is presented where the developer specifies the item model file, the test bank output file, the answer key file, and the generation options. The generation options can include size of the generated item bank, the order of the options, and the number of options for each generated item. Once the item model is specified in IGOR, the final step in Stage 2 is to generate a small sample of items. This sample is created so the SME can review a range of the content combinations included in their generated items. The sample includes 40–60 generated items.

3.3 *Stage 3: Conduct Item and Model Quality Review*

The SME is responsible for evaluating the content and the logic specified in the item model (see Fig. 1). This review is conducted by reviewing the generated item sample from Stage 2. It is also conducted by reviewing the *AIG validation table*

[10]. The validation table is a unique summary for SMEs that is only available with AIG. It is used in Stage 3 to evaluate the structure of the knowledge and skills required to produce the correct and incorrect options for the generated items. It also provides a way to describe and evaluate the target construct for the assessment. The validation table contains a summary of the concept required to produce the correct option. For our Fig. 1 example, the item model is designed to measure students' ability to solve problems involving multiplication and division. This key concept in this model requires the use of the formula [<NUMBER 1> * <NUMBER 2> + <NUMBER 3> / <NUMBER 2>]. Students must be able to implement the division operation specified within the context of the <PRODUCT MATERIAL> and <PRODUCT NAME> variables using the formula in order to produce the correct answer. Students must also be able to differentiate the correct formula from three related but incorrect formulas presented in this model. Each incorrect option contains an error or misconception related to the correct answer. In our example, the first incorrect option uses the correct number plus 1. The second incorrect option uses the incorrect number 2. The third incorrect option uses the incorrect number 3. If the models are correctly specified, then the generated items will reflect the correct combinations of content and logic outlined in models. For the example in Fig. 1, the SME must verify that the main idea, formula, algorithm, and logical outcome are correctly implemented. If the item model is correctly specified, the workflow proceeds to Stage 4. Alternatively, if the item model is not correctly specified, then the workflow goes back to Stage 1 where changes are made to the model. These changes require updates to the database. All updates are tracked so the revisions can be monitored.

3.4 Stage 4: Add Themes and Expand Item Models

Stages 1–3 are implemented to create an accurate item model capable of generating high-quality test items. In Stage 4, the generative capacity of the item model is expanded dramatically in two ways [21]. The first type of expansion requires the addition of themes. In Fig. 1, the example focused on crayons and paper murals. But additional themes can be included to add variation in the context of the generated items for the <NAME>, <PRODUCT MATERIAL> and <PRODUCT NAME> elements. Themes could include a variety of male and female names (e.g., Markus, Damien, Michael, Allison, Cindy, Elizabeth) in different contexts such as books (paperback) in a library, cutlery (spoons) in a kitchen drawer, clothes (sweaters) in a closet, or fruit (apples) in a grocery store. These themes serve as additional content layers in each model, so the names and contexts change in the generated items. The second type of expansion requires the elaboration of the item models so they measure different types of knowledge and skills. For example, the item model in Fig. 1 could be expanded to measure writing and interpreting algebraic thinking using the model <NAME 1> has <NUMBER 2> times as many <PRODUCT MATERIAL 1> as <NAME 2>. Then, <NAME 1> got <NUMBER 3> extra PRODUCT MATERIAL 2> from a friend. Which of the following expressions represents the total number of <PRODUCT NAME 1><NAME 1> and <NAME 2> have? One sample item with content from this model would be: *Leena has 13 times as many crayons as Alex. Then, Leena got 27 extra colors from a friend. Which of the following expressions represents the total number of crayons Alex and Leena have?* By identifying different types of knowledge and skills, different but related types of item models can be created. In sum, Stage 4 is a model expansion step where the SMEs can use their results from Stages 1–3 to elaborate their item models to include different themes and to measure different curricular outcomes. These additions result in a significant increase in the generative capacity of the original Stage 1 item model.

3.5 Stage 5: Generate All Items and Export to Bank

In Stage 5, all of the items from a model or set of related models are generated. The format of the generated items is customized for each content area and grade level. Typically, the IMS Question and Test Interoperability specification (QTI) is used to define a standard format for the generated items. QTI permits a relatively easy exchange of content between authoring and delivery systems, repositories, and learning management systems. The generated items, saved in QTI format, are exported into the ACT Inc. item bank for use on educational tests.

4 Results

A summary of the outcomes from the five-stage workflow is presented in Table 1. Using this workflow between the University of Alberta and ACT Inc., a three-member SME team created ten different mathematics models that generated more than 68,907 items. The number of themes per model ranged from 2 to 5. The number of grade levels covered with a modified model ranged from 2 to 5. The time to implement Stage 1 ranged from 1 h to 1 day; for Stage 2 from 30 min to 2 h; for Stage 3 from 15 min to 1 h; for Stage 4 from 1 to 3 days; and for Stage 5 less than 1 h. Hence, the total amount of time to implement all five stages in order to produce high-quality, content-specific items ranged from approximately 1–6 days per model. In the most recent 18-month development cycle between the University of Alberta and ACT Inc., three three-member SME teams created 54 models that resulted in the generation of more than 651,000 mathematics items across the elementary, junior high, and high school levels.

Table 1 Summary of item modelling results for one team of SMEs using the five-stage process

Item model	Themes	Grade levels	Generated items
1	5	3	>10,000[a]
2	2	2	660
3	4	3	>10,000
4	3	4	3198
5	3	3	>10,000
6	3	3	3468
7	4	4	8521
8	3	4	3060
9	5	5	>10,000
10	4	5	>10,000

[a]Item generation was truncated at 10,000 as an upper bound

5 Conclusions

Testing organizations like ACT Inc. require large numbers of diverse, high-quality items because computer-based tests are now delivered with variable forms at many times during the year using designs where students receive detailed feedback. But the demand for large numbers of test items far out strips the current supply. Test items, as they are currently written, evoke a process that is both time consuming and expensive because each item is written, edited, and reviewed by an SME. AIG is a method that combines content expertise with technological innovation to solve the practical problem of rapidly expanding the item development process. AIG is a method for using models to generate items with the aid of computer technology. The purpose of our study was to describe the workflow in a strategic partnership between researchers at the University of Alberta with content experts at the testing company ACT Inc. The purpose of this workflow is to help transform item and passage development at ACT Inc. from what is currently a manual, labor-intensive, non-scalable process to a specification driven, automated, highly scalable process. ACT Inc. provides the content expertise in the form of SMEs who are responsible for the creative task of identifying, organizing, and evaluating the content needed to create item models. The University of Alberta research team provides the AIG expertise in the form of research outcomes and computer technology required for the generative task of systematically combining large amounts of information in each item model. By merging the outcomes from the content-based creative task in Stages 1, 3, and 4 with the research-based generative task in Stages 2 and 5, automated processes can be used to facilitate and promote a new approach to item development that is characterized by speed, efficiency, quality, and cost-effectiveness. This workflow was recently used to create 54 different items models in mathematics that produced more than 651,000 unique test items across the elementary, junior high, and high school levels. Mathematics was the focal content area in this project presented in our study. But AIG has also been used in other content areas including science, dentistry, non-verbal reasoning, business management, medicine, and nursing to generate high-quality, content-specific test items efficiently and economically. Hence, the workflow we describe in this study can be used by any testing company that requires large numbers of high-quality assessment tasks.

References

1. F. Drasgow, *Technology and Testing: Improving Educational and Psychological Measurement* (Routledge, New York, 2016)
2. F. Drasgow, R.M. Luecht, R. Bennett, Technology and Testing, in *Educational Measurement*, 4th edn. ed. by R.L. Brennan (American Council on Education, Washington, DC, 2006), pp. 471–516
3. R.M. Luecht, Computer-Based Test Delivery Models, Data, and Operational Implementation Issues, in *Technology and Testing: Improving Educational and Psychological Measurement*, ed. by F. Drasgow (Routledge, New York, 2016), pp. 179–205
4. S. Sireci, A. Zenisky, Computerized Innovative Item Formats: Achievement and Credentialing, in *Handbook of Test Development*, 2nd edn. (Routledge, New York, 2016), pp. 313–334
5. S. Lane, M. Raymond, R. Haladyna, S. Downing, Test Development Process, in *Handbook of Test Development*, 2nd edn. ed. by S. Lane, M. Raymond, T. Haladyna (Routledge, New York, 2016), pp. 3–18
6. M.J. Gierl, T. Haladyna, *Automatic Item Generation: Theory and Practice* (Routledge, New York, 2013)
7. R.M. Luecht, Automatic Item Generation for Computerized Adaptive Testing, in *Automatic Item Generation: Theory and Practice*, ed. by M. Gierl, T. Haladyna (Routledge, New York, 2013), pp. 196–216
8. L. Rudner, Implementing the Graduate Management Admission Test Computerized Adaptive Test, in *Elements of Adaptive Testing*, ed. by W. van der Linden, C. Glas (Springer, New York, 2010), pp. 151–165
9. K. Breithaupt, A. Ariel, D. Hare, Assembling an Inventory of Multistage Adaptive Testing Systems, in *Elements of Adaptive Testing*, ed. by W. van der Linden, C. Glas (Springer, New York, 2010), pp. 247–266
10. M.J. Gierl, H. Lai, A process for reviewing and evaluating generated test items. Educ. Meas. Issues Pract. **35**, 6–20 (2016)
11. M.J. Gierl, H. Lai, Automatic Item Generation, in *Handbook of Test Development*, 2nd edn. ed. by S. Lane, M. Raymond, T. Haladyna (Routledge, New York, 2016), pp. 410–429
12. M.J. Gierl, H. Lai, Using automated processes to generate test items. Educ. Meas. Issues Pract. **32**, 36–50 (2013)
13. M.J. Gierl, H. Lai, S. Turner, Using automatic item generation to create multiple-choice items for assessments in medical education. Med. Educ. **46**, 757–765 (2012)
14. H. Lai, M.J. Gierl, C. Touchie, D. Pugh, A. Boulais, A. DeChamplain, Using automatic item generation to improve the quality of MCQ distractors. Teach. Learn. Med. **28**, 166–173 (2016)
15. M.J. Gierl, H. Lai, Using automated processes to generate test items and their associated solutions and rationales to support formative feedback. IxD&A J. N. **25**, 9–20 (2015)
16. M.J. Gierl, H. Lai, K. Fung, B. Zheng, Using Technology-Enhanced Processes to Generate Items in Multiple Languages, in *Technology and Testing: Improving Educational and Psychological Measurement*, ed. by F. Drasgow (Routledge, New York, 2016) pp. 109–127
17. I.I. Bejar, R. Lawless, M.E. Morley, M.E. Wagner, R.E. Bennett, J. Revuelta, A feasibility study of on-the-fly item generation in adaptive testing. J. Technol. Learn. Assess. **2**(3), 1–30 (2003)
18. A. LaDuca, W.I. Staples, B. Templeton, G.B. Holzman, Item modeling procedures for constructing content-equivalent multiple-choice questions. Med. Educ. **20**, 53–56 (1986)

19. M.J. Gierl, J. Zhou, C. Alves, Developing a taxonomy of item model types to promote assessment engineering. J. Technol. Learn. Assess. **7**(2), 1–51 (2008)
20. C.B. Schmeiser, C.J. Welch, Test Development, in *Educational Measurement*, 4th edn. ed. by R.L. Brennan (National Council on Measurement in Education and American Council on Education, Westport, CT, 2006), pp. 307–353
21. M.J. Gierl, H. Lai, J. Hogan, D. Matovinovic, A method for generating test items that are aligned to the Common Core State Standards. J. Appl. Test. Technol. **16**, 1–18 (2015)

Alternative Learning Experiences: An Innovative Project Stimulating Creative Faculty of Humanities Students

Rabeya Binte Habib

1 Introduction

This new millennium, the epoch of globalisation, demands teachers to prepare graduates for workplace readiness and lifelong learning increasing teachers' job in effectively incorporating students at learning. Teamwork, effective communication, shared decision making, sharing information, utilising information technology, innovation, and time management are crucial in today's enterprise. Going beyond the era of imparting basic skill competencies and knowledge transmitted to passive learners, most frameworks: 21st Century Skills Map, 2011; Pacific Policy, 2010; P21, 2007 [1–3] dated back in last 10–12 years label a good number of skills to be conveyed to learners to face twenty-first century demands. The list is long and mentioned in literature review. However, many think-tanks concur on developing the following four critical areas of learning that this present study also focuses exclusively:

- Collaboration and teamwork.
- Creativity and imagination.
- Critical thinking.
- Problem solving.

Hitting the popular search engine Google for the topic of interest 'students teaching students' or 'peer teaching' in researcher's context did not end up much positively. There is hardly any primary source on this topic in the context, despite the same search engine flooding with information for *twenty-first century skills* to be imparted to learners being defined and categorised by many policy Frameworks

R. B. Habib (✉)
Daffodil International University, Dhaka, Bangladesh

A. Reyes-Munoz et al. (eds.), *EAI International Conference on Technology, Innovation, Entrepreneurship and Education*, Lecture Notes in Electrical Engineering 532, https://doi.org/10.1007/978-3-030-02242-6_4

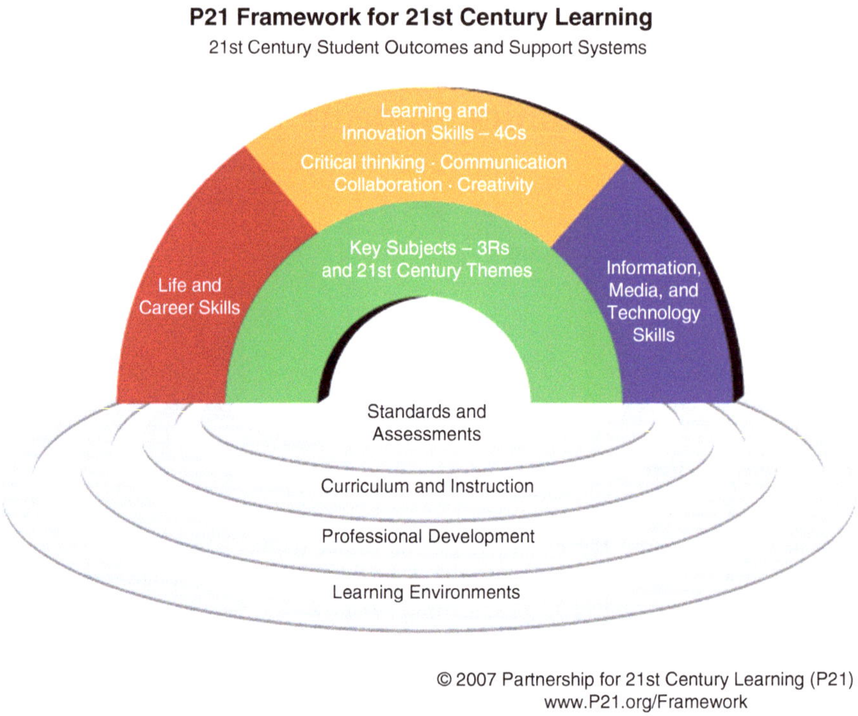

Fig. 1 P21 Framework for 21st Century Learning (Source: p21.org/our-work/p21-framework)

shifting learners' roles as doers in classroom and outside as global resources. *21st Century Skills* as shown in the following figure [1] manifests input taken from educators, education specialists, and business leaders. The framework embellishes skills, subject or content knowledge, and other skills, often defined as soft skills and supports systems students require to succeed as a citizen at work and life (Fig. 1).

Besides, teaching is not an easy cup of tea and it is not all comers affair. Therefore, those who have taken decision or are prompt to venture in it must be well prepared and ready for it. Also, students going in front of the classroom transform and understand more of teacher's jobs having a tactile experience of learning. Finally, engaging students as teachers does not merely mean teaching new tricks to an old horse in a race but to challenge younger foals, along with those older ones, to teach others, and to teach themselves.

Moreover, many novice teachers in private-run organisations in Bangladesh enter into classrooms with less or no previous exposure or training that again contradicts mock teaching being popular trend in teacher requirement contexts. Bangladesh is not any exception [4] in this. Thus, Students Teaching Students (STS), a student centred approach, is introduced and applied with 48 students of two core courses coded ENG-413 and Eng-434 titled Introduction to ELT (English Language Teaching) and English Language Proficiency-1 respectively in English department of Daffodil International University. Presenters in small groups were responsible for thorough analysis of course material as they lead with their chosen topic(s) from linguistics or language teaching or language acquisition (Appendix 1). While setting students on such challenge to become well versed on a particular topic of course material, the present paper also aimed to increase learner's innovation and learning skills.

1.1 Research Objectives

What and how best a teacher can do to help his/her learners for effective learning? The answer may invariably point towards the incorporation of learners in their own learning. Therefore, research focus of this case study were to engage students in active learning through peer-teaching to an advanced level and bring positive self-images among STs.

The present study, thus, aimed to:

1. identify if students were able to cope and adapt well with a student-centric approach enhancing innovation and creativity.
2. pinpoint the impact such approach may have on learners.

2 Literature Review

2.1 Active Learning

"*There is, in fact, no teaching without learning*" [5]—well versed by Freire in 1998 implicating the need to incorporate students in team work being actively engaged in learning which in turn increases level of course satisfaction among students [6]. There is a plethora of research saying in favour of engaging students in learning and more specifically through teaching peers [2, 5–7] resulting in better learning than being taught in traditional methods. The concept of peer teaching can be traced back to Aristotle's use of *archons*, or student leaders, and to the letters of Seneca the Younger [5]. When students are exposed to teach students they are involved in active thinking about materials, critical analysis and applicable solutions and processing the information to be conveyed in their teaching [6] as cited in Samson, 2015.

Active learning shifts learner's role from being sheer passive receiver to a 'maker' and this increases learner's intentness and focal point, interpersonal communicative skills, motivation towards critical thinking skills, learner inventiveness, and many more which eventually helps everyone achieve the course objectives. Numerous research works validate active learning or students' immersion in their own learning [1, 2, 5, 8–11] into and out of a small/large group discussion(s), debate(s), demonstration(s), especially through peer/micro teaching, the approach that the present paper has undertaken. Thirty years back, Chickering and Gamson [10] thought alike stating the implication of student's active engagement in learning:

> Learning is not a spectator sport. Students do not learn much just by sitting in classes listening to teachers, memorizing pre-packaged assignments, and spitting out answers. They must talk about what they are learning, write about it, related it to past experiences and apply it to their daily lives. They must make what they learn part of themselves . . . (p. 3).

2.2 *Twenty-First Century Skills*

Projects that defined skills for first century of third millennium or twenty-first century skills almost in similar manner are many [1–3]. The first and the third, conducted in Washington D.C., are representative sources on this topic. Due to maintain a good size of this present paper, most of these themes are just written in a brief manner except Innovation and Learning Skills, the major focus of this writing that has received utter attention. 21st Century Skills of learning are broadly categorised into key subject that, at first, includes knowledge of and competency in English along with few other subjects incorporating interdisciplinary themes into key subjects. These are:

- Global awareness.
- Financial, economic, business, and entrepreneurial literacy.
- Civic literacy.
- Health literacy.
- Environmental literacy.

Learning and Innovation Skills, which is the main focus of the present study are the most demanding skills to be transmitted to learners in this era of global competitions and excellences, entrepreneurship, effective intelligible communication with multilingual speakers, to name a few. The skills focus of this area includes:

- Creativity and innovation.
- Critical thinking and problem-solving.
- Communication.
- Collaboration.

The first two skills in this sector are supporting the nature of 'changes and demand of inventiveness' while the next sets of skills are the key to knowledge and learning skills addressing the higher demand of today's workplace skills [12].

Thirdly, information, media and technology Skills in today's technology and media driven world require students acquiring functional knowledge in this sector to attain:

- Information literacy.
- Media literacy.
- ICT (Information, Communications and Technology) literacy.

- Succeeding, Life and Career Skills give steering to learners developing their thinking skills, content knowledge, and social and emotional competencies required for life and work. P21's essential Life and Career Skills include:
- Flexibility and adaptability.
- Initiative and self-direction.
- Social and cross-cultural skills.
- Productivity and accountability.
- Leadership and responsibility.

Finally, twenty-first century support system requires an innovative support system engaging learners' through practical skills and knowledge, proper use of technologies, bringing real world to classroom or classroom knowledge in practice and necessary. P21 has identified five critical support systems to make creating that students have acquired twenty-first century competency:

- Twenty-first century standards.
- Assessments of Twenty-first century skills.
- Twenty-first century curriculum and instruction.
- Twenty-first century professional development.
- Twenty-first century learning environments.

In order to draw an end, mastery of learning and innovation skills and proper knowledge and use of ICT and basic subject knowledge are the ones our graduates need to be a well prepared global resource.

The idea of *creativity and innovation* at education can be traced back to 1980s. Many think-tanks refer to innovation as an attempt to bring positive changes in educational settings 'by doing something new or different.' This also aims at bringing changes at pedagogic by reducing teacher-talks for enhanced student participation or creating a student centric learning environment through task based language teaching [13]. This millennium has also seen changes in teaching materials, technological advancements like the incorporation of Facebook, Twitter, computer-assisted language learning, or assessment through e-portfolio and many others. The main aim of innovation and creativity is to keep the pace of ever-changing and always demanding world outside the classroom and to create an atmosphere of creative activity for building learners of today to be tomorrow's global resources.

Considered to be the new basics of twenty-first century learning: *communication*, *critical thinking*, and *problem-solving skills* are today's catchwords of demand in global market. Experts with cognitions, complex logical thinking, and communicative competence are much needed. Critical thinking relates to the application of knowledge learned utilising critical thinking, problem solving approach through communication. The concept of critical thinking, that is, to think on thinking has shattered years-old taxonomy of learning: knowledge, comprehension, application, analysis, synthesis, and evaluation, to a revised one. Now, it goes with how learners *remember*, *understand*, *apply*, *analyse*, *evalu-*

ate, and *create* (http://www.celt.iastate.edu/teaching/effective-teaching-practices/revised-blooms-taxonomy). A learner usually requires critical thinking skills in order to analyse, interpret, evaluate and synthesise their content knowledge and apply the achieved result to solve a problem at further.

Being able to *communicate* in English is also considered to be the key of survival of the fittest at many workplaces. Knowing to communicate effectively in English at entry-level interviews to almost all official communications, writing official papers and correspondences, giving presentations and reports are to name a few.

Lai [14] mentioned more than 35 literature references favouring '*collaboration*' beneficial towards positive effects on student learning. Collaborative strategies in learning often involve 'peer learning', 'group work', 'active learning', and so on. The idea of collaboration gained much interest in last century promoting the sense of community learning in classroom and enhanced group work skills. Both are triggered to develop student-centred learning environment reducing teacher-centred traditional learning. Tibbetts and Hector-Mason in 2015 [15] illustrate the implication of such approach "revolving the needs and abilities of students" as part of planning, teaching and assessment. Collaboration can specially have positive effects in low-achieving students.

2.3 Graduate Job Market in Asia and Bangladesh

Graduate employability is today's catchword and has attracted many literatures within education contexts though much less in Asian contexts [8]. Socioeconomic development, industrialization, and technological development have led to a global demand in the educational sector to create graduates with occupation readiness.

Employability refers to a graduate's ability to demonstrate occupation readiness towards what the eyes of employers are looking for: skills, knowledge, attitudes, and international competitiveness, graduates who are willing to take initiative, are creative, flexible, and adaptable. Although graduate employability has received most research attention in western rather than Asian context, global economy, tertiary education keening towards outcome-based learning, industrial and social development are making it gradually an issue deserving examination in Asia (Ahmed and Crossman [8], Erling [16]).

An employable graduate should have both hard and soft skills incorporating as many as 108 soft skills suggested by George Allen's research work, mentioned in Matin et al. [17] in 2003, is known as Skill-based approach which is utilised to address problems of employment. Other two popular approaches to address employability issues are Graduate Identity Approach that claims a learner will create his/her identity in a given context. The third approach is the USEM (understanding, subjective specific and generic skills, efficacy beliefs and metacognition) Model. Graduate's sense of self is the focal point of this approach (Fig. 2).

Employability factors in Bangladesh is well-reflected on a report published in a popular online news portal titled as *Graduates mismatch with job market*

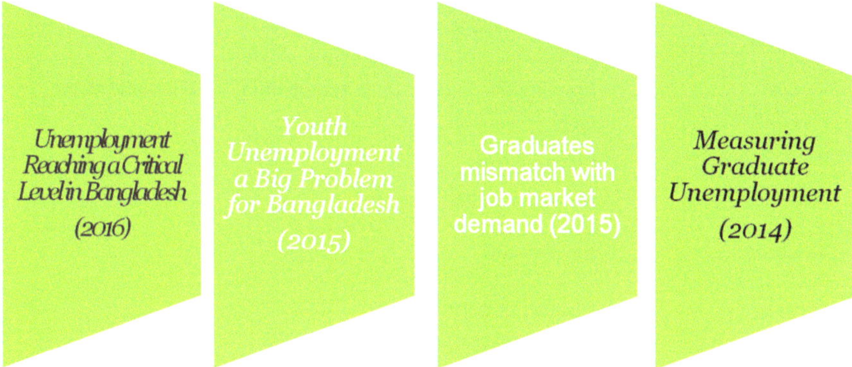

Fig. 2 Newspaper reports on graduate (un)employability in Bangladesh

demand mentioned education system as incongruous in the job market setting. It has been found that "*applicants have low skills in English literacy, computers, communications, problem solving and entrepreneurship*" [18].

2.4 Wagner's Seven Survival Skills

Wagner [19] 2008 criticised that teens leave schools in the USA being prepared for jobs having less significance in the economy and are hardly prepared for the changes due to globalisation and technological spreads at large. He argues that even best schools fail to cope with the ever-changing demand of workplace. Hence, he points out seven skills that should inevitably be imparted to learners at school for survival at life. These are;

- Critical thinking and problem-solving.
- Collaboration.
- Agility and adaptability.
- Initiative and entrepreneurialism.
- Effective oral and written communication.
- Accessing and analysing information.
- Curiosity and imagination.

2.5 Task-Based Language Teaching

Candlin and Murphy's seminal collection of papers in 1987 aroused researchers' and educators' interested towards task-based language teaching (TBLT) [20, 21]. Nunan claimed that TBLT shifts the learning focus from 'what' to learn from

Phase	Examples of options
Pre-task	* Framing the activity (e. g. establishing the out come of the task) * Planning time * Doing a similar task
DuringTask	Time pressure
Post-task	* Number of participants * Learner report * Consciousness-raising * Repeat task

Fig. 3 A framework for designing task-based lessons by Ellis [20]

'how to learn' involving learners as the driving force, navigating and controlling their own routes to learning [20].

In general, most TBLT lesson designs contain phases as mentioned in Ellis [20]. The first phase considers activities that teachers and students undertake while planning through demonstrations of similar at classroom for the actual task, may or not, following a time frame in second phase. As a follow-up activity, learners may be asked to reflect on their tasks conducted at previous phases during the post-take activity. In all senses, TBLT requires learners to carry out tasks performing for language development (Fig. 3).

3 Research Design and Methodology

3.1 This Study and Its Implication

As an attempt to fill up the research gap in context, this case study following a qualitative task-based teaching methodology was designed to investigate students reactions towards a student-centric approach and the implications that may have on them. Though scarcely found, this case study was the course teacher's own custom-design coursework to help learners shun their habit of rote-learning in a form-focused examination-mad teaching context. Active learning based on active participants' tailor-made lesson plans exploiting the extensive use of authentic materials using English as a medium of communication covers many aspects of content, learning and innovation skills mentioned in P21 framework.

Each ST's group was assigned for teaching their juniors for 25–30 min on their topic(s) based on their lesson plan (Appendix 1) that exhibits a wide range of topics incorporating participant from STS-1 teaching their topic(s) focusing

Table 1 ST's demographic data analysis

Population	Male	Female	Age	No. of groups	Nationality
16 (STS 1)	3 (18.75%)	13 (81.25%)	21–24	5 (1 foreign students' group containing 4 members)	Local
11 (STS 2)	4 (36.36%)	7 (63%)		4 groups (3 groups of 3 member and a groups of 2)	3 Turkish
21 (STS-3)	6 (28.57%)	18 (85.71%)		7 groups	3 Turkish
Total: 48	13 (27%)	45 (72.91%)		16	42 local 6 Turkish

Multiple Intelligence (MI) given by (http://multipleintelligencesoasis.org/) Howard Gardner, 1983 in his book Frames of Mind. STS-2 emphasised student teacher(s) to develop lesson(s) focusing on proficiency skills (listening and speaking) while the third STS had its focus largely on innovation and critical thinking in teaching through varied materials: *Realia*, titling it as 'Teaching beyond Lecture'.

In total, 48 student teachers (Table 1) taught, during the trimesters of 2016, more than 100s of the department under a continuous effort named as Students Teaching Students (henceforth STS). This study was conducted in three task cycles: planning (pre-task), action and observation (during task), and finally reflecting (post-task). Usually, it took two months of preparation and planning to come up with the final lesson plan (Appendix 2). Post STS presentation, students had to reflect their experience of teaching analysing different teaching learning situations at their Notes, handwritten or electronic, submitted by email or at Google Classroom page. Reflection was also required while answering an ideational examination paper (Appendix 3). The question paper required them to evaluate and re-evaluate their learning to their active teaching practice while writing answer for six descriptive questions out of ten options. Their performance in final examination was praiseworthy.

Observation data were collected from two observant groups: 107 student observers who were actively engaged in (ST)'s classroom and our four experts. SO's observation report consists of an explorative write-up sharing their overall experience and further suggestions. Four Expert Observers (henceforth EO) who graced the active learning sessions with their presence in different occasions of STSs were sent a semi-structured questionnaire and the return rate from both were 60 and 75%.

All four of our experts have extensive research and a great deal of teaching experience and belong to the esteemed group in educational settings being Professors in the respective field. All three of EOs responded while one did not return the questionnaire duly filled in. EO's Observation Report contains four questions following two close-ended questions, ranging from five for excellent to one for poor, further subdivided in three sections to reflect and evaluate the presence of effective preparation and overall presentation skills, language and twenty-first century learning and innovation skills as observed in STSs. The result varied from the highest scale to the second highest only differing in responses given by one of the experts on the matter of clarity of expression and conclusion. He thought there ought to be more clear conclusions as well as clarity of expression by scaling three to these two factors. They also wrote answer to two open-ended questions sharing their overall experience of attending STSs with further valuable suggestions and comments (Table 2).

Third of its kind, an explorative Reflective Note was collected from STs and scrutinised for answering Research objective no. 2 in order to determine their attitudes towards active learning, more specifically STS.

The varied lesson topic exhibited through innovative teaching practices utilizing authentic materials support that students' were imparted with more than content knowledge. ST's made use of varied teaching aids incorporating songs, video clips,

Table 2 EO's observation towards the initiative and their active engagement in student teachers' classroom

Enjoyable, entertaining and participatory. Students not only learn from each other, they gain apprenticeship of pleasures of teaching if they choose to be one. They have shown creative talents and thoughtfulness in planning their lessons and executing them.
It was a very good experience to attend the student-centered approach initiated to develop students' innovative skills, critical thinking and utilize their creative faculty in an effective way. I hope this will also enable learners to initiate independent learning in which learners learn to learn and express and use their learning in a productive way.
Great initiative! A very innovative way of language teaching!

puzzles, riddles, self-made games, customised board games, and contextually customised Google Map and many more to teach, assess, and re-evaluate the learning of their lessons in their classroom. They incorporated SOs in their classroom dragging them to board for lessons like Affixation or Cultural Differences, all around the classroom in giving directions as part of a speaking lesson to multiple head knocks where they were engaged in groups solving a custom-made game on collocation as part of a lesson on Lexical Relationships or knock out of room to check on their visual literacy under a lesson on Learning Styles.

STs' Notes were received through screen shots, email, or handwritten report state that many were at first terrified of such innovative practices that they were never put into. At first, they felt *nervous/low/anxious* which gradually turned into a *huge* experience that they considered essential in their further endeavour in life and at work.

'Best Endeavour to Teaching Award' sponsored by the course instructor motivated them to win and was helpful to bring in a sense of healthy competition among groups pushing away anxiety, nervousness, and low self-esteem considered to be the barriers to learning. The winning team was decided by SO's on the spot observation report along with general observation from our EO.

Generally, SO wrote their observation report on any form they found advantageous and at their exposure including immediate observation report sent through email, handwritten journal, mobile screenshots, and also audio recordings. These were considered as spot observation report which they submitted within 2/3 days of STS they attended. Their report showed increasing interest of learning beyond lecture being engaged in fun filling learning environment.

4 Summary of Findings

It is critical to share with students the reason active teaching is incorporated into their courses. Students are more familiar with traditional lecture, so active learning comes to them as something actually frightening. Therefore, Student Teachers were asked to write on their STs' understanding of the significance of active learning and whether they can even mention ten benefits of such practice in determining research objective no. 1. Students here are supposed to mention ten comments on the implication of active learning writing comments such as *active learning helps us learn and understand the key concepts.*

> *We, the maximum students feel very shy and nervous while speaking before unknown/less-known people. The rate of nervousness increases with the number of audience. The best part of STS was to get us out from our comfortable zone & make us to stand up before a huge audience & present ourselves.*
>
> *Because they are another form of learning so we can study or understand in a different way.*
>
> *To help different people learn better incorporating different ways.*
>
> *It has got to be some new technique that teachers are trying to experiment with. It is to help get our involvement in the active learning process and keep our brains on task.*

For effective teamwork in a given time.

To hear ideas from our peers/classmates.

So the material we learn is reinforced and repeated in order to catch the concepts better.

To help us learn better and to open our minds in other activities other than just having to study our notes.

To help us to be ready for job market.

To increase our confidence, self-beliefs and motivation.

To help us learn and let us see how much we know. Interacting with fellow department mates, learning from others.

They are incorporated to help us learn the material better. It is supposed to make us learn it and it sticks with us. I like it a lot. It changes things up so we are not just taking notes.

It helps one actually participate so besides just the listening in lecture one has to think about the topics. It developed our critical thinking or thinking outside the box.

When considering the variety of learning activities, students were found interested in working with own handmade context-specific teaching aids that incorporated 'Key to Reveal about You' as an warm up activity to "Throw the Ball in Right Basket labelled with Appropriate Motivation" following a Q/A activity in a lesson to teach about Motivation (Table 3). These well versed STs' ability to think outside the box and their improved critical thinking and imagination supporting Research Objective No. 2 (Table 4).

Majority of SOs showed their interest in variety of learning activities that were made at their exposure in our classroom. The data so far collected from SO's Observation Report showed an exemplary sample stating STS that at first seemed quite a *drag* turned into to an *splendid experience/fun* of learning from their seniors who looked like professionals due to the *clarity of their delivery*, body language, *dress up*, and excellent presentation though the topic for discussion was a boring one, 'preposition' (Table 5).

The findings also show that learners' motivation and confidence level were increased, after the successful completion of the task, though highly stressed in the beginning. Many of them were even reluctant to take part and/or were losing consistency, had little persistence to stick to the task, instructor had to act like a classroom catalyst being empathetic towards their state(s). However, students showed positive attitude towards innovative teaching practices and problem-solving critical thinking skills that they were exposed to which answers both of the central research questions positively.

The researcher observed students' urge to do something new than repeating the similar task while preparing for second STS under English Language Proficiency-1 course with almost same students (16 from STS 1) of Introduction to ELT from Spring 2016. Students, at first were quite reluctant to go for the similar challenge but when they were told that the focus of second STS is developing/helping SO's with listening and speaking proficiency, they were ready to do it. This clearly reflects to the fact that today's students want to take novel risks and prefer not to experience the same boring mechanical process of learning.

Table 3 SOs' Active Participation Triggering Much Success of ST's Teaching

Pictures on left are of classroom picture with students actively participating (up) in class and at board (middle) and a warm-up activity 'Key to Reveal about You' from Learning Styles lesson Pictures in right are showing students being occupied in giving direction (up) and busy in group to solve a custom-made game on collocation (middle) and 'Throw the Ball on Different Kinds of Motivation Box based on Q/A game'	

(continued)

Table 3 (continued)

Table 4 ST-A and ST-B's reflective journals

ST-A and -B's reflective journal	Keywords
In STS, I was a performer in this program. Basically, it was a journey of huge experience. Today I am going to share my experience. *From the very beginning when I heard that I have to perform in STS I was so nervous. I thought that I couldn't do that. But I would like to thank our respective Rabeya mam for her great effort and devotion for this journey. We practiced for many times before our presentation and the most shocking news for us was our dean sir will be there and we have to present in front all of the faculty members Before the day, we decorated our auditorium. It was so fun to decorate the room for our own presentation with friends. And the presentation day was like a day of joy. All friends were in formal dresses and the whole day we presented one by one. It was the first day of my life that I presented something before huge audience. I was nervous though our dean sir and Head Sir and other faculty they showed their interest that provided us courage to perform better* *At the end of the presentation our dean sir had given us certificate and prize for the winning team. Again I want to thank Rabeya mam to make that happen with me. I was really very confused. But the experience I gained is huge. I learned so many things about performing as a presenter. The most important this is my intention is to be a teacher so I think this presentation will help me in my future plan* [ST:A from STS -1 at an email] *STS was held on Dec 3, 2016 under the course ELT. The journey of STS was quite challenging and fascinating, too. We ... got plenty of time to improve ourselves as a participator. Our course teacher... took presentation at every class segmentally. She talked about different methods and styles of teaching and told us how to apply those in the classroom... At first, mam said to choose topic... we had exchanged (*she meant changed) our topics several times... the positive comments of madam encouraged us to take risk... we made worksheets, different funny and interesting games. We were nervous, anxious, but as we started we went on.* [ST:B's Journal from STS-3]	• Performer of learning demonstration • Nervous at the beginning • Practiced many times before (Pre-task) • Fear over expert observation • Day of joy (Task phase) • Decoration • Encouragement • Huge experience • Lifelong career plan • Challenging and fascinating task • Plenty of preparation time (Pre-task) • Teaching the basics and theories beforehand • Preparation, decoration • Final demo competition • Enhanced motivation at post-task phase

4.1 Impact of Findings on My Teaching

A continued practice of letting students explore the significance of active learning in such courses will be continued. The researcher will also try to modify this assessment tool for better or try to explore something even better than STS that suits the course objectives and student interest. Before getting any adequately alternative, a variety of active learning strategies will be used.

Table 5 SO A, B, and C's Observation report

SO A	Keywords
In my opinion STS is such a great platform for proving oneself. For future students teacher. They already arranged so many activities. They all are creative 1. *Physical activities* 2. *Some interesting games like related to the subjects* 3. *For making it more interesting and informative they arranged some short quizzes* 4. *The student teaches were confidence* 5. *What they gave us lessons were informative*	. . . so many activities Creative, confident teachers Interesting games related to lessons Assessment: Short quizzes
SO B *Which have appealed to me the most is that the student-teachers have used many innovative techniques in their lectures. They tried their utmost to hold the audiences' attention, at which they have succeeded beautifully. By using music, videos and several forms of fun games they have made the teaching process less stiff and more fascinating for the audience. The fact that, they too are students and thus have a clear understanding of how to motivate the Student-audience reflected in their approach*	Innovative techniques Successful at holding audience attention Varied activities based on lesson
SO C *To be honest, when I first heard of STS program, I thought it was going to be a total drag. We were like what would students teach us that would be of any help to us? We were not much interested in attending it. So in an unwilling manner, we walked into the auditorium . . . But as soon as we stepped in we were awestruck by the amazing decoration. The decoration alone lifted our sprit just like that . . . student teachers used so many innovative techniques that it graced everyone's attention. For instance, preposition was taught through songs and students were asked to prepare songs with preposition and sing along*	Total drag . . . Splendid work

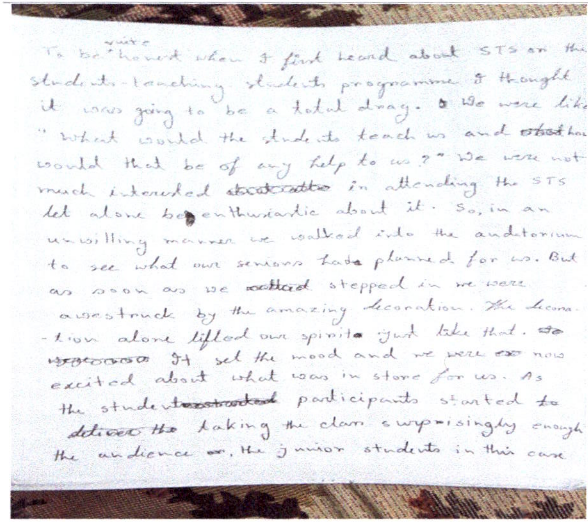 To be honest when I first heard about STS or the students-teaching students programme I thought it was going to be a total drag. We were like "What would the students teach us and how would that be of any help to us?" We were not much interested in attending the STS let alone be enthusiastic about it. So, in an unwilling manner we walked into the auditorium to see what our seniors had planned for us. But as soon as we stepped in we were awestruck by the amazing decoration. The decoration alone lifted our spirits just like that. It set the mood and we were now excited about what was in store for us. As the student participants started taking the class surprisingly enough the audience, the junior students in this case	*At first, thought it . . . .to be a total 'drag' awestruck after entering the classroom seniors were not only all dressed like professionals but also teaching like professionals used so many innovative ideas and materials to teach*
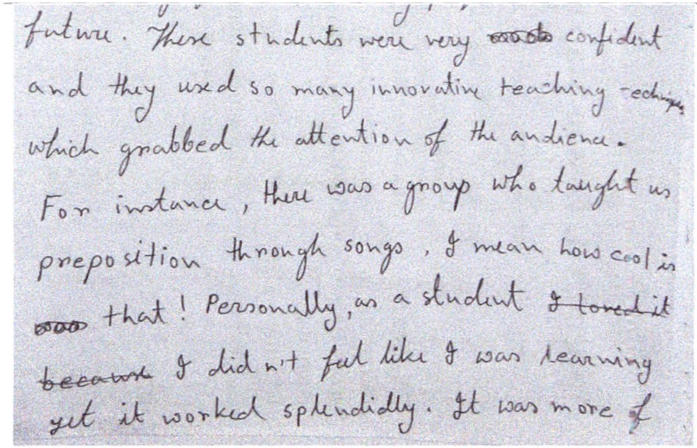 future. These students were very confident and they used so many innovative teaching techniques which grabbed the attention of the audience. For instance, there was a group who taught us preposition through songs, I mean how cool is that! Personally, as a student ~~I loved it~~ ~~because~~ I didn't feel like I was learning yet it worked splendidly. It was more of	*Preposition was taught through songs, not textbooks or worksheets* *Greatest opportunities for students who want to be a teacher splendid work*

It is great to discover from the findings that active learning is helping students push away the barriers to learning such as anxiety, low self-esteem, and fear of facing audience. One SO group's Immediate Observation Report added extra spirit to the researcher's intention for carrying out such practice even further. The report expresses these student observers' interest to be an ST while they wrote in email that they "*are also looking forward to doing the same way like our fellow seniors*" (Table 5).

4.2 *Conclusion and Future Implications*

A small number of students being part of STS have recently graduated and attended convocation in March 2017, among them some are already teaching in different sectors. Few are planning or have started doing their higher education intending to be a teacher. This research does not only focus on student teachers' teaching readiness but also inclination towards any profession they would fit exhibiting wide range of twenty-first century skills and through their persistence to display collaborative problem-solving skills and inventiveness to make a cornerstone role towards their life and at work.

Since the teaching context is exam-mad and tight-syllabi focused, this type of practice is a new start for the students and the instructor, too. The research framework for this kind of work is hardly found, especially in researcher's context. Hence, the study faced many theoretical hitches and methodological glitches at first. Collaborative action research in researcher's context or input from employers of these participants or influence of active learning for enhanced self-efficacy may be conducted as further research scopes of this study.

Appendix 1. Lesson Map of STS-1, STS-2, and STS-3

Topics	Warm-up	Materials	Skill focus
Personality factors in second language learning	Video clips showing several personality factors	Video clips from movies, custom-made worksheet	Intrapersonal intelligence; speaking and writing
Conditionals	Popular superstitions using conditionals	Video clips, song, and custom-made game to complete conditional clauses	Logical and mathematical intelligence; writing skill

Topics	Warm-up	Materials	Skill focus
Affixation	Children nursery rhyme: "if you are happy and you know it"	Story books; board-game: complete the flower with correct petals using prefixes and suffixes	Linguistic intelligence
Asking for direction	Video clip from an English TV series on the topic	Video clips, engaging game activity in class, and worksheet	Visual and spatial intelligence
Preposition	English song and tailor-made worksheet	English song; custom-made worksheet to assess the use of preposition	Musical intelligence; basic proficiency
Transitional words	Asking about the 'how to' or a process like 'how to make a paper plane?' with random student in class revealed the topic	Creative game: how to make a paper boat? relating the lesson topic	Bodily kinesthetic intelligence; writing skill
Cultural differences between Bangladesh and Turkey	Small talks	Board game, worksheet, and video clips	
Lexical relationship	Picture puzzles	Worksheets, video puzzles, and a game in group on collocation	
Learning styles	'Key to explore about you'	Tactile: picture puzzle, kinesthetic: toy making, visual: toy puzzle to observe the changes, auditory: English and Bangla karaoke, to name a few	Various
Word formation process	Jokes	Worksheet and video clips	Writing
Giving direction	Riddles	Self-made Google map for giving direction to DIU, UC from Rajlaxmi campus	Speaking
Motivation	Video clip of a little girl climbing upon a pony on the 17th attempt	Video clips, custom-made worksheets, and so on	Various
Personality traits	Whole class discussion on general related to the topic	Worksheets, role play, game in class	Various

Appendix 2

Daffodil International University

Uttara Campus
Faculty of English Department

Lesson Overview

Course	Psycholinguistics
Teacher / Facilitator	Farhana Yesmin, Roksana Yesmin
Aim of Lesson	This lesson is aimed to help students evaluating themselves by learning personality factors and finding their own.
Class Description	Third year, first semester student
Date	9th April, 2016
Lesson Topic	Personality Factors in Second Language Learning
Activity	Self-evaluation test
Skill Development at the Discourse Level	Intrapersonal skill through writing and speaking skill.
Learning Resources	Video clips: Movie Clip—"Tare Jamin Par" Mr. Bean—The Exam Power Point Presentation Printed Material: task sheet
Learning Intelligence	Intrapersonal Intelligence
Teaching Method	CLL and CLT

Lesson Plan
Topic: Personality Factors in Second Language Learning
Time: 25 minutes

Stages	Teacher activity	Student activity	Time
1.	The teacher will warm up learners with some funny talk and ask them to draw their feeling by funny emo.	They will be listening and answering teacher's questions and draw emo of feeling.	2-3 min
2.	The teacher will give a preview of last class and discuss with the students.	They will be interacting with teacher	1-2min
3.	The teacher will start new lesson of personality factors and before the detail discussion, teacher will ask them to write about some of their own personal qualities. Then teacher will show them a video and describe the factor related with it. After that, teacher will describe another factor and show a video related with it.	They will write some of their own personal qualities. Then, they will be observing and listening to the lecture and interacting with teacher.	8 min
4.	The teacher will discuss next two factors—introversion and extroversion and describe the differences between these factors and their effects and occasionally ask questions to the students.	Students will be listening to the teacher and interacting with her.	5 min
5.	Teacher will take a quiz test on "introvert or extrovert" and divide them according to their personality. Then, teacher will ask the extrovert students to evaluate their personal quality through writing and the introvert students through speaking.	They will answer the quiz according to their personal preference and afterward, evaluate themselves through speaking and writing.	6 min
6.	The teacher will end the class by give student homework to make their own journal and evaluate their daily activity.	They will interact with teacher to know more about their homework.	2 min

Appendix 3

Department of English
Faculty of Humanities and Social Science (FHSS)
Final Examination, Semester: Spring 2016
Course Code: Eng-413 Course Title: Introduction to ELT
Time: 2 hours Marks-40
Course Teacher's Initials:RBH

Part-A: MCQ for 16 Marks

This part has been intentionally chopped off to maintain length of the submission.

Write brief answer for any SIX of the following. (04 x 06 = 24)

1. Imagine you are at a teacher-parent meeting. One of the parents questions the efficiency of student-centred teaching making students talk more than the teacher. Explain the main principles of this approach and argue for your position.
2. List the five core identified characteristics of the Communicative Approach. What is Realia?
3. What is an information gap activity? Can you design one information gap activity for Linguistic and Literary Terminology course? Define the term "on -the-spot correction" and "delayed correction". Justify their application.
4. What are the most common types of tests? What justifies a good test?-Elaborate with examples.
5. Describe three activities in order to demonstrate different ways of using visuals in the classroom. List three different ways in which you can get feedback from your students.
6. List some questions you would ask yourself before and while planning a lesson. What are the main components of a lesson plan?
7. List a number of possibilities for a teacher to ensure that the students understand his/her instructions (e.g. when setting up an activity). What the following statement means through one specific example:*"Before a freer activity be sure students have a good command of essential language items."*
8. Describe five classroom situations which you feel would justify the use of the mother tongue. What advice would you give to a beginner-teacher on the use of the mother tongue?
9. How would you depict the current situation of ELT in Bangladesh?
10. List some problems which may emerge during pairwork and groupwork. Give ideas for prompting/dealing with each one. Give an overview of different types of published ELT materials.

References

1. Framework for 21st Century Skills, Author, Washington (2007), http://www.p21.org/our-work/p21-framework
2. Pacific Policy Research Center, *Twenty-First Century Skills for Students and Teachers* (Kamehameha Schools, Research & Evaluation Division, Honolulu, 2010)
3. 21st Century Skills Map, Author, Washington (2011), https://www.actfl.org/sites/default/files/pdfs/21stCenturySkillsMap/p21_worldlanguagesmap.pdf
4. I. Mahmud, S. Rawshon, Micro teaching to improve teaching method: an analysis on students' perspectives. IOSR J. Res. Method Educ. (IOSR-JRME) **1**(4), 69–76 (2013)
5. A. Flectcher, Meaningful student involvement. Retrieved from P. Goss, J. Sunnmann, K. Griffiths (2017), Engaging students: creating classrooms that improve learning. Grattan Institute report No. 2017-01 (2017)
6. C. Park, Engaging students in learning process: the learning journal. J. Geogr. High. Educ. **27**(2), 183–199 (2003)
7. M. Fitzsimon, Engaging students' learning through active learning. Irish Journal of Academic Practice **3**(1), 9–18 (2014)
8. R.I. Ahmed, J. Crossman, Significance of employability factors: Bangladesh perspective. Res. Gate, 61–66 (2014)
9. M. Andrawis, Engaging students in learning [PDF document] (2011), http://www.sdstate.edu/sites/default/files/engr/upload/Engaging-Students-ASEE-CoE-3-24-11.pdf
10. W. Chickering, Z.F. Gamson, Seven good principles for good practice in undergraduate education. Appeared in in March 1987 American Association for Higher Coleman, H. (2010). The English language in Development. Report submitted to British Council ISBN 978-086355-638-8. Education (AAHE) Bulletin and later printed in spring as a special report in the Wingspread Journal in Racine, Wisconsin (1987)
11. P. Goss, J. Sonnemann, Engaging Students: Creating Classrooms that Improve Learning (2017), https://grattan.edu.au/wp-content/uploads/2017/02/Engaging-students-creating-classrooms-that-improve-learning.pdf
12. B. Trilling, C. Fadel, 21st century skills: learning for life in our times (2009). https://books.google.co.uk/books?hl=en&lr=&id=VUrAvc8OB1YC&oi=fnd&pg=PR7&dq=innovation+and+learning+skills+at+21st+century&ots=DwaQRsfpdO&sig=dOmn0rePoSQw7Em4ErJaGX9Ad4M#v=onepage&q=innovation%20and%20learning%20skills%20at%2021st%20century&f=false
13. D. Carless, in *Innovation in Language Teaching and Learning*, ed. by C.A. Chapelle. The Encyclopedia of applied linguistics (2013)
14. J. Lai, Collaboration: a literature review (2011), http://images.pearsonassessments.com/images/tmrs/Collaboration Review.pdf
15. J. Tibbetts, A. Hector-Mason, Innovation collaboration in adult education utilizing practices that reflect twenty-first century learning contexts. Published by California Department of Education. Research Briefs (2015)
16. E.J. Erling, *The Role of English in Skills Development in South Asia: Policies, Interventions and Existing Evidence* (British Council, Manchester, 2014)
17. M. Matin, T. Ali, P. Wiebe, Conversation with employers (2003), http://research.brac.net/reports/conversationwithemployers.pdf
18. Graduates Mismatch with Job Market Demand (2017), http://archive.dhakatribune.com/business/2015/apr/21/graduates-mismatch-job-marketdemand#stash.BelFXYsz.dpuf
19. T. Wagner, *The Seven Survival Skills for College, Career, and Citizenship. Adivsor's Corner in the Global Achievement Gap: Why Even our Best Schools Don't Teach the New Survival Skills our Children Need—And What We Can Do about it* (Basic Books, New York, 2008)
20. R. Ellis, The Methodology of Task-Based Teaching (2002)
21. R. Ellis, *Task-Based Language Learning and Teaching* (Oxford University Press, Oxford, 2003)

Teaching Creativity in the Context of a Business School

Andrew Jackson

1 Introduction

Whilst it is acknowledged that creativity is a fundamental element of business success [1, 2] the partnerships between employees in creative roles and business planners and strategists is often uneasy [3]. The mutual suspicion engendered by the differences in attitude embodied by these roles frequently extends into Higher Education, and can act as a barrier in the sharing of experiences between academics working in different disciplines. This is a challenge that has been tackled with mixed success. Whilst there are number of high profile examples of business schools embracing concepts such as "design thinking" (for example the Stanford *d.school*, or "design school" which the *New York Times* described as one of the most highly sought after destinations at Stanford [4]) most business schools, and indeed business students, struggle to get to grips with the teaching of creativity. In response to this problem, this paper demonstrates how teaching practices that deal with creativity can be successfully transferred from the art and design educational environment into a business school context.

My teaching career has been spent teaching students both in a "studio" situation, and in the more conventional university setting of the lecture theatre and seminar room. I have primarily taught in the areas of design, design history, cultural studies and marketing. This followed an 8 year career working as designer with my own practice. In this paper, I want to show how I have used this varied experience to develop my teaching of creative skills to students who are primarily based within the discipline of business and management. I will first review the way that students have been conventionally taught in the "art school" system in the UK, expanding on

A. Jackson (✉)
Canterbury Christ Church University, Canterbury, UK
e-mail: andrew.jackson@canterbury.ac.uk

A. Reyes-Munoz et al. (eds.), *EAI International Conference on Technology, Innovation, Entrepreneurship and Education*, Lecture Notes in Electrical Engineering 532, https://doi.org/10.1007/978-3-030-02242-6_5

the theoretical frameworks that underpin this approach. I will then contrast this with the conventional academic approach found in Business Schools, and describe how I have developed a module that bridges this gap. I will explain how the module is taught, and reflect on the experiences of the staff and students who take part in the module.

2 Pedagogy and Andragogy

The call for papers for this conference asked how we might teach creativity to students by moving from a pedagogic approach to andragogy. Pedagogy is defined simply as the study and theory of the methods of teaching, but also implies the teaching of children, the word deriving from the Greek *pedo*, or relating to children. Andragogy refers more specifically to “adult learning”, and suggests a learner-centric approach to education. Knowles [5] developed four andragogic assumptions, stating that for adults to be successful learners they should be encouraged to move from dependency to self-directedness and draw upon their reservoir of experience for learning. They tend to be ready to learn when they assume new roles and want to solve problems and apply new knowledge immediately.

However, the distinction between the two words also implies striving for a less didactic teacher-centred approach to education, and a move towards student centred learning. This is a problem for traditional universities who tend to base their approach on the lecture/seminar method of teaching—delivering a predetermined, set curriculum to students who are then examined to test whether they have remembered and understand what they have been taught. Biggs [6] has argued against this surface learning approach asserting that education is about conceptual change, not just the acquisition of information in other words, “deep learning”. He believes that deep learning takes place when:

1. It is clear to students (and teachers) what is “appropriate”, what the objectives are, where all can see where they are supposed to be going, and where these objectives are buried in the assessment tasks.
2. Students experience the felt need to get there. The art of good teaching is to communicate that need where it is initially lacking. “Motivation” is a product of good teaching, not its prerequisite.
3. Students feel free to focus on the task, not on watching their backs. Often, attempts to create a felt need to learn, particularly through ill-conceived and urgent assessments, are counter-productive. The game then becomes a matter of dealing with the test, not with engaging the task deeply.
4. Students can work collaboratively and in dialogue with others, both peers and teachers. Good dialogue elicits those activities that shape, elaborate, and deepen understanding ([6], pp. 60–61).

In order to see this approach in action, I now want to turn to the practice of art and design education.

3 Art Schools in the UK

Until the end of the twentieth century (in the UK at least) creative education was primarily carried out not in universities, but in art schools, and was tackled in a quite different way to conventional academic subjects. So why should we take note of the way teaching in art colleges in the UK has been developed? Name any one of the UK's most famous designers, musicians or artists, and they are likely to have studied at an art school at one time or other. One of the most celebrated art school graduates (in the business world, at least) is Sir Jonathan Ive. Ive, who is currently Chief Design Officer at Apple studied industrial design at Newcastle Polytechnic, graduating in 1989. Ive's contribution to Apple has been a key part of its rise to become one the most powerful brands in the world [7]. Looking at the kind of education Ive would have received at Newcastle gives us some insight into why British art schools have had such a strong track record in training creatives.

The art and design faculty at Newcastle Polytechnic (now Northumbria University) began life as an archetypal art school, being founded in 1844 as one of the original Government Schools of Design [8]. By the beginning of the 20th century, almost every region in the UK had its own art school provision, with counties such as Kent having as many as five, each serving their local area [9]. Today, most of these unique institutions have now either closed or have been absorbed into larger universities. In many cases, however, the educational approach inherited by art and design faculties remains unchanged, offering a sharp contrast with the pedagogy employed by the wider universities within which they are now housed.

Art school teaching is founded on the belief that learning is a dynamic process that consists of making sense and meaning out of new information and connecting it to what is already known. To learn well and deeply, students need to be active participants in that process. This typically involves doing something—for example, thinking, reading, discussing, problem-solving, or reflecting ([10], p. 94).The key to the art school system is the "atelier" method. An atelier is an artist's or musician's workshop that is led by a "master" or, in modern parlance, a tutor. In art and design faculties the chief learning spaces frequently follow this model. Within this space, which is designed to replicate a professional design studio, students are required to work independently on briefs that have been set by their tutors; they are autonomously engaged in self-directed problem-solving. Weaver describes this process in the context of architectural education:

> The reason for learning through problem solving is that this activity 'thinking like an architect' can only be learnt through experience; that is, learning to solve problems through the experience of solving problems. It is not for the teacher to tell the student everything, as if knowledge went from one head into another [...] so the essential task in teaching consists of organising the situation so students will have necessary experience [11].

As Belluigi states, "the atelier method places teaching and learning within one-to-one 'studio conversations' between supervisor and student. The espoused theory of this model involves offering constructive criticism on ideas and proposals, and helping the students realise their ideas" ([12], p. 23).

4 Problem-Based Learning

This is an example of "problem-based learning"—although in my experience this term is rarely, if ever used in the art and design faculty. Here, tutors inherit teaching methods from their forebears without naming or necessarily analysing them; the method is regarded as self-evidently the best way to teach. Problem-based learning encourages active participation, student independence and deep engagement. With regard to three-dimensional design and engineering education, Green and Bonollo [13] note: "The teaching of industrial design and product design is usually conducted in an industrial design studio, a place that has developed traditions of learning-by-doing within the traditions of project-based and problem-based education" ([13], p. 269). The objective of problem-based learning is to get students to solve problems they will meet in their professional career. The method is to present them with problems to solve; the assessment is based on how well they solve them. The method is genuinely student centred and tutors guide the students rather than "teach", with new knowledge obtained through self-directed learning. Students frequently work in groups, with a problem forming the basis of their organisational focus, thus stimulating learning, problem solving skills and the cognitive process.

5 Creativity in a Business School

I now want to describe how these methods can be transposed into a business school context by using the example of the Christ Church Business School Creative Campaign development module, which I wrote and have been delivering since 2011. This module is core for students studying BSc Advertising in the Christ Church Business School. The advertising industry places great store by its creative capacity, and is classified by the UK Government as one of the "creative industries". The creativity of advertising agency teams is the key point of difference that separates them from client-side marketing departments [14].

In the UK, advertising courses tend to take one of three distinctive approaches: those based in business schools encompass the wider strategic aspects of advertising and incorporate elements of the general marketing curriculum and management theory; those based in schools of media and cultural studies tend to approach the topic from the perspective of critical cultural theory and sociology; finally those based in departments that were once a part of an art college take an almost exclusively creative standpoint on the subject. The advertising course at Christ Church Business School is modelled on the first type, incorporating many other aspects of management theory—although a small number of students take advertising as single honours course, most combine advertising with either marketing or business studies, and their programme is typically made of general business models that follow a pedagogic format. By the time students reach the third year their primary goal has become transactional, and is to achieve good grades. Their understanding is that this is achieved by producing assignments that demonstrating the recall of a given curriculum and the ability to cite text book marketing theory and business models (examples include the "Boston Matrix", "Porters Five Forces", and the "Marketing Mix"). However, at the beginning of their third-year Advertising students are faced with a very different proposition—they are asked to independently develop creative ideas for advertising campaigns that answer set briefs, and do not explicitly reference theory. They must arrive at answers for themselves, and these answers are not known in advance by the tutor; the work comes entirely from them. For some students who have been used to rote learning this is an alarming prospect. For most it is a pleasurable release from the routine of lecture/seminar/assignment.

Within constraints of the University rooming allocation, the module resembles a simplified form of the atelier system outline earlier in this paper. The course runs for 11 weeks, and is delivered across four hours on a single day. It is effectively the students' sole focus for an entire day, and they work together in a room with computer access presentation facilities and meeting tables. This very different to a typical module in the business school, which runs over 22 weeks, with a single one-hour lecture and single one-hour seminar each week, often delivered on different days. Like many other academic university courses, the focus of these "long thin" modules is on passing the exam or course work, and accumulating grades. In other words, most students are extrinsically motivated by the need to pass. As a result, weekly attendance at lectures and is poor, with some modules achieving as little as 10% attendance, with occasional students never attending and simply submitting assignments to be marked. In contrast attendance on the Creative Campaign Development module rarely falls below 80%.

In the first four weeks of the module students are introduced to theories of creativity, with strong focus on methods that can be utilised in order solve a set of given problems. These include then concepts of brainstorming [15], lateral thinking [16], concept fans [17], mind mapping [18], and method prompt cards [19, 20]. These are supplemented with educational videos that clearly illustrate advertising creatives at work, showing their techniques and processes [21]. Interestingly, research has shown that very few advertising creatives consciously use creative theory in their work, and there is a well-documented divide between academic and practitioner theory [22], and I return to this issue later in this paper. Initially, students embark on a series of short projects ranging from 1 week to 3 weeks in duration, preferably working with one other partner to generate ideas. This mirrors the advertising agency convention of creative teams working in pairs, where one partner is a copywriter and the other an art director [23]. The final half of the module, typically around five weeks in duration, is spent working on a single larger scale project, normally to an external creative brief available as a national student competition by D&AD (the UK professional body representing designers and art directors). These briefs are given to the D&AD by real organisations who have a problem to solve, and this helps to set the students' academic work in a strong professional context.

Throughout the module, the emphasis is always on process rather than outcome, and students are required to keep a logbook that records their progress. This contains visual research, notes and ideas and records of brainstorming sessions. Students are encouraged not to be precious about these books, and to regard them as tools that help them to externalise and record ideas—the more spontaneous the better, even if this means the books are scrappy and messy (in fact, the messier the better). This is a direct mirror of the sketchbook that all art and design students are encouraged to keep as core part of their creative development. For some students, this approach is hard to adopt; they see the logbook as an "assessable item" rather than a tool for creativity. As a result, they labour over presentation and endeavour to avoid "mistakes". Other challenges include a reluctance to move beyond the first idea, overly literal interpretation of problems and solutions, and an inability to generate a large quantity of possible solutions. Tutorial support and theoretical examination of the nature of creative blocks [24] aims overcome these barriers, whilst always encouraging students themselves to arrive at, and evaluate their ideas.

The module culminates in a pitch that as far as possible replicates the professional pitching situation encountered by advertising creatives, designers, and architects in their day-to-day work. In addition to the pitch, students also have submit a reflective essay that records and critiques the creative process they have used, and refers back to the literature on creativity that they have encountered over the course of the module.

Successful students gradually move from a deductive, convergent approach to more inductive, divergent thinking. For some students, this is a challenge, and the skills of the tutor who, in the atelier system, needs to be an experienced practitioner, are essential in coaching students towards this goal. In planning the module, we have made a point of employing practicing advertising and design creatives who give a day a week from their practice to teach. Interestingly, although these tutors play an essential role in delivering the teaching, it is very rare for any of them to have consciously engaged with the theory that underpins their professional practice, and these lecture-based components of the module are generally delivered by staff who are part of the faculty. The question remains about whether this part of the module is strictly necessary in order for students to achieve the learning outcomes, and there is potential for further research into both this, and the wider academic practitioner divide in the creative industries [22].

6 Conclusion

The atelier style situation combined with the problem-based learning approach creates a situation where most students become intrinsically motivated [25]. In module evaluation feedback students frequently tell us that this was their favourite module and they enjoy being challenged to "think outside the box". The direct connection with "real world" industry briefs is also highly valued by students as preparation for employment.

This meets both Knowles requirements of an andragogic education, and Biggs' formula for deep learning. Students are encouraged to move from dependency to self-directedness, to draw upon their reservoir of experience for learning, and to solve problems and apply new knowledge immediately. Importantly, in ways that the conventional transactional approach to business education finds hard to achieve, this method can develop in students' creative habits such as a tolerance for ambiguity, analogic thinking, or a resistance to closure.

References

1. N. Anderson, K. Potočnik, J. Zhou, Innovation and creativity in organizations. J. Manag. **40**(5), 1297–1333 (2014)
2. G. McCracken, *Chief Culture Officer: How to Create a Living Breathing Corporation* (Basic Books, New York, 2009)
3. C. Bilton, *Management and Creativity* (Blackwell, Malden, 2010)
4. N. Perlroth, Solving problems for real world, using design. *New York Times* (2013), https://search.proquest.com/docview/1471319854
5. M.S. Knowles, *The Modern Practice of Adult Education* (Association Press, Chicago, 1980)
6. J. Biggs, What the student does: Teaching for enhanced learning. High. Educ. Res. Dev. **18**(1), 57–75 (1999)
7. K. Badenhausen, Apple heads the world's most valuable brands of 2017 at $170 billion. *Forbes.com* (2017), https://www.forbes.com/sites/kurtbadenhausen/2017/05/23/apple-heads-the-worlds-most-valuable-brands-of-2017-at-170-billion
8. J.A. Schmiechen, Reconsidering the factory, art-labor, and the schools of design in nineteenth-century Britain. Des. Issues **6**(4), 58–69 (1990)
9. D. Haste, *The Art Schools of Kent* (Werter Press, London, 2013)
10. E.F. Barkley, *Student Engagement Techniques* (Jossey-Bass, San Francisco, CA, 2010)
11. N. Weaver, *The Atelier Principle in Teaching*. Symposium on New Directions of Architectural Education - The Necessity of a Cultural Paradigm Responsive to the Majority (Delhi, India, 1999)
12. D.Z. Belluigi, Constructions of roles in studio teaching and learning. Int. J. Art Design Edu. **35**(1), 21–35 (2016)
13. L.N. Green, E. Bonollo, Studio-based teaching: History and advantages in the teaching of design. World Trans. Eng. Technol. Edu. **2**(2), 269–272 (2003)
14. M. Dahlén, S. Rosengren, F. Törn, Advertising creativity matters. J. Advert. Res. **48**(3), 392–403 (2008)
15. A.F. Osborn, *Applied Imagination*, 3rd rev. edn. (Scribner, New York, 1963)
16. E. De Bono, *Lateral Thinking* (Harper Perennial, New York, 1990)
17. E. De Bono, *Serious Creativity* (Schäffer-Poeschel, Stuttgart, 1996)
18. T. Buzan, B. Buzan, *The Mind Map Book* (BBC, London, 2001)
19. B. Eno, P. Schmidt, *Oblique Strategies* (Self-published, 2001)
20. IDEO, *IDEO Method Cards* (IDEO, London, 2003)
21. ZKK for Channel 4 (Producer), S. Foster (Director), (2009), Advertising uncut programme 2: Where do ideas come from? [Video/DVD]
22. G. Nyilasy, L.N. Reid, The academician–practitioner gap in advertising. Int. J. Advert. **26**(4), 425–445 (2007)
23. A. Tibbs, *Advertising: Its Business, Culture and Careers* (Routledge, London, 2009)
24. J. Adams, *Conceptual Block Busting* (Perseus Books, Massachusetts, 1986)
25. R.M. Ryan, E.L. Deci, Self-determination theory and the facilitation of intrinsic motivation, social development, and well-being. Am. Psychol. **55**(1), 68–78 (2000)

Enhancing Student Retention Within Higher Education

Kam Gill

1 Introduction

The UK is often regarded as a country with relatively high barriers to university entry. There has been debate about access to higher education traditionally the focus on social group differences [1]. This seems more apparent with the more prestigious UK universities such as the 'Russell Group' the fact that more prestigious universities tend to be those in which social groups with historically low participation rates are least well-represented students from lower social class backgrounds, besides being hugely under-represented in higher education generally in the UK [2, 3], found in Boliver [4], whilst this is well-established that those from lower social class backgrounds, it can be seen certain ethnic minority groups are starkly under-represented in the UK's more prestigious universities.

2 Factors that Influence Retention

Universities have been seen as:

- Communities dedicated to the learning and personal development of their members, especially students.
- Sources of expertise and vocational identity.
- Creators, testers, for the evaluation and application of new knowledge.
- Important contributors to society and nations.

K. Gill (✉)
Coventry University, Coventry, UK
e-mail: kam.gill@coventry.ac.uk

A. Reyes-Munoz et al. (eds.), *EAI International Conference on Technology, Innovation, Entrepreneurship and Education*, Lecture Notes in Electrical Engineering 532, https://doi.org/10.1007/978-3-030-02242-6_6

3 Reasons for Early Withdrawal

I have categorised in Table 1 a series of reasons why students have withdrawn from university, although the reasons seem quite broad, there are underlying concerns that are inherent earlier on the term.

In addition to statements from Table 1, when students were interviewed common causes for early withdrawal were also:

- Preparation for study at the institution and their previous expectations
- Academic study at the institution earlier on the term
- Personal reasons for leaving

In short other researchers shared the same view and stated that 'the purpose of higher education is to give students of all backgrounds and abilities a platform to develop themselves individually, both academically and socially', this was described by Tremblay et al. [5, pp. 16–19]. This statement was further supported by British Council for Higher Education 'The shape of things to come: higher education global trends and emerging opportunities to 2020', pp. 4–6 (2012).

To support this further, 'International Trends in Higher Education' [6] published their mission that for 'all universities offers equal and fair treatment to all students'. Such statements and views have a strong contemporary resonance to widen participation to a wider audience.

More recently, HESA [7] found that the UK alone the student numbers show that in 2015–2016, there were 992,125 entrants to 162 higher education institutions in the UK. These figures show growth and estimation of approximately 260 million students at worldwide universities by 2025. These changes have affected countries in many different ways certainly in terms of governance perspective. Changes that have been directed by the government in particular with the widening participation strategy, demographically and economically, this viewpoint has been shared by many others [8], suggested that the mission for all universities is shaped by serving all segments of society through open-access admissions teaching and learning that offers equal and fair treatment to all students.

The challenges that face higher education have alluded to a number of questions. How will universities accommodate for an increasingly diverse student population walking through their doors along with diverse teaching, learning strategies and curriculum offerings. The QAA [9, pp. 4–8] research showed that on 16 institution

Table 1 Reasons for early withdrawal

Statement
Personal
Transfer to another college/institution
Medical
Inappropriate course
To seek employment
Financial

across the UK during 2012–2013 the students had high expectations and perceptions of higher education. Students want 'value-for-money' for their education, this was one of many expectations.

Many authors have supported voice of the student, but interpreted in different ways [10]; interpretation was student as stakeholder or representative, or students as consumers or a customer.

Addressing the voice of student was explored by Healey et al. [11], all suggested that students can be collaborators in learning. Therefore, this can be seen as understanding consumer needs, this is of paramount importance to the governance and university. This is further supported more recently with the White Paper, BIS (Department for Business Innovation and Skills) [12] which further reinforces the discourse of students as consumers with the establishment of a Teaching Excellence Framework (TEF), as may also a new governance model in the form of the Office for Students (OfS).

Table 2 above shows student expectations at the beginning of their course, the comments referring to the "*poor support from tutors and academic staff* "were made by the greatest proportion of students. Not all students were critical of the academic support they had received; however, the comments were from students who said that they had found the support offered by the institution staff had been helpful to them [13].

4 Engagement Through an Institution

Engagement from the student and institution through partnership has produced a substantial number of research studies on retention. During this time, universities have undergone major changes in demographics, including evolving from a traditional to a more non-traditional and diverse student body on many campuses. According to the universities in UK (2015), the number of students from disadvantaged backgrounds studying full-time for a first degree has also risen by 42%. With this evolution come significant challenges in retaining these new and growing student populations. Student retention is one of the more interwoven and intricate issues of modern higher education from the student perspective, involuntary

Table 2 Student expectations at the beginning of term

Statement
I did not feel that I was given helpful academic support by my department
I wasn't given enough time and detailed feedback to help me improve my grades
I found the transition from other prior study to higher education work very difficult
I found the assessment tasks very much harder than I expected
I did not know where to go to seek academic help or advice
I was disappointed by the marks I had been getting

withdrawal because of academic failure or inability to cope with the demands of the educational system lowers self-confidence and self-esteem and likely represents a negative economic impact.

It can be seen that universities are now becoming knowledge economies and the shift from manufacturing-based economies to the transition of universities, taking a hold on the importance of physical capital and massive production, whilst knowledge and knowledge-intensified innovations have more strength to be the source of present (and future) wealth. The expansion of higher education in the UK, and how it has changed from elite to a mass system. The changes to the socio-economic state and the impact it has on the role and purpose of higher education. Indeed, the rising demand for more access to higher education has become more apparent and visible that the existing higher education system has become overloaded causing problems such as high student drop-out rates leading to poor retention rates as it falls into the growing dilemma between the obligations to offer more seats at universities as discussed by Keohane and Petrie [14].

Although there appears to be no shortage of research on varying aspects of retention as it relates to students in general. Studies that focus on retaining students particularly in the UK appear to be limited. Limitations appear not to consider expansion, and response to these changes, that may consider issues such as retention which has grown in significance with the expansion of the HE system. These issues can be seen to highlight the problems of a system with elite instincts and traditions, challenged by an agenda of inclusion and widening participation.

5 What Causes Low Student Retention Rates?

Reasons for student withdrawal have been discussed by Yorke and Longden [15]. Their work concluded that there is rarely a single reason why students leave. In most cases, the picture is complex and students leave as a result of a combination of inter-related factors, such as personal reasons, lack of integration, and dissatisfaction with course/institution. Thus, it is important that students have easy access to help outside the classroom through extra tuition or personal academic support as discussed by Thomas [16]. Thomas felt that if students view a course as having little value, their motivation to attend classes will decrease, by beginning higher education study for the first time can often require a significant adjustment in lifestyle for a student.

Thomas, discussed further that first-year students who may find it difficult to concentrate on their studies when demands outside of study are causing them distractions, such as family problems, the breakdown of relationships, paid employment as having negatively affected their study.

Vossensteyn [17] discussed that students who are not only failing to attend classes but are also not engaging with the institute on the most basic level are even more of a concern. This may include students who do not check their email for notifications provided to them by the institute, do not log on to student management systems to check their grades after release or do not collect marked assignments from tutors. All of the above actions would indicate an 'at risk' student who is not likely to complete their qualification, particularly if this pattern occurs early during the course.

Vossensteyn [17] discussed further that students who are struggling are embarrassed to admit that they are not performing academically and will not ask for help, thus it is the education provider's responsibility to establish communication and provide struggling students with information regarding resources that they can use to improve their grades. Similarly, admission strategy is important here—if you admit weak or unmotivated students you will get a higher drop-out or failure rate. Without guidance to deal with these problems, students may feel as though they are not supported at the institution and choose not to return.

6 Significance of Student Retention in Higher Education United Kingdom

According to Moore et al. [18] to have a university education and to gain educational experience can be a fundamental enabler of social mobility in the UK. This experience can open doors to a wide range of professional careers, university education is now much more widely available than it was in the last 50 years. Moore et al. [18] discussed the expansion of higher education provision since the 1990s has seen the sector in England subject to such considerable pressure for change. This has been further discussed in the Government's White Paper (2012) titled, 'Higher education: students at the heart of the system' this document calls for 'radical reform' to put higher education on a 'sustainable footing'.

7 Sources of Help

On the government's 2016 white paper BIS [12] the key recommendation was to improve the student experience. In terms of student empowerment and autonomy, the paper also discussed increasing the social mobility provision for the widening participation agenda for all universities.

The UK Quality Code for Higher Education is the important governing body who sets the standards within the UK, this is a starting point to ensure the education provision is comparable and consistent across the UK. The framework has seven core values which put all students at the heart of the education system below sets the minimum expectation for an intuition.

- Treating students with fairness, dignity and respect
- Providing all students with opportunities to learn
- Informing students about relevant matters
- Having transparent and robust policies and processes
- Exercising responsible governance
- Ensuring sufficient external scrutiny
- Supporting staff effectively

It's important that universities have a better understanding of the student expectation and perception as a minimum. This is of a significant factor as argued by Thomas [16]. Her research found that around 40% of students in the UK think about withdrawing during their first year. Thomas's research also discussed that the important factor for early student withdrawal was a student who felt a need for self-belonging. Undergraduate self-belonging is a very broadly studied topic within HE, much early research goes back to America by Tinto [19]. His work focused on self-belonging and student integration, which are important factors and critical for success. His research later found that students come into an institution with a variety of backgrounds, intents and commitments. Understanding, the source of differences in first-year students 'and highlighted the benefits of retention and completion rates, such as meaningfulness and belonging play important roles in meeting psychological needs and driving student engagement. However, many authors such as Thomas argued for many years earlier that for sense of belonging is a psychological factor focusing on students subjective feelings of connectedness or cohesion to the institution. Thomas's work explored a set of factors associated with sense of belonging, such as membership in religious and social groups, concluding that these were essential contributors to student retention.

8 Conclusion

This research has focused on studies in institutions with similar characteristics. Many researchers have found the first-year student retention is one of the most important factors of any success initiatives. To better understand student retention it is important to take a holistic view of previous research undertaken to date. Although research on student retention is relatively new, a significant amount of literature is available as a starting point for student retention initiatives. According to Seidman, research in student retention is only about 50 years old (2010). Understanding retention took place even more recently, in the past 25 years [20]. These findings have enabled the researcher to conduct an early intervention questionnaire and in-depth semi-structured interviews with the targeted respondents. Moreover, these findings present an interpretation and reflection of the quantitative and qualitative results. Furthermore, comparative and related studies have been introduced, which has led to investigating the current best practices and experience of retention implementation. This will allow evaluating and measuring the student perception of retention strategies and propose and implement a new framework for supporting the student's voice in retention and include testing for appropriateness in the field of education.

References

1. C. Crawford, *The link between secondary school characteristics and HE participation and outcomes, CAYT research report* (Department for Education, Runcorn, 2014)
2. R. Blackburn, J. Jarman, Changing inequalities in access to British universities. Oxf. Rev. Educ. **19**(2), 197–215 (1993)
3. J. Blandin, S. Machin, Educational inequality and the expansion of UK higher education. Scott. J. Political Econ. **51**(2), 230–249 (2004). https://doi.org/10.1111/j.0036-9292.2004.00304. Cited 8th February 2017
4. V. Boliver, *Why aren't there more black British students at elite universities? The conversation* (Durham University, Durham, 2014)
5. K. Tremblay, D. Lalancette, D. Roseveare, *Assessment of higher education learning outcomes: feasibility study report, Vol 1. Design and implementation* (OECD, Paris, 2012)
6. International Trends in Higher Education (2015) https://www.ox.ac.uk/sites/files/oxford/International%20Trends%20in%20Higher%20Education%202015.pdf
7. HESA, Introduction - higher education statistics for the UK 2015/16. https://www.hesa.ac.uk/data-and-analysis/publications/higher-education-2015-16/introduction. Cited: 14th April 2017
8. P.G. Altbach et al., Internationalization and global tension: lessons from history. J. Stud. Int. Educ. **19**(1), 4–10 (2015)
9. Quality Assurance Agency for Higher Education, *UK quality code for higher education, Part B: ensuring and enhancing academic quality, chapter B5 student engagement* (QAA, Gloucester, 2012)
10. B.J. O'Neil, M.A. Wyness, Student voices on an inter professional course. Med. Teach. **27**(5), 433–438 (2005)
11. M. Healey, A. Flint, K. Harrington, *Framework for partnership in learning and teaching in higher education* (Higher Education Academy, Heslington, 2014), p. 4. https://www.heacademy.ac.uk/system/files/resources/hea_framework_for_partnership_in_learning_and_teaching.pdf. Cited 23rd March 2017
12. BIS (Department for Business Innovation and Skills), *Success as a knowledge economy: teaching excellence, social mobility and student choice* (Dandy Booksellers Ltd, London, 2016). https://www.gov.uk/government/publications/higher-education-successas-a-knowledge-economy-white-paper. Cited: 24th June 2017
13. K. Gill, Student academic motivation & engagement strategy to identify early leavers. Int. Educ. Res. J. **2**(8), 76–78 (2016)
14. N. Keohane, K. Petrie, *On course for success? Student retention at university* (Social Market Foundation, Westminster, 2017). http://www.smf.co.uk/publications/course-success-student-retentionuniversity. Cited: 16th July 2017
15. M. Yorke, B. Longden (eds.), *Retention and student success in higher education* (Open University Press, Maidenhead, 2004)
16. L. Thomas, *Building student engagement and belonging in higher education at a time of change: final report from the what works? Student retention and success programme* (Higher Education Academy, Heslington, 2012)
17. J.J. Vossensteyn, *Widening participation in the Netherlands, report submitted to CFE* (CHEPS, Sydney, 2015)
18. N. Moore, M. Christopoulos, K. Dimitriou, J. Hutchinson, *Researching the Aim higher Greater Manchester graduate officer programme* (International Centre for Guidance Studies, London, 2011). https://www.heacademy.ac.uk/system/files/graduate_officer_prog_gtm.pdf. Cited 19th April 2017
19. V. Tinto, *Taking student retention seriously* (Syracuse University, New York, 2007)
20. J.M. Braxton, A.S. Hirschy, S.A. McClendon, *Understanding and reducing college student departure: ASHE-ERIC higher education report, volume 30, number 3* (Wiley, Hoboken, 2004)

Educating Responsible Innovators-to-Be: Hands-on Participation with Biotechnology

Amalia Kallergi and Robert Zwijnenberg

1 Introduction

The emerging popularity of the term "Responsible Research and Innovation" (RRI) marks an increased awareness towards innovation products and processes that respect the needs and values of contemporary societies. Leaving aside the question why isn't all innovation responsible to begin with, one can no longer dismiss the fact that the ethical and societal implications of technological innovations must be carefully considered for any technological innovation to succeed. Given also environmental, economic and humanitarian challenges all around the globe, it is urgent that our technological innovations are conceived, developed and deployed with an eye to the future. RRI emphasizes our "collective commitment of care for the future" [1], both in terms of reducing negative consequences and in terms of promoting positive change. Obviously, if all of our innovations are to be implemented responsibly, then all of our innovation *education* is to be replaced by an education on responsible innovation [2]. But how does one go about educating responsible innovators? This paper recommends a humanities-informed education for responsible innovation and proposes novel forms of pedagogy that go beyond existing models of ethics instruction.

A. Kallergi (✉) · R. Zwijnenberg
Leiden University Centre for the Arts in Society (LUCAS), Leiden University, Leiden, The Netherlands
e-mail: mail@kallergia.com; R.Zwijnenberg@hum.leidenuniv.nl

A. Reyes-Munoz et al. (eds.), *EAI International Conference on Technology, Innovation, Entrepreneurship and Education*, Lecture Notes in Electrical Engineering 532, https://doi.org/10.1007/978-3-030-02242-6_7

According to Von Schomberg's [3] frequently cited definition, RRI is *"a transparent, interactive process by which societal actors and innovators become mutually responsive to each other with a view on the (ethical) acceptability, sustainability and societal desirability of the innovation process and its marketable products (in order to allow a proper embedding of scientific and technological advances in our society"*. An overview of recent definitions of RRI is provided in [4]. Noticeably, most of these definitions emphasize the acceptability of innovation products and/or enforce a collective accountability for both the processes and the outcomes of innovation. Ethical acceptability and social desirability constitute a reasonable response to concerns about the negative consequences of technological innovations while democratic and inclusive processes constitute a legitimate way to achieve this acceptability. For the purposes of this paper, we define responsible innovation as an innovation process that presupposes shared responsibility regarding the societal needs and values that drive a technological intervention. As such, responsible innovation would account for both of the (closely related but not interchangeable) concepts of science *for* society and science *with* society [5]. It should be clear that core to our understanding of responsible innovation is the articulation and negotiation of ethical and social values.

Our work starts with the assumption that higher education institutions can contribute to responsible innovation. Universities are, after all, acknowledged as one of the essential structures in the triple helix of innovation [6] and it is only reasonable to assume that they maintain a similar role also in the case of responsive innovation. Nevertheless, our work focuses on the (somewhat traditional) function of higher education institutions to train capable and well-equipped professionals or, in this case, ethically capable and socially responsible innovators. The role of higher education institutions as a source of ground-breaking knowledge or as responsible entrepreneurial entities themselves is beyond the scope of this paper. Furthermore, our educational objectives and derived interventions share more with the existing landscape of applied ethics education (particularly engineering ethics education) than with innovation education per se; educational practices aiming to enhance the creative capacities of future innovators, such as creativity training techniques for ideation, innovation education, or design thinking, to name a few, will not be further discussed.

Since the Hastings Centre report on teaching engineering ethics, the importance of ethics education in science and engineering curricula has been steadily established.[1] While there is plenty to be done before ethics education is compulsory to all science and technology programs [8], its relevance is undisputed and a number of high-quality educational initiatives are implemented worldwide. Note that the normalization of ethics in science and technology education precedes the discourse on RRI. The disciplines of philosophy and engineering appear to have found a perfect match in the domain of applied ethics, where a large number of the examined cases are directly derived or, at least, intensified by scientific and technological

[1] For a brief history of the major developments in engineering ethics education, see [7].

developments. Such tendencies suggest that the humanities (as the study of ethics and philosophy) are, by now, a respected partner in educational initiatives regarding (socially responsible) technological innovation. At the same time, educators urge for a broader understanding of engineering ethics [9] while others highlight the limitations of the predominant modes of instruction, such as the case method or codes of ethics induction [10]. Emphasis on macro-ethics [11], attention to meta-ethics and calls to better incorporate STS (science, technology and society) concepts in applied ethics curricula are some of the suggestions towards a more meaningful collaboration between the domains of engineering and philosophy in education.

We welcome the aforementioned developments in the domain of applied ethics education but further observe that the exact nature of RRI as a value-laden process amplifies the identified need for a more rigorous contribution of the humanities to science and technology education. In other words, the challenge of RRI introduces a need and an opportunity for a more pluralist integration of the discipline of the humanities. We expand on this thesis in Sect. 2 and further propose a pedagogy that would best serve our educational objectives, namely hands-on participation. Section 3 documents a case study on RRI education, namely the module "Ethics, culture and biotechnology", part of the minor Responsible Innovation, and the educational activities we implemented along the lines of hands-on participation. Our experiences and observations from three subsequent editions of the course are discussed in Sect. 4.

2 A Humanities-Informed, Hands-on Pedagogy for RRI

The contribution of the humanities to RRI and to RRI education can and should span beyond ethical analysis of (already implemented) technological innovations. Consider as an example the successful synergies between science, technology and humanities in the domain of sustainability. The well-established field of environmental humanities is acknowledged as an essential resource in coping with our present environmental crisis [12, 13], while transdisciplinary approaches are urged with regards to sustainable development. The urgency of tackling RRI problems from a multiplicity of perspectives is not missed by several theorists of RRI who suggest that multidisciplinary teams are crucial in implementing RRI. We believe that a valuable component of RRI education programs is to be found in humanities-related content. Successful RRI education programs should strive to integrate a humanities-attentive consideration of RRI, e.g. one that recognizes the socially constructed nature of the values that guide responsible innovation.

Secondly, and in order to fully prepare responsible innovators-to-be for the negotiation of values that characterizes RRI, successful RRI education programs should strive to sharpen and sensitize the anticipatory, reflective and inclusive capacities of their students. As implied in Stilgoe et al.'s [14] framework of RRI, RRI can only be truly responsible if it is performed in an anticipatory, reflective, inclusive and responsive manner. These attributes imply a capacity to productively converse with different stakeholders and a readiness to consider critically one's own role and position, qualities that are hard to find in mature professionals, let alone young professionals in training. Similar elements can be traced in the practical skills that are considered core but non-trivial to integrate in RRI education. Pavie et al. [2] observe that one of the difficulties in educating responsible innovators (particularly engineers) lies in familiarizing them with the multiplexity and fluidity of the design space as well as with the complex dynamics of including multiple stakeholders. Hollander et al.'s [15] list of required practical skills highlights the need to comprehend various stakeholder perspectives and to identify value conflicts. Successful RRI education should, thus, prepare students to actually cope with multiple perspectives (as opposed to only understanding them) and to continually reflect upon their own perspective.

To facilitate the aforementioned objectives, we propose and implement a form of pedagogy that is highly interactive, experiential and embodied, i.e. one that requires students to participate hands-on in practical exercises with a tangible character and a moral edge. Our rationale behind this approach is two-fold. Firstly, active learning is a more engaging and more effective way of learning that should better support our students in comprehending the ethical and societal issues at stake. Secondly, we expect that inviting our students to go beyond their comfort zone should enhance their appreciation of the complexity of RRI and promote a more considerate and open attitude towards the perspectives of different stakeholders.

For the remainder of this paper, we focus on the domain of biotechnology as a case study that typifies the complexities of RRI as well as the challenges of RRI education. Biotechnology, as the use of living things to make useful products for human use, is a booming industry distinguished by rapid scientific developments and rapid commercialization. It is also an industry that unavoidably touches upon fundamental questions surrounding human values, from the definition of life to the definition of human nature. The values that drive the acceptability of biotechnological innovations are often influenced by emotions and unarticulated assumptions and further problematized by the inaccessibility of the subject matter to the general public. To support and implement a hands-on participation with the issues in an educational context, we introduce two practices distinguished by a unique level of interactivity with and exposure to the materials and methods of biotechnology, namely DIY (do-it-yourself) biology and bioart.

2.1 *Tactic 1: DIY Biology*

DIY biology is an emergent practice that advocates, promotes and facilitates informal experimentation with biotechnology [16–18]. As specified in [19], DIY biology is "the pursuit of biology outside of scientific institutions by amateurs, students, 'hobbyists' ". Amateur DIY biologists all around the world, often organized in informal local groups or maker spaces, experiment with the materials and methods of biotechnology, actively performing a variety of biotechnological protocols such as DNA extraction or DNA sequencing [20].

The DIY biology movement is, arguably, an interesting model for innovation in the life sciences and one worth introducing to future innovators-to-be. For example, DIY biologists are routinely praised for developing innovative, sometimes ingenious but always cheaper, alternatives to standard laboratory technologies. Furthermore, it is expected that wide and pluralistic access to biotechnology will allow for novel biotechnological solutions to emerge and/or stimulate viable and competitive start-ups in an industry that is largely dominated by big players.

Its economic potential aside, DIY biology can be a relevant educational activity for responsible innovators, thanks to both its embodied nature and its collective character. Obviously, DIY activities are distinguished by a high degree of active learning: students touch, use, handle and experiment with the materials and methods of biotechnology; they perform experiments and get directly exposed to the concepts they would normally only theorize about. Furthermore, the sheer availability of these protocols outside of an institutional context raises questions about self-regulation: how far are we willing to go with the technologies at our disposal when we are the ones responsible for their use? Notably, DIY biology groups are self-regulating groups that tend to operate non-hierarchically and to adhere to the ethics of community, access and collaboration. As such, DIY biology communities can become spaces where values and ethical frameworks are negotiated in an equalitarian manner. In effect, staging or enabling a DIY biology initiative becomes a direct invitation to question, challenge and negotiate how far we are willing to go with the technologies at our disposal.

2.2 *Tactic 2: Bioart*

Bioart refers to artistic practice that engages with the methods of the life sciences and/or employs living material as its medium [21–23]. Much of bioart literally comes out of the laboratory for it is "created in test tubes, using the laboratory as the art studio" [24]. By definition, bioartists are directly implicated with the materials and methods normally ascribed to science, from organisms such as genes, cells and bacteria to apparatuses such as electrophoresis gels, microscopes and incubators.

Bioart is typical of the ongoing interest of artists in emerging technologies and exemplary of the potentials of art–science collaboration programs [25]. Art–science collaborations are frequently hailed as spaces of cross-contamination that can be beneficial for both research and innovation. For example, interdisciplinary collaborations involving artistic capacities and expertise can be valuable for creativity and innovation. In the case of art that engages with sciences, art can often assume a mediator role between the sciences and the public. More importantly, however, bioart works can become spaces for valuable encounters with the ethical issues surrounding biotechnology, particularly due to the material dimension of these works [26].

3 Case Study: "Ethics, Culture and Biotechnology"

3.1 The Minor Responsible Innovation

The minor Responsible Innovation is a joint initiative of three Dutch Universities (Leiden University, Delft University of Technology and Erasmus University Rotterdam). Within the (Dutch) bachelor degree organization, minor programs are coherent units of education (30 ECTS) available as an elective to bachelor students in their third year. Minors are, in principle, open to students from a wide range of backgrounds. In our case, a multidisciplinary body of students is bound to be the case as students are admitted in equal proportions from all three participating universities. In fact, a defining aspect of the program is the combination of expertise and the merging of academic traditions from three different universities.

The aims and organization of the minor are further described in [27]. The program is interdisciplinary by nature and characterized by a problem-based approach to education, with students assigned real-life cases to work with. Parallel to their project work, students attend a number of standalone modules delivered by instructors from all three participating universities. The minor Responsible Innovation is an innovative and valuable experiment on how to structure an educational program for responsible innovation in its entirety. However, our discussion here is limited to the module "Ethics, culture and biotechnology" (organized by Leiden University), i.e. the module in which we explicitly implement our hands-on pedagogy.

3.2 Ethics, Culture and Biotechnology

"Ethics, culture and biotechnology" is a 6-session long module, scheduled at the second half of the minor Responsible Innovation. As discussed in Sect. 2, the course focuses on biotechnology as a relevant and challenging case study for RRI: Contemporary biotechnology is testing accepted ethical and aesthetic values and, as such, challenges us to seek new approaches to the ethical, cultural, juridical and economic issues relating to biotechnological practices. Table 1 summarizes the learning objectives of the course, while Table 2 provides a generalized course schedule.

"Ethics, culture and biotechnology" builds upon the instructors' expertise in teaching and coordinating educational projects at the intersections of art, science and the humanities. The Leiden University Honours class "Who Owns Life? Ethical, Juridical, and Artistic Encounters with Biotechnology" [28], organized since 2007, is one of the signature courses of the instructors distinguished by both a hands-on perspective and a devotion to the voice and agency of artists. Often under the supervision of invited artists, students from a multiplicity of backgrounds explore the ethical, legal and societal implications of biotechnology by engaging with the materials and methods of the life sciences inside the biology laboratory.

Table 1 Learning objectives as published in the e-guide

Learning objectives
Describe key ethical issues in biotechnology and its products
Identify individual and social barriers that play a role in the application of biotechnological innovations
Identify various perspectives and values in the public debate surrounding biotechnology
Demonstrate debating skills and critical reading skills

Table 2 Course overview (some variations across editions apply)

Course overview
Introduction and DIY DNA extraction workshop
Lecture "Why Art?"
Debate 1: human enhancement
Debate 2: patents
Artist workshop in the lab
Roleplay CRISPR and human germline modification technologies
Skill lab: board game design

It should be noted that our partnering university Delft University of Technology has a substantial tradition in engineering ethics education. Core material on responsible innovation, applied ethics and value-sensitive design is introduced to the students during the module "Introduction to RI" (organized by Delft University of Technology). Our efforts towards a hands-on RRI education should be understood as complementary to this module.

4 Our Approach: Hands-on Participation

Across the three editions of the module "Ethics, culture and biotechnology" (academic years 2014–2015, 2015–2016 and 2016–2017), we implemented several educational activities along the lines of hands-on participation, namely DIY biology workshops and bioartist workshops. Note that the boundaries between bioart and DIY biology are not always clear-cut, as several bioartists make use of DIY biology practices and/or relate to a DIY biology ethic.

4.1 DIY Biology Workshops

DIY DNA Extraction DIY DNA extraction is a simple activity widely used for both science communication and DIY biology workshops. It is a straightforward, inexpensive exercise with concrete results: DNA is extracted from organic matter

Fig. 1 Results of the strawberry DIY DNA extraction as performed by the students of "Ethics, culture and biotechnology" (academic year 2015–2016). Image credits: A. Kallergi

and is visualized in a clear layer of alcohol (cf. Fig. 1). The protocol can be performed using only household supplies and items available in one's kitchen or local supermarket.

We conducted DIY DNA extractions using strawberries in two subsequent editions of the course, as both an opening exercise and an introduction to biotechnology, in general, and genetic research, in particular. These extractions took place in a regular classroom and were led by the course instructors (both non-life scientists). Generally speaking, our protocol yields clear results without much need for guidance or accuracy. Regarding the visceral experience of a strawberry DIY DNA extraction, the process is messy but harmless. Plastic bags and coffee filters are likely to break during the process but strawberries are a very friendly and forgiving material: they smell nice and are not disgusting to clean.

The DIY DNA extraction was popular among our students. It is a surprising and fun activity and one not readily associated with a university classroom setting. Looking at strings of DNA for the first time is a rewarding experience and students happily take (and post) photos of their results. Obviously, the protocol succeeds in turning an abstract concept into something visible and tangible. As a conversation starter, it raises discussions mostly on the potential of DIY biology. What can one do with the extracted DNA? And is this a credible scientific procedure? Such discussions are often stimulated by the presence of science students who tend to reject the activity as relevant for educational or illustration purposes only.

Building a DIY Sterile Hood For the third edition of the course (2016–2017), bioart pioneer Oron Catts introduced out students to the technique of tissue culture.[2] Tissue culture (or cell culture) is the practice of growing living cells outside of an organism. The students first visited the tissue culture facilities of the Faculty of Science, Leiden University, where they were introduced by the lab responsible to the techniques, laboratory equipment and biological agents, i.e. living cells, involved in tissue culture. Oron introduced the students to his artistic use of the technique and guided them to the making of a DIY sterile hood (a component that allows growing cells in sterile conditions). A DIY hood should allow one to perform tissue culture outside of the laboratory, as Oron and his colleagues have successfully demonstrated at various settings.

The exercise of building a sterile hood was presented to the students as an engineering problem. Students worked in groups, were given a list of materials and a schematic diagram of a sterile hood and were asked to devise their own designs/ solutions. As such, the exercise was rather appealing to our students, particularly the ones with an engineering background, who took to the task and competed on their designs; one group took their construction home with them. Oron contextualized this exercise as not only a means to perform DIY tissue culture but also as an activity that requires one to think carefully about conditions of sterility and the artificial environment of the laboratory. Yet, it remains unclear whether this critical message was evident to the students, who seemed to struggle with connecting the wetware, living part of tissue culture to the hardware part of it.

4.2 *Artist Workshops*

General Setup For each edition of the course, we invited a practicing bioartist to lead a 1-day session with our students inside the biology lab. Invited artists were encouraged to draw from their own artistic practice and develop practical exercises that relate to the topics of the course (Fig. 2). Other than a couple of consultations with the instructors, the artists were free in the type of activities they wished to perform with the students. The workshops were hosted at the Faculty of Science, Leiden University, with the kind support of the Junior Science Lab.

Adam Zaretsky: Do-It-Yourself Biolistics For the first edition of the course (2014–2015), internationally renowned bioartist Adam Zaretsky invited our students to participate in his DIY biolistics protocol/art performance. A biolistic particle delivery system or gene gun is a device used to introduce foreign genes in the genome of an organism by "shooting" DNA-coated micro-particles to the target cells [29]. Adam's workshop attempted to replicate the functionality of a gene

[2]While strictly speaking an artist workshop (cf Sect. 4.2), we classify this activity as a DIY exercise due to the staging of the workshop, the form of participation assumed by the students and the straightforward nature of the activity.

(a) (b) (c)

Fig. 2 Student activities inside the biology laboratory (**a**) Building a DIY sterile hood (**b**) DIY DNA extraction the Adam Zaretsky way (**c**) Making breast milk butter. Image credits: A. Kallergi

gun using gold nanoparticles—a substance typically used in commercially available gene guns (cf. Helios gene gun [30])—and a powerful jackhammer. The workshop entailed the extraction of DNA from various food sources (an exercise frequently performed by Adam Zaretsky in public and art spaces) and Adam's subsequent attempt to use his makeshift gene gun with the extracted DNA. For the record, a variation of this performance was delivered a few days later in Kapelica Gallery, Ljubljana [31].

Much can be said about Adam's artistic practice in general (often described as "bioethics in action" [32]) and his DIY mechanical genetic engineering experiments in particular. Here, it should suffice to say that Adam's practice problematizes the discussion surrounding genetic engineering by, among others, amplifying the grotesque in the existing scientific methods and techniques. In that respect, Adam's workshop was a staged performance intended to merge scientific terminology, laboratory equipment, biological agents and oddity. The quirkiness of his procedures was not missed by the students who played along with Adam's instructions but would readily admit that Adam is mocking or circumventing scientific procedures rather than enacting them.

Alice Vandeleur-Boorer and Heath Bunting: Survival Food Tech For the second edition of the course (2015–2016), Alice Vandeleur-Boorer and Heath Bunting organized an artist talk and a number of short indoor and outdoor exercises motivated by the notion of food and food technologies. During the laboratory exercise, the students produced butter from human breast milk[3]; the practicum was inspired by Alice's interest in food politics and her latest project Vaghurt (2013–2015) [34]. In the public space nearing the laboratory facilities, Alice and Heath demonstrated a number of survival techniques such as making fire and producing ice in the wild. These outdoor activities related to Heath's long-lasting practice of public space, cross-border, activist interventions as well as the duo's recent joint practice. In the artists' own words, "they teach domesticated artists and visionaries to nurture and protect themselves in order to release them back into the wild" [35].

Survival Food Tech triggered extreme responses by our students, at both ends of the spectrum. The artists take a clearly anti-corporate stance on issues of biotechnology and globalization and tend to take personal risks in their projects. These attributes were unexpected and outlandish to our students who, in turn, responded to the subsequent exercises, particularly the outdoor ones, either with irritation or with eagerness. When looking for a spot to make a fire, students were asked to climb walls or walk through vegetation in what was, in fact, a search for wildness and invisible borders in the public space. Students were free to withdraw any time for the activities and several of them actually did.

5 Discussion

A word of caution: our short descriptions of the artist workshops are bound to fall short to the experiences undergone by our students. To make things worse, our limited coverage fails to properly address the artistic value of the artworks directly or indirectly referenced in the workshops. While these limitations are crucial for a full appreciation of the educational activities organized, it is not our intention here to provide a fully fledged analysis of the artist portfolios; neither do we wish to assess the artistic value of the workshops as artworks themselves. Instead, we will focus on what we observe to be relevant points in understanding and evaluating these activities as hands-on participatory experiences for the sake of RRI education.

5.1 Course Evaluation

The module has consistently received very good evaluations by the students and is highly appraised for its interactive character. Unfortunately, course evaluations were organized independently to the module instructors and varied enormously across the three editions of the minor. As a result, we are particularly limited in the quantitative data at our disposal. Nevertheless, the 2016–2017 edition received

[3]For a short commentary on this activity, see [33] (in Dutch only).

the best quality rating across all modules (3.75 out of 5) and the second best overall rating (3.69 out of 5). For the 2015–2016 edition of the course, over 84.2% of the students gave a quality rating of 5 or higher in a scale of 1–7, making our module the best rated module for block 2 (no data for block 1 are available to us). Overall, students appreciate the high level of interaction ("Truly an interesting course. the small groups and interaction was very well organized", "fun outdoor activity") and, reportedly, find value "in the opportunity of learning from people outside the university or from other fields". Our own, informal in-class evaluations also testify that students do appreciate the course and approach taken[4]; some difficulties with connecting the course to RRI were reported but such comments declined progressively across editions.

We are pleased with the quality of the course and believe to have delivered a curriculum that successfully implements an active and experiential form of learning. Still, what can we say about the impact of our educational activities and their relevance for RRI education? Did our implemented activities meet our expectations and assumptions regarding the potential of DIY biology and bioart as a pedagogically relevant hands-on participation? Given our experiences and observations from the course, we suggest that the assumed potential of our hands-on pedagogy materialized in the form of two concrete contributions: (1) Hands-on encounters with the materiality of biotechnology and (2) Hands-on encounters with a vastly different knowledge system.

5.2 *Encountering the Material: Repurposing the Materials and Methods of Biotechnology*

One of the main contributions of our hands-on pedagogy is to be found in the tactile, tangible, embodied encounters with the materials and methods of biotechnology. As it was hypothesized, DIY biology activities were successful in bringing abstract or unfamiliar concepts within reach. More importantly, they allowed an equalitarian consideration of these concepts and facilitated a discussion that was not dominated by the expert (scientific) discourse. In the case of DIY DNA extraction, the subject matter became something that everyone was eligible to talk about, raising questions about sterility, purity, and the underlying quest of science to study life in isolation. Both Oron's and Adam's workshops were characterized by a clear intention to break the corresponding technologies down to their core elements, exposing the usually concealed mechanisms of controlling and manipulating life.

While successful as embodied and equalitarian experiences, the full potential of DIY biology may have been underrepresented in our current implementations. More

[4]While we have no means to specify which (out of the multiple) interactive activities are most appreciated, our informal evaluations suggest that both the DIY biology activities and the artist workshops make a lasting impression to the students.

specifically, a missed opportunity is to be found in the collective possibilities of DIY biology. Our activities lacked a practical follow-up that would allow the group to feel as part of a DIY project/community and to, consequently, take ownership of the process. As a result, the possibilities of DIY biology as a space of moral negotiation remained hypothetical.

5.3 *Encountering Another: The Artistic Perspective*

While DIY biology exercises tend to be a new but comfortable territory for our students, artist workshops aren't. Surprisingly, it seems to be the case that simply bringing our students in contact with the practices, idiom and ways of working of artists is quite a striking experience. Our students seem to be affected by the vast differences in worldviews and methodologies that were revealed during their encounters with artists at work. It can be argued that encounters with (any) real stakeholders could train the anticipatory and inclusive capacities of our students. Still, their encounters with artists were unique in challenging the authority of the students' existing and assumed knowledge systems. As such, encountering the artistic perspective not only enriches the anticipatory and inclusive capacities of the students with additional viewpoints but also confronts them in a way that requires them to reflect upon their own certainties.

Of course, there is and should be much more to art than coming to terms with alternative ways of knowing. Successful bioart works are platforms where moral dilemmas are enacted and explored, a quality that may or may have not been present in our current implementations.

5.4 *Limitations and Additional Recommendations*

Despite our encouraging results, a limitation of our approach may be the choice of subject matter. Throughout the paper, we reasoned over the relevance of biotechnology as a case study for RRI and RRI education. Nevertheless, we are aware that RRI education spans across a variety of technological subjects. The tactics chosen as the means to implement our pedagogic stance on hands-on participation (DIY biology, bioart) may be unique to the domain of biotechnology and may not be directly applicable to other technological subjects. Furthermore, it might be the case that these activities were made possible only thanks to our long lasting involvement with the field and our existing professional network. We acknowledge these limitations but are confident that a humanities-informed approach which values active learning and respects the artistic perspective should be relevant and applicable to several technological subjects.

As a humanities-informed hands-on pedagogy finds its way in RRI education, some practical risks and recommendations must be taken into consideration. First and utmost, motivated instructors should be attentive to the fact that their students are about to embark on a challenging territory. A hands-on participation with potentially confusing and/or morally charged situations should by no means put the wellbeing of students at risk. Sufficient preparation and full disclosure on the voluntary nature of the activities are some of the ways to address this risk. Furthermore, inviting students to go beyond their comfort zone requires some good will from the participants and can only happen in a safe learning environment. To our experience, workload pressure (unrelated to our module) and stress from minor-wide related issues can be detrimental to student motivation. As it is often the case, a clear articulation of the course objectives and of the relevance of the course to RRI is essential.

On a similar tone, we insist that hands-on participatory exercises are properly contextualized. DIY biology activities can be easily reduced to science communication activities or engineering tasks while the idiom of art can be particularly confusing to students with minimum to no prior exposure to contemporary art. The experiences of students may remain unarticulated or, even worse, get dismissed if not properly contextualized, embedded, discussed and theorized. This need for contextualization is yet another reason why we believe that interactive exercises are not enough: hands-on participation requires a form of instruction that is both experiential and humanities-informed.

6 Conclusion

This paper contributed a pedagogic stance for RRI education that emphasizes a humanities-informed, hands-on participation with the complexity that defines RRI. We suggested that active and experiential learning that brings students in close contact with the material aspects of the issues at stake and exposes deep differences in stakeholder perspectives can be an engaging way to train the anticipatory and reflective capacities of future responsible innovators. We implemented such a hands-on participatory pedagogy using the tactics of DIY biology and bioart, two practices distinguished by a unique materiality and a potentially challenging moral edge. Our experiences with three subsequent editions of the module "Ethics, culture and biotechnology", minor Responsible Innovation, are particularly encouraging and motivating. Our observations have also enabled us to separate our original notion of hands-on participation into two potentially promising dimensions, namely the encounter with the material and the encounter with vastly different knowledge systems. It remains to be seen whether hands-on participation can be implemented also according to our original motivation, i.e. as a hands-on engagement with actual moral dilemmas. And it remains to be seen whether the experiences of our students can translate to personal changes as well as to changes in the institutional and professional settings of RRI practice. Still, we hope to have encouraged RRI educators to consider and implement their own versions of hands-on participation with the fundamental aspects and challenges of RRI.

Acknowledgements The authors wish to thank all students of the "Ethics, culture and biotechnology" module for their insights and participation. Also, our colleagues at the minor Responsible Innovation for realizing such an ambitious and innovative program.

References

1. R. Owen, J. Bessant, M. Heintz, *Responsible Innovation: Managing the Responsible Emergence of Science and Innovation in Society* (Wiley, Chichester, 2013)
2. X. Pavie, V. Scholten, D. Carthy, *Responsible Innovation Within Research and Education* (World Scientific, London, 2014)
3. R. Von Schomberg, A vision of responsible research and innovation, in *Responsible Innovation: Managing the Responsible Emergence of Science and Innovation in Society* (Wiley, Chichester, 2013), pp. 51–74
4. B.J. Koops, *The Concepts, Approaches, and Applications of Responsible Innovation* (Springer, New York, 2015), pp. 1–15
5. R. Owen, P. Macnaghten, J. Stilgoe, Responsible research and innovation: from science in society to science for society, with society. Sci. Public Policy **39**(6), 751–760 (2012)
6. H. Etzkowitz, *The Triple Helix: University-Industry-Government Innovation in Action* (Routledge, London, 2008)
7. C. Mitcham, E.E. Englehardt, Ethics across the curriculum: prospects for broader (and deeper) teaching and learning in research and engineering ethics. Sci. Eng. Ethics 1–28 (2016), https://link.springer.com/journal/11948/onlineFirst/page/5
8. H. Zandvoort, I. VanDePoel, M. Brumsen, Ethics in the engineering curricula: topics, trends and challenges for the future. Eur. J. Eng. Educ. **25**(4), 291–302 (2000)
9. E. Conlon, H. Zandvoort, Broadening ethics teaching in engineering: beyond the individualistic approach. Sci. Eng. Ethics **17**(2), 217–232 (2011)
10. A. Colby, W.M. Sullivan, Ethics teaching in undergraduate engineering education. J. Eng. Educ. **97**(3), 327 (2008)
11. J.R. Herkert, Ways of thinking about and teaching ethical problem solving: microethics and macroethics in engineering. Sci. Eng. Ethics **11**(3), 373–385 (2005)
12. G. Palsson, B. Szerszynski, S. Sörlin, J. Marks, B. Avril, C. Crumley, H. Hackmann, P. Holm, J. Ingram, A. Kirman, Reconceptualizing the 'anthropos' in the anthropocene: integrating the social sciences and humanities in global environmental change research. Environ. Sci. Pol. **28**, 3–13 (2013)
13. S. Sörlin, Environmental humanities: why should biologists interested in the environment take the humanities seriously? BioScience **62**(9), 788–789 (2012)
14. J. Stilgoe, R. Owen, P. Macnaghten, Developing a framework for responsible innovation. Res. Policy **42**(9), 1568–1580 (2013)
15. R.D. Hollander, C.R. Arenberg, National Academy of Engineering. Center for Engineering, Ethics, and Society, *Ethics Education in Science and Engineering* (National Academies Press, Washington, 2009)
16. A. Delgado, DIYbio: making things and making futures. Futures **48**(0), 65–73 (2013)
17. J. Keulartz, H. van den Belt, DIY-bio – economic, epistemological and ethical implications and ambivalences. Life Sci. Soc. Policy **12**(1), 1–19 (2016)
18. G. Seyfried, L. Pei, M. Schmidt, European do-it-yourself (DIY) biology: beyond the hope, hype and horror. Bioessays **36**(6), 548–551 (2014)
19. T. Landrain, M. Meyer, A.M. Perez, R. Sussan, Do-it-yourself biology: challenges and promises for an open science and technology movement. Syst. Synth. Biol. **7**(3), 115–126 (2013)
20. D. Grushkin, T. Kuiken, P. Millet, Seven myths and realities about do-it-yourself biology. Report, Woodrow Wilson International Center For Scholars (2013)

21. J. Hauser, Bioart-taxonomy of an etymological monster, in *Hybrid: Living in Paradox* (Hatje Cantz Verlag, 2005), pp. 182–187
22. A. Kallergi, Bioart on display: Challenges and opportunities of exhibiting bioart. http://kallergia.com/bioart/docs/kallergi_bioartOnDisplay.pdf (2008). Last accessed: 22 Aug. 2017
23. R. Zwijnenberg, Preface, in *Art, the Life Science, and the Humanities: In Search of a Relationship* (Springer, New York, 2009), pp. xiii–xxix
24. M. De Menezes, *The Laboratory as an Art Studio. The Aesthetics of Care*, pp. 53–58 (2002)
25. S. Wilson, *Information Arts: Intersections of Art, Science, and Technology* (MIT Press, London, 2002)
26. R.E. Mitchell, *Bioart and the Vitality of Media* (University of Washington Press, Seattle, 2015)
27. Leiden University, Responsible innovation, 2016–2017, e-prospectus, Leiden University. http://studiegids.leidenuniv.nl/en/studies/show/4903/responsible-innovation (2016). Last accessed: 22 Aug. 2017
28. Leiden University, Who owns life? Ethical, juridical, and artistic encounters with biotechnology, 2016–2017, e-prospectus, Leiden University. http://studiegids.leidenuniv.nl/en/courses/show/65703/who-owns-life-ethical-juridical-and-artistic-encounters-with-biotechnology (2016). Last accessed: 22 Aug. 2017
29. S.K.V.R. Edwards, J.H. Bauer, M. Olianas, P. Onali, High-velocity microprojectiles for delivering nucleic acids into living cells. Nature **327**, 7 (1987)
30. Bio-Rad Laboratories Inc., Helios gene gun systems, life science research, Bio-Rad. http://www.bio-rad.com/en-us/product/helios-gene-gun-system (2017). Last accessed: 22 Aug. 2017
31. A. Zaretsky, Mutafelch – methods of transgenesis: Genegun (biolistics). http://www.kapelica.org/index_en.html#event=993 (2016). Last accessed: 22 Aug. 2017
32. J. Zylinska, *Bioethics in the Age of New Media*, vol. 1 (MIT Press, Cambridge, 2009)
33. M. Groot, Nano supermarket, vaghurt en moedermelk. Article Food for Thought. Blikken op kunst en eten(15), 10–14 (2006)
34. A. Vandeleur-Boorer, Vaghurt. http://www.alicevandeleur-boorer.co.uk/vaghurt_2013-2015.html (2013–2015). Last accessed: 22 Aug. 2017
35. A. Vandeleur-Boorer, Alice and heath school. http://www.alicevandeleur-boorer.co.uk/alice_and_heath_school2014-ongoing.html (2014). Last accessed: 22 Aug. 2017

Smart Enhanced Context-Aware for Flipped Mobile Learning: SECA-FML

Fatima Ezzahraa Louhab, Ayoub Bahnasse, and Mohamed Talea

1 Introduction

Today the development of information and communication technologies has changed the way of teaching and learning [1]. The mobile technologies use has given learners the opportunity to test new learning methodologies. And especially the use of mobile devices that offer much more benefits compared to the fixed devices [2].

In parallel with this development, the learning methods have known a great change, starting from distance learning (D-learning) [3], E-learning, which is defined by Urdan and Weggen [4] as "the delivery of content via all electronic media, including the Internet, intranets, extranets, satellite broadcast, audio/video tape, interactive TV, and CD-ROM," and finally arriving at mobile learning (M-learning). Figure 1 illustrates the different forms of distance learning.

1.1 Mobile Learning

Mobile learning is a kind of e-learning training adapted to the mobile uses of learners. It has recently appeared and enables the delivery of remote training on other media than computers. Thus, thanks to the various applications developed, the learner can continue his training, wherever he is by his mobile device, whether it is a smartphone, a multimedia player or a mobile tablet. The technical capacities of

F. E. Louhab (✉) · A. Bahnasse · M. Talea
LTI Laboratory, Department of Physics, Faculty of Science Ben M'Sik, Hassan II University, Casablanca, Morocco
e-mail: louhabfatimaezzahraa@gmail.com

A. Reyes-Munoz et al. (eds.), *EAI International Conference on Technology, Innovation, Entrepreneurship and Education*, Lecture Notes in Electrical Engineering 532, https://doi.org/10.1007/978-3-030-02242-6_8

Fig. 1 Distance learning forms

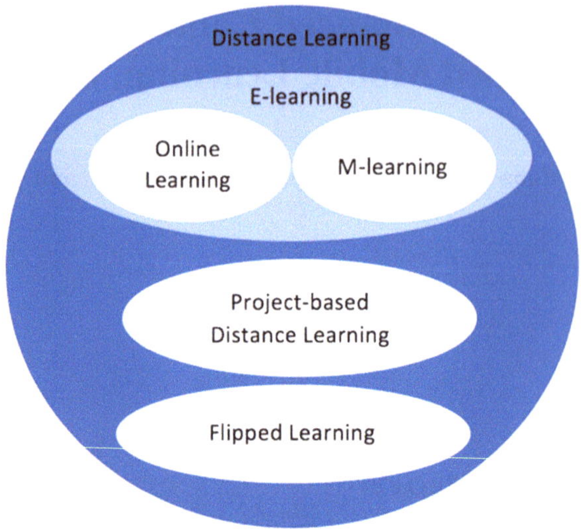

these supports make them a remote learning tool since they are generally able to read most of the media used in e-learning such as texts, videos, sounds, or images.

In the literature, m-learning is defined as a method that allows for acquiring knowledge and skills using mobile technologies anywhere and anytime throughout the day [5]. It can be distinguished "by rapid and continual changes of context, as the learner moves between locations and encounters localized resources, services and co-learners" [6]. Mellow [7] explains that m-learning is "a means to enhance the broader learning experience, it is not a primary method for delivering courses/distance learning." The main difference between e-learning and m-learning is that the first one takes place in front of a computer or in Internet labs, while the second takes place at any location [8].

1.2 Flipped Classroom

The Flipped Classroom (FC) or flipped learning is a learning strategy that reverses the roles of the teacher and the learner. Students are equipped with the learning content before class. According to Bergmann and Sams [9], the flipped classroom is a method created to provide online classes outside the classroom and move homework into the classroom, where students can discuss the subject in depth.

Figure 2 shows a comparison between the traditional classroom and the inverted or flipped classroom.

As the flipped classroom offers several advantages for learners and teachers, it also has disadvantages (Table 1):

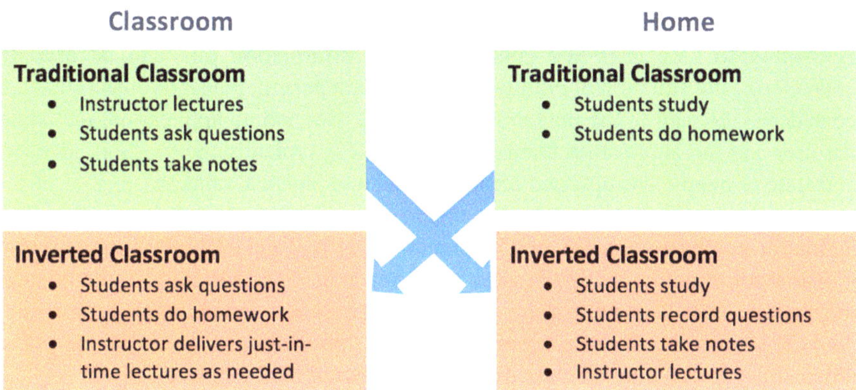

Fig. 2 The traditional classroom and the inverted classroom

Table 1 The advantages and disadvantages of flipped classroom

Learners		Teachers
Advantages	Learners learn at varying speeds Learners are provided opportunities for review Materials are ready and prepared for learners who are absent or sick Parents can view lessons and better assist learners Learners are actively working with their peers	Teachers focus on being the "Guide on the Side" not the "Sage on the Stage" Teachers spend more time supporting learners with practice Teachers are involved with learners learning rather than lecture Teachers spend less time on classroom management of learners' behaviors Teachers collaborate with peers in creating materials
Disadvantages	Learners without devices can feel left out of flipped mobile classroom Higher risk of inappropriate use by learners during school hours due to lack of IT control	More preparation time (Technology, designing tasks . . .) Needs motivated learners More need for monitoring and assessment

1.3 Contextual Learning

In the last few years, learning outside the classroom has become widespread. As a result, an increase in demand for adapted educational content is perceived [10]. We all agree that no single pedagogical strategy is the best for all learners. Learners will be able to achieve the learning objectives more effectively when learning procedures are adapted and personalized to their individual differences and characteristics [11].

This issue leads us to talk about a new aspect of learning that is obviously the contextual learning. This type of learning takes into account the learner context in order to allow him to access content adapted to his needs, characteristics, environment, etc.

When we talk about contextual learning, we have to mention the term "context." According to Dey [12], the context is "any information that can be used to characterize the situation of entities (i.e., whether a person, place, or object) that are considered relevant to the interaction between a user and an application, including the user and the application themselves. Context is typically the location, identity, and state of people, groups, and computational and physical objects."

According to the researchers, the three types of information to be collected to define the context are places, people and things [13]. To classify the context types, a distinction must be made between its dimensions. [12, 14] classify the context dimensions into two categories: external or extrinsic, and internal or intrinsic. As [15] groups the dimensions into two types: physical or logical. The physical dimensions describe the context that can be detected by the device sensors such as light, proximity, etc. While the logical dimensions represent the user context such as his objectives, his knowledge, etc.

Kofod-Petersen and Mikalsen [16] have identified the task context that captures the user activities and goals, the social context that describes the user relationships and roles, the personal context that encompasses the user mental and physical properties, the spatiotemporal context which represents concepts such as time and location and finally the environmental context that deals with the user environment.

Schilit et al. [12, 17] divide the context into three categories:

- *Computing context*: Network connectivity, communication costs, communication bandwidth, and nearby resources such as printers, displays, and workstations.
- *User context:* The user profile, location, people nearby, and social situation.
- *Physical context*: Lighting, noise levels, traffic conditions, and temperature.

Chen and Kotz [18] have added time as the fourth category of context. Schmidt et al. [19] have added the task category and have defined the following dimensions: User, user social environment, tasks, location, infrastructure, physical conditions, and time.

Zimmermann et al. [20] list individuality, activity, location, time and relationships as fundamental categories of context. Individuality is subdivided into four elements: Natural entity, human entity, artificial entity and group entity.

Based on these works, we present a taxonomy of the different context elements in Fig. 3.

In the flipped mobile learning context, there are certain limits that can influence the content quality delivered to the learner, such as the learner's learning style, its equipment and the environment in which the course is delivered. Most of the works carried out dealt with the context problem in the traditional mobile learning environment, this was an inspiration for us to adapt them in the flipped classroom context in which the mobile device used and the environment for consulting the course are unpredictable and uncontrollable.

Our aim, therefore, is to exploit the different context elements in order to give the learners a course content format adapted to their environments in a mobile situation

Fig. 3 The context elements

and in the context of flipped classroom. And this by presenting our approach called Smart Enhanced Context-Aware for Flipped Mobile Learning SECA-FML.

The rest of this paper is organized as follows: In the second section, we will present the various recent works that have been carried out in relation to the mobile learning, the flipped classroom, and the contextual learning. In the third section, we will situate our contribution in relation to the works already cited. Through the fourth section, we will describe our proposed approach while explaining how it works. Finally, we will conclude in the last section.

2 Related Works

In this section, we will present the works carried out in the mobile learning, the flipped classroom, and the contextual mobile learning.

Hung et al. [21] developed e-library activity worksheets that assisted the students concentrate on their outdoor ecology observation activities. The e-library delivered resources to clarify their observed descriptions, while the automatic scoring and feedback systems were helpful in supporting the students' persistent effort.

Shih et al. [22] carried out a study at the Peace Temple of southern Tainan with the inquiry-based mobile learning system. They used pre- and post-questionnaires along with observations and focus group interviews.

Reynolds et al. [23] developed a web-based museum tracks for university level students to access mobile devices in the Victoria and Albert Museum in London. The tracks were used in multiple ways to explore the museum environment and collections.

Sharples et al. [24] conducted an evaluation of MyArtSpace which is a combined mobile phone and web-based service to support learning between schools and museums. The study showed that MyArtSpace had a positive impact on school museum visits, and recognized areas for improvement in the both technical and educational aspects of the service.

Flipped classroom approach has become a popular pedagogy in many education institutes around the world. The basic notion of flipped classroom approach is to deliver the teacher's lectures before class through online resources, in order to free-up the in-class time for active learning and problem-solving activities [25].

The work of Bergmann and Sams [26] offers significant guidance when commencing a flipped classroom study. Nevertheless, it should be noted that the real strength of this pedagogical approach is not the instructional videos but the in-class time that is left to be redesigned and evaluated. This time allows many more opportunities for active, experiential learning to test higher-order cognitive skills.

Hamdan et al. [27] offer a comprehensive review of FC studies conducted in the USA, exploring the applications of the FC in the USA in compulsory schooling and higher education. The authors state that the FC features four pillars: a flexible learning environments that are conducive to active learning strategies, a shift in learning culture from a teacher-centered classroom to a student-centered one, an intentional design of content in order to allow students to develop as much as possible in a variety of skills and competencies, and professional educators who can cope with the demands of entirely active classrooms and mastery-based learning.

Lage et al. [28] discuss the FC as a method to allow lectures, experiential learning, collaborative learning and independent study to all occur simultaneously without having to increase staffing fourfold. When comparing the FC and the traditional classroom, Lage et al. explain that in order to fully realize the potential for staff-student interaction within the FC, a low cohort size may be required. The logistical problems caused by facilitating groups of any real size by either fully experiential methods or completely individualized experiences would potentially be significantly challenging. Additional space and resource requirements would need to be considered in order to allow large groups to work in a truly flipped manner.

Chen and Chao [29] propose an approach for developing a paper-based learning support environment, which the traditional books on mobile phones and the web discussion forum are integrated to improve the student's ability to acquire knowledge. The learners receive contextual messages based on their learning status.

Petersen and Markiewicz describe the PALLAS system [30]. This system offers a personalized language learning on mobile devices. It takes into account the dynamic and statistical parameters of context in order to provide learners with personalized resources. The contextualization is ensured by the use of the learner profile (age, skill, level, native language, interests, courses were taken, and user group) and his environment (location, time and day, mobile device used, and weather).

Ghadirli and Rastgarpour [31] present the adaptive and intelligent tutor by an expert system. This system promotes the learning abilities of learners by considering their learning styles, in order to recommend the appropriate content for their situation.

The research work of Tortorella [32] proposes an adaptation and personalization approach for mobile learning. This approach is based on a combination of the learner's learning style and its environment context, as location, movement, brightness, and proximity, in order to provide the learner with the most appropriate course content format (Text, Presentation, Audio, and Video).

3 Positioning of Our Contribution

The contributions made at the contextual mobile learning are related to context awareness in a traditional environment. However, several forms of learning or teaching have recently emerged, as already mentioned in the flipped classroom.

In the FC context, certain limits may influence the course or the quality of the content delivered to the learner, such as the learner's learning style, its equipment and the environment with which the course delivered is primarily concerned:

- *Learning style:* Constitutes a range of competing and contested theories based on a common concept that learners would differ in how to acquire their knowledge.
- *Mobile device*: The software components supported or installed on the learner's device.
- *Environment*: The situation of the learner at the time of learning and the conditions that surround him.

Most of the work carried out deals with the problem of context in the traditional mobile learning environment, this was a motivation for us to adapt them in the context of the flipped classroom in which the mobile device used and the environment for consulting the course are unpredictable and uncontrollable.

Our contribution called Smart Enhanced Context-Aware for Flipped Mobile learning is to take into account the different dimensions of the context (learner, equipment, and environment) in order to provide the learners with a course content format adapted to their needs within the context of flipped classroom.

4 Smart Enhanced Context-Aware for Flipped Mobile Learning SECA-FML

4.1 Approach Presentation

The proposed model is interposed between the mobile learning platforms and the different learning methods (Project-Based Learning, Flipped Classroom, ...). As part of our contribution, we focused on the flipped classroom or flipped learning method (Fig. 4).

The course content formats proposed (Text PDF, presentation PPT, video MP4, and audio MP3) require the capabilities of the mobile device (technical and software features) to be taken into account. Consequently, an improvement of the preceding approaches is proposed by adding another context element, namely the device context, in order to meet the limitations imposed by the latter on the display, connectivity, battery, etc.

In the context of the flipped classroom, the problem of the course content format is strongly posed, because not all learners have the same methods of acquiring knowledge. In addition, learners' device features may not support the same course content formats. Also, a specific format may not be adequate in a well-defined location (Audio in a conference room or text in a dark place).

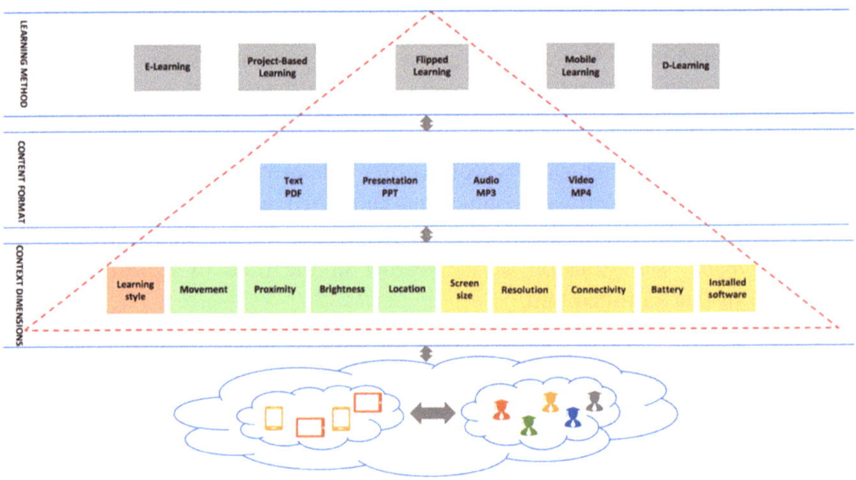

Fig. 4 SECA-FML architecture

4.2 Approach Architecture

Figure 4 shows the architecture of our approach called Smart Enhanced Context Aware for Flipped Mobile learning SECA-FML.

Our approach is based on three layers: Learning methods, content formats, and context dimensions.

1. *Learning methods:* This layer is made up of the different learning methods, namely: E-learning, Project-Based Learning, Flipped Learning, Mobile Learning, D-Learning, etc. Our approach is therefore concerned by the method of Flipped Learning or Flipped Classroom.
2. *Content formats:* In this layer, we find the different course content formats available for learners according to their contexts: Text PDF, presentation PPT, video MP4, and audio MP3.
3. *Context dimensions:* This layer contains the different context dimensions supported by our approach. Each dimension is assigned an appropriate score. After the combination of the different scores, the format with the highest score will deliver to the learner:

    (a) *Learning style*: The learning style of the learner is based on the Index of Learning Style (ILS) developed by Felder and Silverman [33]. This index classifies the learner's learning styles on four dimensions: Active/Reflective, Sensing/Intuitive, Visual/Verbal, Sequential/Global. For each style, we assigned a score that matches the result of the ILS questionnaire passed by the learner.

    (b) *Movement*: Detects the movements of the learner during the learning process. It is measured using the mobile device accelerometer. For example, if the device has detected that the learner is running, in this case, the use of audio format is favored.

    (c) *Proximity*: In this dimension, the proximity sensor is used to detect the proximity of the device with the user. If, for example, the device is held close to the face, the audio format would be favored.

    (d) *Brightness*: The ambient light sensor detects the amount of ambient light in which the mobile device is located. For example, with a very bright or dark environment, the audio format is preferred.

    (e) *Location*: The GPS sensor is able to read the position of the device in terms of its geographical location and returns the longitudinal and latitudinal coordinates. The learner can define by default a specific format for a specific place (Text format in a library).

    (f) *Screen size*: Calculated based on the mobile device diagonal. The use of a mobile device with a small size disadvantages the printed formats.

    (g) *Resolution*: The screen resolution is the number of pixels displayed horizontally and vertically on the screen. It impacts the quality of the visible content display.

(h) *Battery*: Represents the value of the battery charge. Detecting a low battery disadvantages the use of formats that consume more energy (Video).
(i) *Connectivity*: In this parameter, there are two types of Internet connection: limited or unlimited. Detecting an unlimited Internet connection disadvantages the use of formats that consume much more Internet connection.
(j) *Installed Software:* This element is responsible for verifying the software installed on the mobile device, which is required to read the proposed formats. After calculating the final score for each format, we verify the existence of the software responsible for reading this format. If the software not installed, we choose the format of the score ranked the second and so on.

4.3 Approach Functioning

The proposed approach has been implemented in the format of a mobile Android application. It consists of two spaces: teaching space and learning space.

Figure 5 illustrates the use case of the application for the learner and the teacher.

1. *The learner*: To consult the list of courses available on the application, the learner must create an account so that it can authenticate. The creation of an account is done in two steps: The first is the part of the personal information (Name, first name, level, . . .) and the second part corresponds to the phase of the response to ILS questionnaire. Learner information is stored in his space and can be changed at any time. Once the account is created, the learner can select a lesson, each lesson is divided into chapters. The list of lessons is not the same for all learners, it varies according to the level of each learner. By clicking on a lesson's chapter, the learner will have this chapter in the format most appropriate to its context.
2. *The teacher*: Through its space, the teacher can manage learners and courses by removing, modifying, and adding each course in four formats (PDF, PPT, Audio, and Video). The teacher also has the opportunity to manage the various tests associated with the lessons.

5 Conclusion and Perspectives

The concept of context awareness in mobile learning has become increasingly critical view that learners learn in a mobility context. However, and with the arrival of the flipped classroom, this notion has not yet taken its place in this new method of learning.

Through this paper we have proposed the Smart Enhanced Context-Aware for Flipped Mobile Learning approach SECA-FML. This solution allows learners to have a course content format adapted to their situations in the context of flipped classroom.

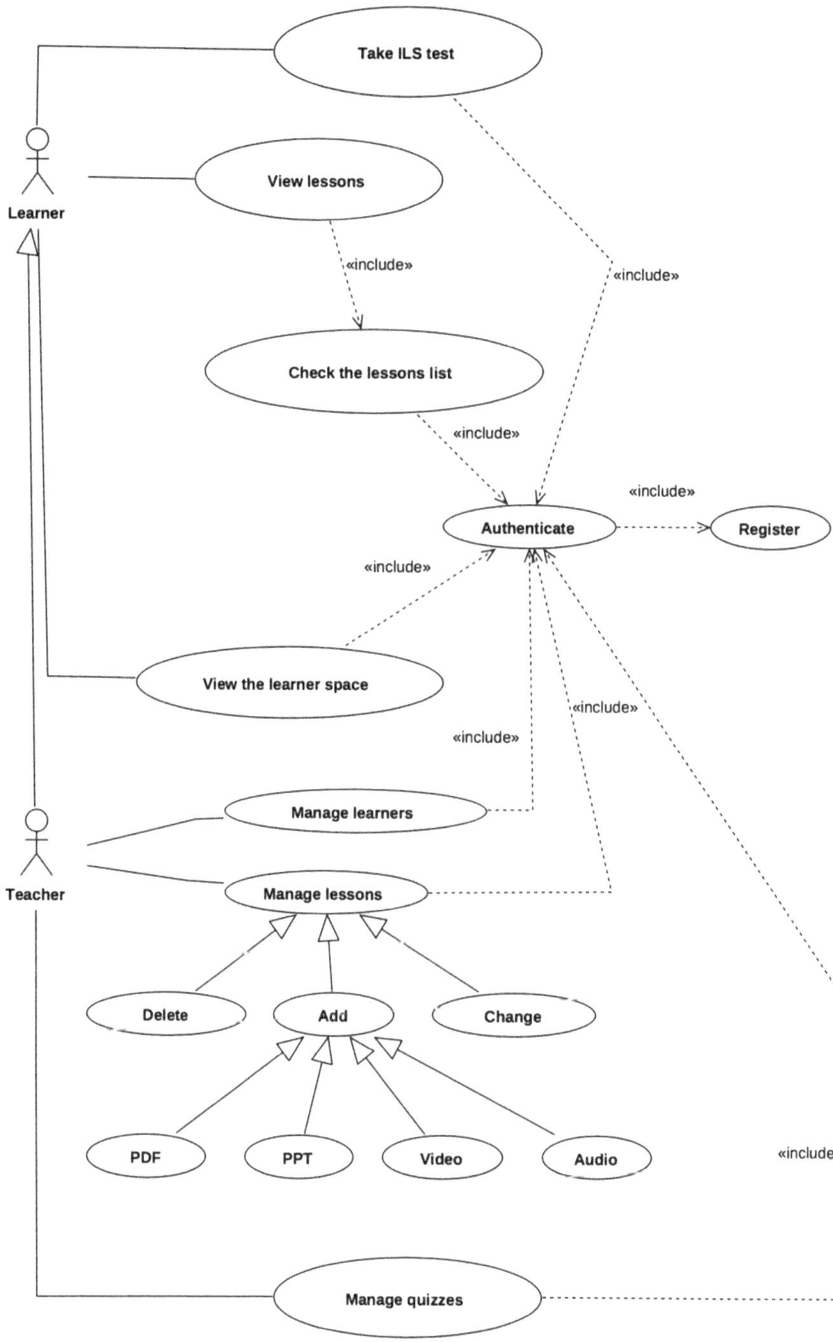

Fig. 5 SECA-FML use case diagram

To validate our model, we have developed an Android mobile application which is in test phase and the results will be the subject of our next paper.

References

1. P. Bentsen, J. Schipperijn, F.S. Jensen, Green space as classroom: Outdoor school teachers' use, preferences and ecostrategies. Landsc. Res. **38**, 561–575 (2013)
2. F. Ozdamli, H. Uzunboylu, M-learning adequacy and perceptions of students and teachers in secondary schools. Br. J. Educ. Technol. **46**, 159–172 (2015)
3. T.H. Brown, *The Role of m-Learning in the Future of e-Learning in Africa*. 21st ICDE World Conference (Hong Kong, 2003), pp. 122–137
4. C.C. Weggen, T.A. Urdan, *Corporate E-learning: Exploring A New Frontier* (WRHAMBRECHT CO Equity Res, San Francisco, 2000)
5. Y. Liu, H. Li, C. Carlsson, Factors driving the adoption of m-learning: An empirical study. Comput. Educ. **55**, 1211–1219 (2010)
6. T. Chan, M. Sharples, G. Vavoula, P. Lonsdale, *Educational Metadata for Mobile Learning*. 2nd IEEE International Workshop on Wireless and Mobile Technologies in Education IEEE Press (Taiwan, 2004), pp. 197–198
7. P. Mellow, The media generation: Maximise learning by getting mobile. Ascilite. **1**, 469–476 (2005)
8. S.K. Sharma, F.L. Kitchens, Web services architecture for m-learning. Electron. J. E-Learn. **2**, 203–216 (2004)
9. J. Bergmann, A. Sams, How the flipped classroom is radically transforming learning. Dly. Riff. **4**, 1–3 (2012)
10. I.J. Chen, C.C. Chang, J.C. Yen, Effects of presentation mode on mobile language learning: A performance efficiency perspective. Australas. J. Educ. Technol. **28**, 122–137 (2012)
11. C.K. Chiou, J.C. Tseng, G.J. Hwang, S. Heller, An adaptive navigation support system for conducting context-aware ubiquitous learning in museums. Comput. Educ. **55**, 834–845 (2010)
12. A.K. Dey, G.D. Abowd, D. Salber, A conceptual framework and a toolkit for supporting the rapid prototyping of context-aware applications. Hum. Comput. Interact. **16**, 97–166 (2001)
13. H. Byun, K. Cheverst, Utilizing context history to provide dynamic adaptations. Appl. Artif. Intell. **18**, 533–548 (2004)
14. P. Prekop, M. Burnett, Activities, context and ubiquitous computing. Comput. Commun. **26**, 1168–1176 (2003)
15. R.M. Gustavsen, *Condor–An Application Framework for Mobility-Based Context-Aware Applications*. The Workshop on Concepts and Models for Ubiquitous Computing (2002)
16. A. Kofod-Petersen, M. Mikalsen, Context: representation and reasoning. Representing and reasoning about context in a mobile environment. Revue d'Intelligence Artificielle **19**, 479–498 (2005)
17. B. Schilit, N. Adams, R. Want, *Context-Aware Computing Applications*. The First Workshop on Mobile Computing Systems and Applications (California, 1994), pp. 85–90
18. G. Chen, D. Kotz, *A Survey of Context-Aware Mobile Computing Research*. Technical Report TR2000-381 (Department of Computer Science, Dartmouth College, 2000)
19. A. Schmidt, M. Beigl, H.W. Gellersen, There is more to context than location. Comput. Graph. **23**, 893–901 (1999)
20. A. Zimmermann, A. Lorenz, R. Oppermann, *An Operational Definition of Context*. International and Interdisciplinary Conference on Modeling and Using Context (Denmark, 2007), pp. 558–571

21. P.H. Hung, Y.F. Lin, G.J. Hwang, Formative assessment design for PDA integrated ecology observation. Educ. Technol. Soc. **13**, 33–42 (2010)
22. J.L. Shih, C.W. Chuang, G.J. Hwang, An inquiry-based mobile learning approach to enhancing social science learning effectiveness. J. Educ. Technol. Soc. **13**, 50–62 (2010)
23. R. Reynolds, K. Walker, C. Speight, Web-based museum trails on PDAs for university-level design students: Design and evaluation. Comput. Educ. **55**, 994–1003 (2010)
24. M. Sharples, P.J. Meek, P. Rudman, G.N. Vavoula, *An Evaluation of MyArtSpace: A Mobile Learning Service for School Museum Trips*. 6th Annual Conference on Mobile Learning (Melbourne, 2007), pp. 238–244
25. C.K. Lo, K.F. Hew, A critical review of flipped classroom challenges in K^{-12} education: possible solutions and recommendations for future research. Res. Pract. Technol. Enhanc. Learn. **12**, 4–26 (2017)
26. J. Bergmann, A. Sams, *Flip Your Classroom: Reach Every Student in Every Class Every Day* (International Society for Technology in Education, 2012)
27. N. Hamdan, P. McKnight, K. McKnight, K.M. Arfstrom, A white Paper Based on the literature review titled: A review of flipped learning. Flip. Learn. Netw. 1–15 (2013)
28. M.J. Lage, G.J. Platt, M. Treglia, Inverting the classroom: A gateway to creating an inclusive learning environment. J. Econ. Educ. **31**, 30–43 (2000)
29. G.D. Chen, P.Y. Chao, Augmenting traditional books with context-aware learning supports from online learning communities. Educ. Technol. Soc. **11**, 27–40 (2008)
30. S.A. Petersen, J.K. Markiewicz, *PALLAS: Personalised Language Learning on Mobile Devices*. The Fifth IEEE International Conference on Wireless, Mobile, and Ubiquitous Technology in Education (IEEE Press, Beijing, 2008), pp. 52–59
31. H.M. Ghadirli, M. Rastgarpour, An adaptive and intelligent tutor by expert systems for mobile devices. ArXiv Prepr. ArXiv13044619 (2013)
32. R.A. Tortorella, S. Graf, Considering learning styles and context-awareness for mobile adaptive learning. Educ. Inf. Technol. **22**, 297–315 (2017)
33. R.M. Felder, L.K. Silverman, Learning and teaching styles in engineering education. Engl. Educ. **78**, 674–681 (1988)

Designing Interactive Tools for Learning in the Digital Age

Andre Campos, Mário Rodrigues, Marlene Amorim, and Alberto Signoretti

1 Introduction

A growing number of voices are calling for the renewal in the competences and skills of individuals, as well as for deep changes in the required learning paradigms and tools that are needed for the competitiveness and sustainability of modern economies. Learning in today's knowledge-based society will become a lifelong activity, spanning the spheres of private and public life and work. Traditional formal learning will need to be increasingly blended with other forms of education and informal learning.

Digital competences have been put forward as key for personal development, active citizenship, employment, and inclusion [1–3]. This will be not only about developing basic literacy on information and communication technologies but rather about enabling individuals to achieve higher-order skills to live and work

A. Campos (✉)
DIMAp - Computer Science and Applied Mathematics Department, Federal University of Rio Grande do Norte – UFRN, Natal, Brazil
e-mail: andre@dimap.ufrn.br

M. Rodrigues
ESTGA/IEETA, University of Aveiro – UA, Aveiro, Portugal
e-mail: mjfr@ua.pt

M. Amorim
GOVCOP/DEGEIT, University of Aveiro – UA, Aveiro, Portugal
e-mail: mamorim@ua.pt

A. Signoretti
DI - Computer Science Department, State University of Rio Grande do Norte – UERN, Natal, Brazil
e-mail: albertosignoretti@uern.br

A. Reyes-Munoz et al. (eds.), *EAI International Conference on Technology, Innovation, Entrepreneurship and Education*, Lecture Notes in Electrical Engineering 532, https://doi.org/10.1007/978-3-030-02242-6_9

in a digitalized and networked society. This paper describes a tool that has been developed for enabling the learning experiences for acquiring competences on coding in an engaged, flexible and creative manner—piBook. The paper offers a brief overview of the future of work and competences as well as of the associated drivers of change that are shaping the future of societies as a conceptual background to frame the motivation and purpose of piBook. The tool is described together with a short discussion about its applications and contributions it can provide for the emerging learning society.

2 The Future Demand for Competences

Recent debates have been calling for a fundamental transformation of education and training approaches to meet the demand for new skills and competences that are required for competitiveness and sustainability in modern economies [2–4]. Among the various drivers of change that have been put forward, a core importance is attributed to the generalized adoption of digital technologies, supported by an unprecedented evolution in computational capabilities, that are enabling the emergence of smart machines, consumers and systems [5]. These drivers will have profound impacts in the economy and in the organization of business and relationships, creating new jobs and professions, and challenging the ability of organizations, and individuals to keep the pace for update their skills and competences on a continuous manner. According to the recent study of the World Economic Forum [3] "in many industries and countries, the most in-demand occupations or specialties did not exist 10 or even 5 years ago, and the pace of change is set to accelerate." In this context, the reskilling of the current workforce will also be a key domain of effort and investment of companies and core concern of public policies.

2.1 Drivers of Change and Evolution in Competence Demand

The task of fore sighting what competences will be relevant for the future of work, and for the development of societies and economies as a whole, cannot be done without a preliminary analysis of the determinants of change.

Among the trends that are shaping the future of work, and demanding for new competences and learning approaches, three drivers are particularly prominent: the digitization of economies; the increasing obsolescence of knowledge; and the growing volume of available information. The generalized adoption of information and communication technologies, and the development of computational capacity are paving the way for the emergence of smart consumption and production patterns that will call for rapid product development cycles and flexible production capabilities. Such change in the production and consumption landscape that is commonly labelled as a new industrial revolution, also known as industry 4.0, is extensively supported by the newly enabled possibilities for interaction and communication between humans and technology (also referred to as cyber-physical-systems). In this approach, the production systems will be able to deliver personalized mass products, by relying on flexible processes, modularized product design and requiring efficient integration between supply chain players [6]. Such manufacturing systems have reconfiguration properties which allow for the definition of case-specific outputs enabling companies to adapt to changing production requirements in a cost-efficient way. On an employee level, the upcoming Industry 4.0 reality will call for workers that are able to increasingly address creative, innovative and communicative activities, as routine activities [7]. The upcoming occupational landscape will move towards knowledge and skills intensive jobs [8]. In other words "we will be entering into a new kind of partnership with machines (...) resulting in a new level of human-machine collaboration and codependence" [1]. Overall, the growth in communication, computation, and connectivity are being extensively acknowledged as the dominant determinants of change. As a result from the inclusion of technology with connection and processing capabilities into everyday objects and contexts, we are witnessing an unprecedented potential for data creation, extraction, and sense making. The emergence of a world that is "computational and programmable" will be a natural consequence of that. Likewise, the future of work, consumption, and management of individual affairs (health, finance, etc.) will become intrinsically dependent on our abilities for using such data for decision making [1]. Altogether, these changes are creating urging needs for the qualification and reskilling of individuals.

Another dimension of change that cannot be disregarded of this analysis has to do with demographics. Recent data on European demographic trends show that in the near future the older segment of the workforce and the elderly will be a major proportion of the population. The forecasts are for the numbers of those aged over 65 in relation to those aged 15–64 will increase from 26% in 2008 to 38% by 2030 [2]. These changes will have major impacts on the characteristics of the people and the skills that will be available. The number of people graduating from universities is expected to decline, and the working age population is actually on a downward trend (the prospects are for a reduction in the order of 6 million, in the period between 2007 and 2020). This scenario is leading to pressures for increases in productivity and in labor market participation, notably for women as well as for the reintegration/requalification of unemployed and immigrant populations.

Overall, the future of work will demand for new competences, while also calling for serious changes in what concerns the policies and practices of continuous education and reskilling of individuals, notably digital and computational competences, and communication skills and adaptability to changes in work organization. The future of Europe is foreseen as far from economic models based on cost and price competition, but rather on the ability to qualify people for the production of higher quality products and services.

2.2 Framing the Idea of Competences and Skills

According to the WEF [3], in less than 5 years from now, over one third of what is considered important in today's workforce will have changed. The adoption of Industry 4.0 principles will call for professionals who are technology literate, able to deal with complex problems in a creative manner while holding the ability to judge, manage, and coordinate with other people. Many of the foresighted competences fall beyond the scope of existing education and training offers, and demand quick responses from universities and policy makers.

A professional profile can be defined as a set of competences that are often proposed to be organized along three dimensions: knowledge, abilities, and attitudes. Knowledge is related to the information that can be acquired and structured by individuals. Abilities are defined as the capability of making productive use of the acquired knowledge, and this can be improved by means of training and experience. Attitude refers to social and affective aspects related to work, resulting in the predisposition towards the adoption of a specific action. The literature offers a variety of definitions for the term "competence." In broad terms, a competence can be addressed as the ability to identify, select, and combine a set of resources (materials, knowledge, etc.) to perform a task, solve a problem, or conduct a project [9]. The idea of competence centered in the professional's reaction towards complex and challenging situations from work. Professionals need to develop competences to take on responsibilities for addressing new and unexpected situations with adequate performance, that is, professionals are expected to mobilize the knowledge to deliver results that meet the required deadlines and quality specifications [10]. In this vein the idea of competence cannot be reduced to that of knowledge, it rather needs to be represented in a context where competences are materialized [11].

2.3 A Prospective Look at Competence and Skills Demand

Several recent reports have been highlighting that the expected core competences and skills for future occupations and citizenship will be quite different from the ones that individuals hold today. Moreover, the same studies also question the ability of the currently existing education and training systems to provide adequate approaches to prepare individuals for such a future [2, 3].

Among the advanced priorities, the future demand for skills related to complex problem solving and systems skills (including systems analysis and judgement and decision making) both stand among the ones with more projected demand across business sectors [3]. Content skills (involving ICT literacy and active learning), cognitive abilities (involving creativity and mathematical reasoning) and process skills (such as active listening and critical thinking) are expected to become part of the key skills requirements across different industries. The generalization of internet technologies and systems connectivity in general will result in a generalized growth in the level technology literacy even in occupational domains like sales, installation, or maintenance.

The research and prospective work expressed in various reports of the European Commission reports point out the importance of fundamental skills such as communication, digital skills, along with aspects related to learning to learn, social and civic responsibility, entrepreneurship, and creativity [4].

In this scenario, coding is acknowledged as a key future skill, and praised for its potential for contributing to the development of many of the aforementioned skills. Individuals who learn to code are believed to benefit from the enhancement of creativity levels, as well as to develop capabilities for working together across physical and geographical boundaries and to communicate in a universal language [12]. According to the recent EU report "A concept paper on digitisation, employability and inclusiveness [13]" "individuals in the workplace will need to engage more comprehensively with machines as part of their everyday activities (. . .) will need to acquire new skills that will be in demand in the new automation age (. . .) the ability to work within the digital environment, and the capacity to go beyond the logical reasoning at which computers excel (. . .) digital skills and complementary skills such as creativity, high-level cognitive and interpersonal skills will be needed". A growing number of professions will require computing notions, something that will be useful in the management of many aspects of individuals' affairs. In other words, we are observing a shift from a general need for knowing how to use technology to the need for understanding the elementary principles of computing based on the belief that it enables the development of skills involved in resolving a wide range of problems [12]. These beliefs, along with a projects shortage in ICT-skilled workers at putting at the forefront of the agendas the need to include coding in school curricula (in some countries starting at elementary schools levels) and leading to a multiplication of efforts to develop coding skills across individuals of all ages and backgrounds. More importantly, coding or computing are increasingly being integrated into different education subjects privileging a cross-disciplinary and cross curricular approach [12]. In this context a top concern is related to the design of effective learning approaches for developing coding skills, that is, figuring out what activities and pedagogical approaches will be more appropriate to reach different individuals and meet their heterogeneous backgrounds, motivations, ages, interest, and capacities.

3 Learning and Teaching in the Digital Age

The prevalent orientations for future qualification and development of individuals are shaped by the ubiquity of information and communication technologies and the need to deploy lifelong learning strategies. Digital technologies will accelerate the reconfiguration of production and consumption systems, leading to an acceleration in the renewal of jobs. In this context individuals will experience frequent professional changes throughout their lives while extending their permanence in the workforce due to the reported demographic changes. The requalification of individuals and the need for continuous skills updating will be a dominant trait of the future economies.

3.1 Updating Learning Contexts to Meet Digitization Demands

The developments in digital technologies will increasingly challenge the contexts of learning and placing multiple calls for updates in curriculum and in pedagogical practices. Despite these calls, education is still ranked as a weak technology-intensive domain [14]. The demands for improving the infusion of technology in learning contexts is also aligned with the concerns for promoting more rapid and more seamless transitions from school to work, and for facilitating the reentrance paths into active labor, given that the working contexts will be marked by the ubiquity of such technologies. The vision for future guiding principles for organizing learning and teaching contexts includes the need to involve technology contact along with personalization, collaboration, and informalization. Generic and transversal skills will become increasingly important as they are understood as a key for helping citizens to become lifelong learners, who can engage in the development of their competences in a proactive manner.

Innovative learning environments are therefore likely to be fostered, in the sense that they fully embed the potential of ICT to change, innovate and modernize learning and teaching practices [15]. The desired innovative approaches in learning refer to the possibility of enabling individuals for developing their thinking and skills and learn in new, flexible and creative manners [16].

3.2 Blending Technology and Learning

The demands for new competences and skills that are being fostered by our evolution towards a knowledge-based society, thoroughly infused with information and communication technologies is followed by a call for change in the learning approaches. There is growing awareness for the need to develop an innovative blend between information and communication technologies and learning in order to meet the new forms of work and the new ways individuals are making sense of their (digitalized) world. Technology can assume different roles in learning contexts. Several benefits have been associated to the employment of computer technologies in learning strategies [17] including the ability to allow for greater flexibility for participants to adjust their own pace. Likewise, the use of software in learning contexts has proved to be useful for leveraging pedagogical approaches [18, 19].

Also important is the fact that technologies have the potential to make learning more pleasant and therefore more attractive (personalized, and media-rich virtual environments, etc.) and more prone to generate emotional engagement (through fun and connectivity, etc.).

4 A Flexible Tool for Blending Learning, Creativity, and Coding

Technology holds the potential to facilitate the acquisition of new competencies by allowing a controlled introduction of new challenges, smoothing this way the learning curves, and removing barriers. However, technology by itself is not a guarantee that people will remain engaged in the learning process. Technology actually prevents some groups to learn or to feel compelled to engage the learning process. This makes important to design tools that introduce the required concepts in a non-disruptive way, while coping with the multitude of topics relevant to the potential users. The approach followed here was to build a story-telling tool—everyone can create its favorite story—while maintaining advanced features for users to slowly experiment and improve their own stories while experimenting coding, and making them reactive and interactive.

4.1 Describing the piBook Tool

piBook is an authoring application to produce programmable interactive books, piBooks [20]. A key feature in piBook is to allow the production of content using ICT skills comparable to the creation of contents using a content manager like WordPress or blog and, at the same time, seamlessly being very flexible by allowing for the usage of tools that involve advanced ICT skills such as computer programming tools. By letting content producers to choose how they want to use it allows the introduction of those concepts at each person rhythm, and thus being more appropriate for wider audiences to acquire such advanced ICT skills. Having the content producers working on topics of their own choice while new concepts are introduced helps to foster a productive and goal oriented environment [21].

4.2 Understanding piBook Operation

The piBook application is composed of several elements that are equivalent to physical book elements, sections, and pages, or metaphors of online content managers such as html editors and gadgets. piBooks are composed of sequences of

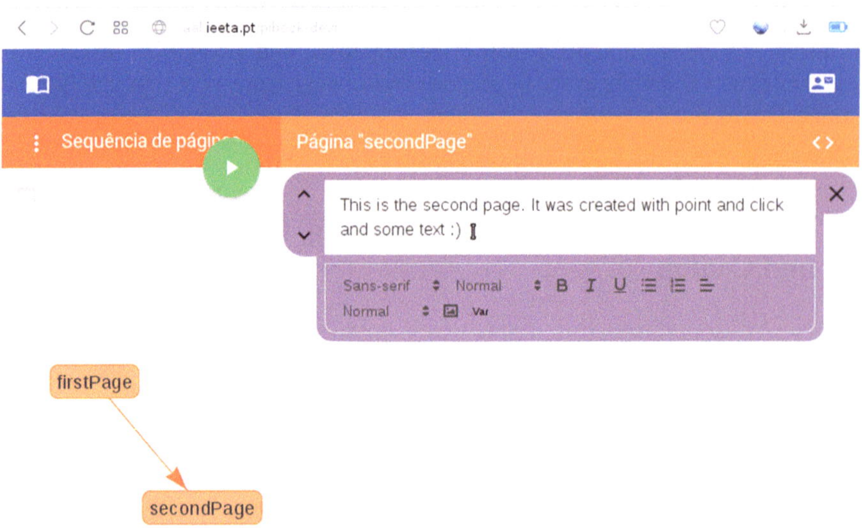

Fig. 1 Producing text content. On the left-hand side is visible a graph that represents the sequence of pages, and on the right-hand side the rich text editor for the pages with some text already produced

linked pages that contain other elements. The book flow is defined by the sequence of links between pages. If the default behavior is used, every new page is placed after the last page available. Figure 1 presents the interface with two pages already created and with the rich text editor to produce content.

The tool offers three page templates: (1) blank page; (2) page with multiple choice question; and (3) page with question and open answer. The blank page template creates pages as the one rendered in Fig. 1. It is the most suitable for authors that just want to produce content as it does not require, but allows, the usage of advanced features as redefining the page sequence graph or custom code. It is also the most suitable template for people getting familiarized with the tool.

The second page template, page with multiple choice question, is suitable for the creation of content such as quizzes. It has a variable that can assume a value from a set of values, and that can be used to define the next transition. Every value can lead to a distinct transition. Figure 2 presents a multiple-choice page where, in the left-hand side, is visible that the multiple-choice node of the graph has three distinct transition possibilities. In this example, each variable value is directly associated with a distinct transition.

The third page template, the page with a question and an open answer, is a page that renders a question to the reader and asked for some text input. The input text will be bound to a variable and can be used programmatically.

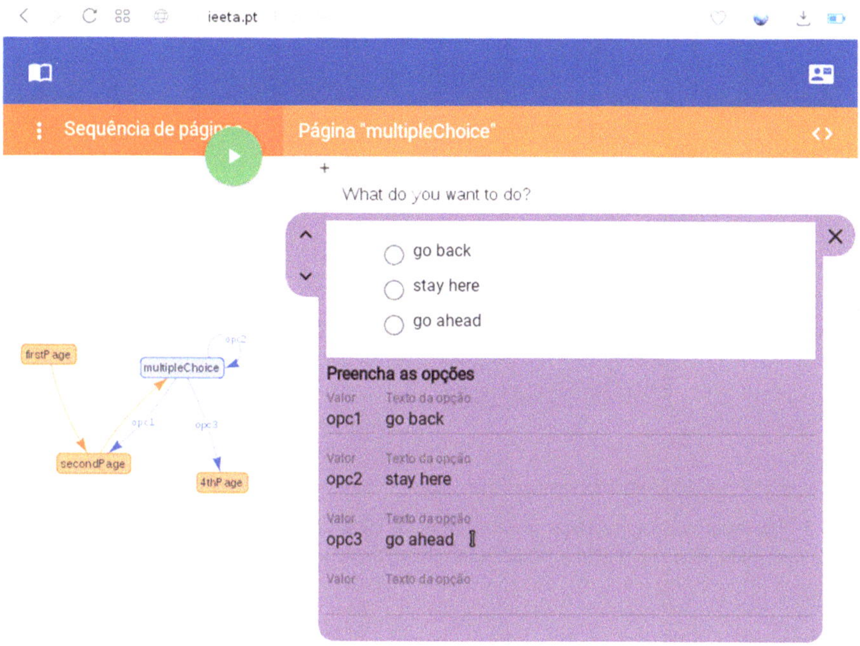

Fig. 2 A multiple-choice page. Every option can lead to a distinct transition. In the left-hand side is visible that the purple node of the graph has three distinct transition possibilities which, in this case, are directly associated with the option that the reader will select

Independently of the page template selected, the content producer is free to manipulate the graph with the mouse cursor and to use the page programming mode using the Blockly interface or Javascript code (see Fig. 3). Five types of blocks are available: page events, page transitions, logical, values, and variables. Without entering in too much detail, the blocks available allow to control every feature of the books.

A book can be created with any possible mix of page templates, graph definitions, and custom code, including not using any of those features by just using the blank pages template.

4.3 Innovativeness and Contributions of the piBook Tool

This tool has the potential to reach audiences that are not the typical ICT profiles. Using storytelling has the potential of engaging people that prefer other areas of knowledge such as humanities. Its interface is suitable to be used on tablets and the colorful environment helps capture the attention of younger public. Also, its

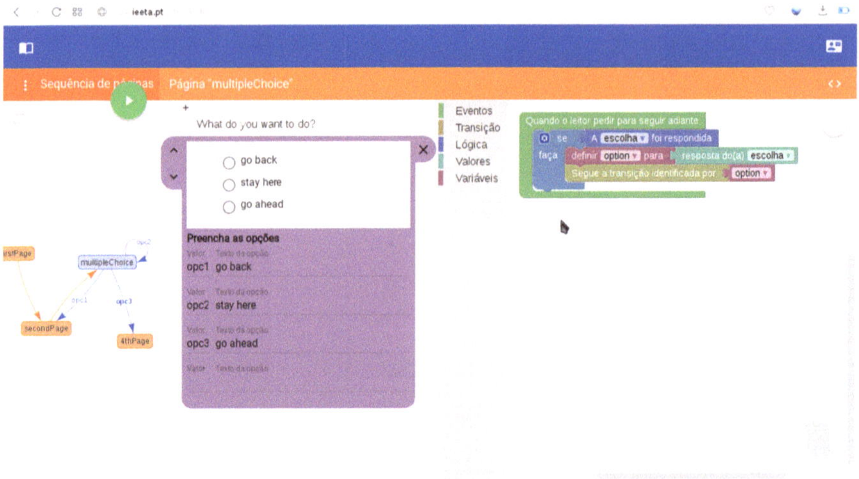

Fig. 3 The programming interface. It is visible in the right-hand side a set of blocks that represent the source code. Is also possible to select a Javascript mode instead of a Blocky interface

programming interface is, by default, based on Blockly which has a good track record when used by children [22].

The availability of this and of other tools of this type can provide a good contribution for introducing key and advanced ICT concepts, as well as logical reasoning, to vaster audiences thus helping prepare our societies for this new digital era [23].

References

1. A. Davies, D. Fidler, M. Gorbis, *Future Work Skills 2020*, in Institute for the Future for the Future (The University of Phoenix Research Institute, Palo Alto, 2011)
2. European Commission, *New Skills for New Jobs: Action Now*. A report by the Expert Group on New Skills for New Jobs (European Commission, Luxembourg, 2010)
3. World Economic Forum, *The Future of Jobs* (World Economic Forum, Geneva, 2016)
4. European Commission, Recommendation of the European Parliament and of the Council of 18 December 2016 on key competences for lifelong learning. Off. J. Eur. Union, 10–18 (2006)
5. W. Bauer, M. Hämmerle, S. Schlund, C. Vocke, Transforming to a hyper-connected society and economy–towards an "Industry 4.0". Procedia Manufactur. **3**(2015), 417–424 (2015). https://core.ac.uk/download/pdf/82257745.pdf
6. G. Schuh, C. Reuter, A. Hauptvogel, C. Dölle, Hypotheses for a Theory of Production in the Context of Industrie 4.0, in *Advances in Production Technology*, (Springer, New York, 2015), pp. 11–23
7. H. Kang, J.Y. Lee, S. Choi, H. Kim, J.H. Park, J.Y. Son, S. Do Noh, Smart manufacturing: Past research, present findings, and future directions. Int. J. Precision Eng. Manufact.-Green Technol. **3**(1), 111–128 (2016)

8. C. Redecker, M. Leis, M. Leendertse, Y. Punie, G. Gijsbers, P. Kirschner, S. Stoyanov, B. Hoogveld, *The Future of Learning: Preparing for Change* (Publications Office of the European Union, Luxembourg, 2011)
9. A. Stoof, R. Martens, J.J.G. van Merriënboer, The boundary approach of competence: A constructivist aid for understanding and using the concept of competence. Hum. Resour. Dev. Rev. **1**, 345–365 (2002)
10. T. Hoffmann, The meanings of competency. J. Eur. Ind. Train. **23**(6), 275–286 (1999)
11. D. Mesquita, R.M. Lima, M.A. Flores, C. Marinho-Araújo, M. Rabelo, Industrial engineering and management curriculum profile: Developing a framework of competences. Int. J. Ind. Eng. Manage. **6**(3), 121–131 (2015)
12. A. Balanskat, K. Engelhardt, *Computing Our Future: Computer Programming and Coding* (European Schoolnet, Brussels, 2015)
13. European Commission, Directorate-General for Communication Networks, Content and Technology, *A Concept Paper on Digitisation, Employability and Inclusiveness* (The Publications Office of the European Union, 2017)
14. D.L. Lowther, F.A. Inan, J. Daniel Strahl, Does technology integration "work" when key barriers are removed? Edu. Media Int. **45**(3), 195–213 (2008)
15. European Commission - Joint Research Centre Institute for Prospective Technological Studies, *Innovating Learning: Key Elements for Developing Creative Classrooms in Europe* (Joint Research Centre of the European Commission, Seville, 2012)
16. R. Cachia, A. Ferrari, K. Ala-Mutka, Y. Punie, *Creative Learning and Innovative Teaching* (Joint Research Center -Institute for Prospective Technological Studies, Seville, 2010)
17. R. Wegerif, The role of educational software as a support for teaching and learning conversations. Compu. Edu. **43**(2), 179–191 (2004)
18. P. Chera, C. Wood, Animated multimedia 'talking books' can promote phonological awareness in children beginning to read. Learn. Instr. **13**, 33–52 (2003)
19. J.P. Akpan, T. Andre, Using a computer simulation before dissection to help students learn anatomy. J. Comput. Math. Sci. Teach. **19**, 297–313 (2000)
20. A. Campos, A. Signoretti, M. Rodrigues, *An Interactive Book Authoring Tool to Introduce Programming Logic in Schools*. 9th International Conference on Computer Supported Education (Porto, 2017)
21. B. Chan, D. Churchill, T. Chiu, Digital literacy learning in higher education through digital storytelling approach. J. Int. Edu. Res. **13**(3), 1–16 (2017)
22. N. Fraser, *Ten Things We've Learned From Blockly*. IEEE Blocks and Beyond Workshop (Blocks and Beyond, 2015)
23. Y. Kafai, Q. Burke, *Connected Code: Why Children Need to Learn Programming* (MIT Press, Cambridge, 2014)

Semantic-Based Search Engine System for Graph Images in Academic Literature

Sarunya Kanjanawattana and Masaomi Kimura

1 Introduction

An ontology specifies the representation of a conceptualization. Recently, ontologies have been recognized as an important feature of information retrieval systems owing to their ability to link knowledge in different areas of the ontology. Ontology-based search engine systems can acquire more useful information than traditional search engine systems because users can find not only particular concepts obtained by a given query but also other related concepts. However, for the practical usage of ontologies, an application must be developed to present the results from an ontology query because the realization of ontologies alone is extremely difficult for the average end user. Typically, to input a query via ontology, it is necessary for the user to be skilled in a query language; in addition, a specific ontology realization will be required. This creates problems for a general user who would then avoid using the complex system. Therefore, via constructing a web-based application with user interfaces, we introduce a handy and capable search system that does not require any computer skills.

In recent years, there has been a substantial amount of research on information retrieval. There are many types of data that are regular targets for search systems, including text [1] and image [2] data. In particular, image-based information retrieval is a growing topic in several study fields (e.g., computer vision and knowledge-based information retrieval) because methods for extracting data from images are more complicated than those for extracting data from text. Hence, researchers require particular methods for extracting information from an image.

S. Kanjanawattana (✉) · M. Kimura
Shibaura Institute of Technology, Tokyo, Japan
e-mail: NB14503@shibaura-it.ac.jp; masaomi@sic.shibaura-it.ac.jp

A. Reyes-Munoz et al. (eds.), *EAI International Conference on Technology, Innovation, Entrepreneurship and Education*, Lecture Notes in Electrical Engineering 532, https://doi.org/10.1007/978-3-030-02242-6_10

For example, a system of image-content extraction could analyze an image's low-level features [3] such as colors.

Images in academic literature contain essential information that may not be described in any other part of the paper. Moreover, graphs in academic literature always present statistical data, relations, or comparisons that are different to a generic image such as a photo. With the greater use of infographics, it is very useful for users if they are able to acquire implicit information from a graph and thus obtain new knowledge. Hence, not only the textual information but also the graphical content should be utilized for queries in search engine systems. However, dealing with both the graphical content and the linguistic information in the graph can result in a semantic gap problem, which characterizes the difference between linguistic and graphical representations. If the gap is large, it is likely that the information presented in the combined graphical and linguistic representations will be misunderstood. Therefore, one solution is to use ontologies or semantic relations to narrow the gap. In the past, we have attempted to extract graphical information from a graph, including graph components such as axis titles and legends [4] and graph data information such as data located in the axes of the graph [5, 6]. We constructed an ontology and stored the information extracted from the graphs and their descriptive contents (i.e., captions and cited paragraphs) in the ontology. The objectives of the currently proposed system are therefore to utilize an ontology based on the structure designed in our previous study [5] and to propose an ontology-based search engine system. An ontology matching may be an important integration to enhance search engine systems [7]. It helps to bridge search terms which contains a partial relation.

In the past several decades, there have been several studies that proposed semantic search engine systems. Li et al. [8] developed a fuzzy search by allowing mismatches between query keywords and answers. To freely explore data, this fuzzy search accepted keywords even in the presence of minor errors. Based on this existing study, we realized that results or knowledge provided by a conventional system are limited because they are solely dependent on given keywords. In the most recent decades, studies on information retrieval have advanced to semantic systems utilizing ontology concepts that enhance and extend the obtained knowledge based on user specifications. Jayalakshmi et al. [9] proposed a semantic search engine system that depended on inverse document frequency and text mining. The proposed search engine system created the search indexing using the contents of the files to retrieve the relevant document from a computer. Another existing study constructed a scalable semantic search for geospatial data [10] in which an application layer and a search service that provided a specific search functionality inspired by resource description frameworks was introduced.

In this study, we propose an ontology-based search engine system that utilizes the ontologically stored textual and graphical information that is extracted from images containing graphs, including their captions and paragraphs. In our proposed system, users can select questions and assign required settings to complete their queries. Depending on the selected questions, the search engine system provides relevant graph images and extended information.

2 Methodology

We utilize graph component extraction [4], graph information extraction [5], and optical-character-recognition-error (OCR-error) correction [11] to obtain the graph information. This graph information is then added to an ontology with the structure designed in the previous study [5]. Note that the size of graph image dataset is 600 images.

2.1 *Database and Ontology Designs*

We used a relational database to store data that was related to our target data (i.e., a collection of graph images, including their captions and text paragraphs) and other necessary information such as captions, text paragraphs, and graph profiles. A database design is presented in Fig. 1.

Five tables were used to record the following information from each graph: graph, contained, description, text_types, and ocr_text_extraction. The "graph" table collected the profiles of graph images, such as the graph name. The "description" table contained the graph's captions and the paragraphs that referenced the graphs. The "text_types" table contained the different types of the graph's descriptions (e.g., caption, *X*-title, and legend). To acquire the graph components, we used OCR to first recognize and convert them into digital data. These data were stored in the "ocr_text_extraction" table.

The ontology used in this study was based on the structure design in [5], but we redesigned it to be more applicable and to meet the requirements of the search engine system analyzed in this study. Herein, we describe the updated parts that differ from the previous version of the ontology. Note that the previous ontology was simply applicable to singular data in plot, line, and bar graphs but could not handle multiple data. Thus, we have since added a few relations that allow the ontology to

Fig. 1 Database diagram for storing the graph information

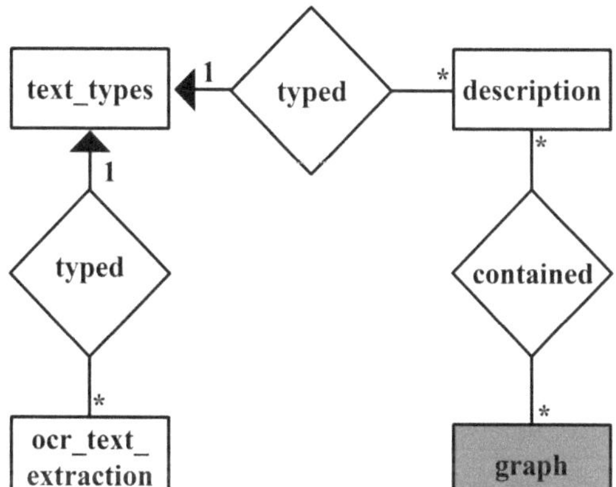

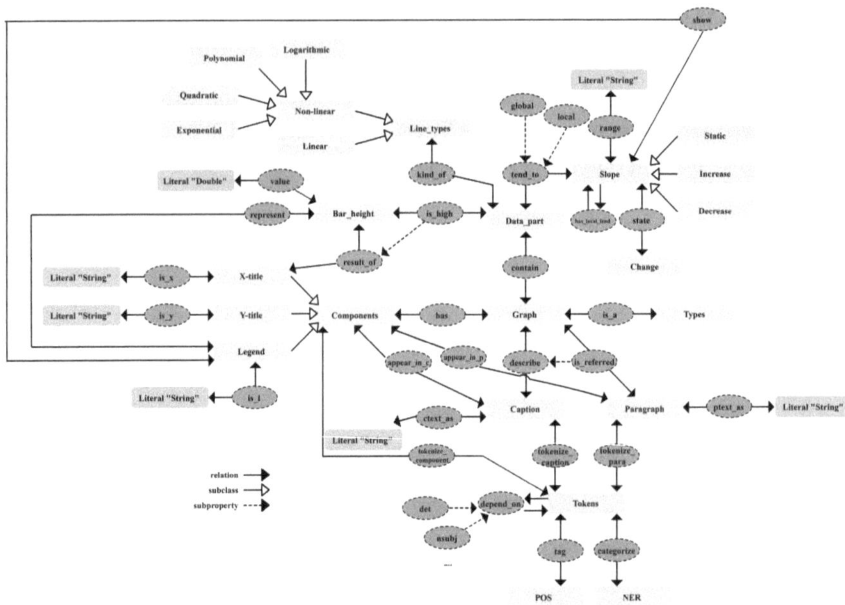

Fig. 2 Ontology design with new relations

be applicable to bar graphs containing multiple data labeled in a legend. A relation named "show" now connects the Legend and Slope classes because it describes the data tendency of each data label that belongs to different categories in the X-title class. A "represent" relation indicates the height of each data. Figure 2 illustrates the ontology design including the two new relations described above.

Note that there were a total of 636 graph images collected from academic literature that were separated to 138 images relating to biology and 498 images relating to computer science. We focused on two types of graphs such as 170 bar graphs and 466 two-dimensional charts (2Dcharts), which included line and plot graphs. Information on these graphs was extracted and added to the ontology.

2.2 *System Implementation*

An ontology-based search engine is a search engine application that utilizes ontologies to fulfill user inquiries. Through the use of ontologies, it is possible to obtain relevant results to a query as well as to obtain new extended knowledge. Such a search engine system helps users to retrieve images of graphs that contain information they require, such as a relation between the X- and Y-axis labels and a comparison of each bar in a bar graph. The users can easily comprehend the graph and its descriptive details because the search engine system precisely provides

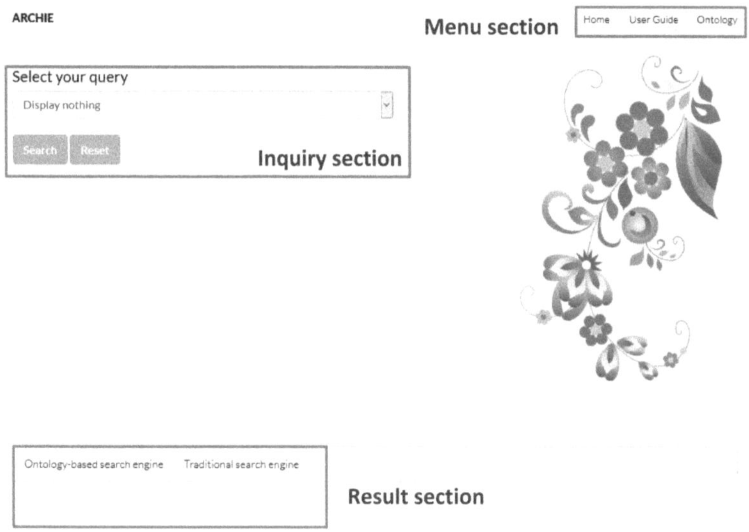

Fig. 3 Illustrating a user interface of a search mode with three sections

detailed information such as main ideas and tendencies. The system allows the users to select specific questions for inquiring; moreover, some settings must be accepted to restrict the amount of obtained results.

We implemented the described search engine system in a search application. The developed application can query the search engine system by specifying some keywords and specific questions. The relevant results are then returned and displayed in a search page. This system was designed to support simplicity and immediate availability. To that end, only necessary functions such as the query settings are shown on the web page. Three sections such as menu section, inquiry section, and results section are presented on the main search page, as shown in Fig. 3.

The menu section contains three tabs, namely home, user guide, and ontology. The home tab is the default screen when the system is launched in the search mode. The user guide is a page that briefly explains the system and its components, including a guideline of the system process and examples of system simulations. The ontology tab displays the ontology schema that is used in this system.

In the inquiry section, the users can select questions and input some required settings. The acquired relevant results are displayed on the results section. In addition, the question option can be selected by the users. We offer a few options that can help the users to filter unnecessary results (i.e., conditions, as shown in Fig. 4, and features, as shown in Fig. 5).

The condition box contains five conditions. First, the users can restrict the graph type. We distinguish the graph types into two types such as bar graphs and 2Dcharts. Second, the users can select results that belong to a specific group. Third, the results shown in the results section can be filtered based on a specific regression type. (For

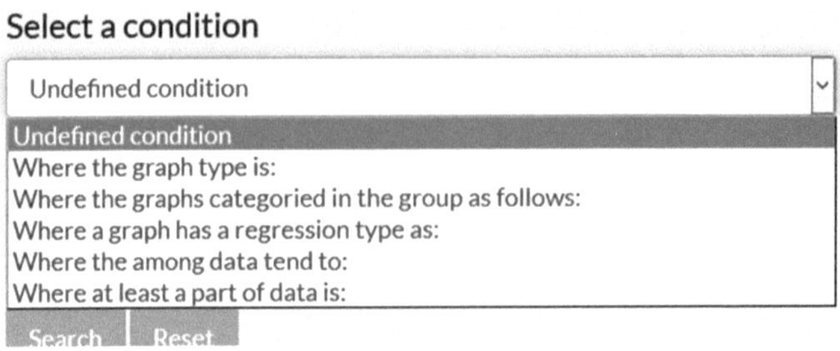

Fig. 4 Selectable conditions for results filtering

Fig. 5 Selectable features that can be presented in the results section

example, a user might need only graphs that have linear regression.) Fourth, the users can select the results with a specific tendency such as increasing or decreasing. Finally, for the line and plot graphs, a local tendency is also a significant option, because changes in the graph might identify essential information. Thus, users can filter the results based on the data variation. The feature box was created to cover the needs of all users because additional information such as the graph caption or X- and Y-labels might be required by a user.

In the results section, results from our ontology-based search engine systems are presented. Depending on the user's system, a user can independently choose a tab to examine the results.

Herein, we discuss the questions that are included in the system and the settings that must be entered by the user. There are six queried questions, described as follows:

- Question 1: Display the graphs involving the following keyword(s).
- Question 2: Display the graphs involving following keywords and their main idea of captions.
- Question 3: Display the graphs involving following keywords and their maximum and minimum values of graphs.
- Question 4: Display the graphs relationship extracted from axis titles.
- Question 5: Display the relationships between two different tokens.
- Question 6: Display the comparison of bar values on different X-categories but same data label.

The first question is the most basic because it is similar to a keyword-based search engine system (e.g., Google search). There are a few settings that are required and must be completed by the user. For example, a user may need graphs that feature

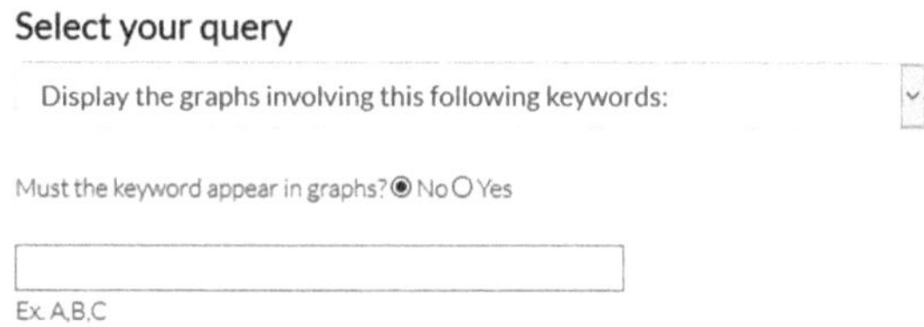

Fig. 6 A query form of Question 1

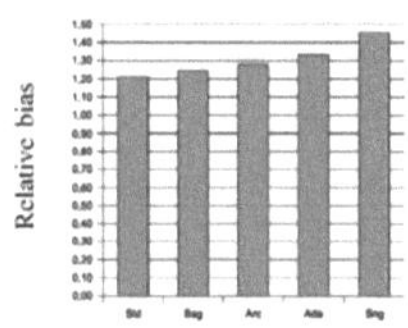

Graph name	graphdata.PMC200026_figure5d_1
Caption	Comparison of relative bias as defined by Kong and Dietterich [90] for the ten data sets. A value of 1 has been chosen for the estimation obtained for the cooperative ensemble bias and we represent the relative bias of the rest of the ensemble methods to the cooperative ensemble. The y axis represents this relative bias. A dataset is glass.
Main idea	• Comparison of relative bias as defined by Kong and Dietterich [90] for the ten data sets. • A dataset is glass.

Fig. 7 Example results for Question 2

a specific inputted token in the graph for deep discussion on a particular topic. An example query form for Question 1 is illustrated in Fig. 6. The user simply inputs at least one keyword to the text box separated by commas (for example, the string "data, test, accuracy" can be inputted to the text box). Moreover, the user can specify whether the keyword(s) must appear in the graph's components (e.g., X-label, Y-label, or legend) by choosing either the "yes" or the "no" radio button above the text box. There is an optional text box that asks the user's intention for the query; however, to complete the evaluation, the user should describe their intention for their query. For example, a user may input keywords such as "neural network, accuracy, image," and the intention would be to obtain graph images relating the accuracy of neural networks when dealing with images.

The second question requires only keyword(s) from the users to produce the relevant results. Moreover, the question asks for the main idea of the graph descriptions (e.g., the caption). Therefore, an extra feature has been added to the results section (Fig. 7), that presents the main idea. Sentences containing the main idea are selected by analyzing the appearance of keywords and the first sentence of the paragraph. A user can use this question to summarize information to realize the underlying concept of a graph.

The third question is similar to the second question, and it requires keyword(s) to be set. The bar height and local trend features are initially selected and displayed in the results section, including the highest and lowest values identified (Fig. 8). However, for a bar plot containing multiple data, it is difficult to identify the highest and lowest values; thus, a comparison between each bar height and the legend are displayed instead. A user may use this question to analyze statistical data to compare to their results.

In general, there are significant relations that are established in any given graph. The fourth question is used to indicate which tokens are a part of a graph's relationships. For this question, the user inputs keyword(s), just as in the previous questions, and the relevant results are presented, including some tokens related to

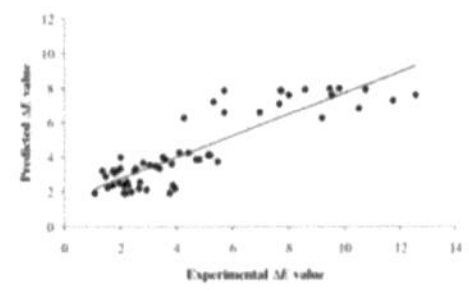

Graph name	graphdata.PMC200081_figure6_1
Bar heights	NONE
Local trends	The status of tendency at (1.46, 2.91) is increase which is slope_change **<- MIN** The status of tendency at (11.30, 10.11) is static which is slope_change **<- MAX**

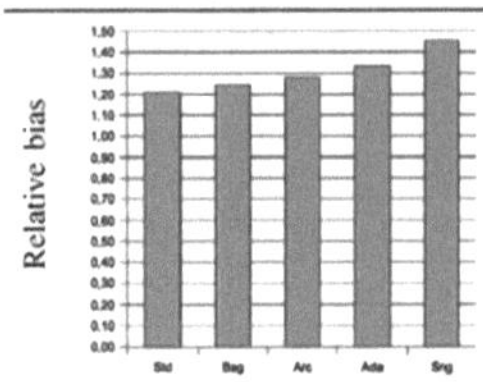

Graph name	graphdata.PMC200026_figure5d_1
Bar heights	Ada = 14.043478260869565 **<- MAX** Std = 13.217391304347826 Sng = 2.869565217391304 **<- MIN** Arc = 14.043478260869565 Bag = 13.608695652173912
Local trends	NONE

Fig. 8 Example results for Question 3

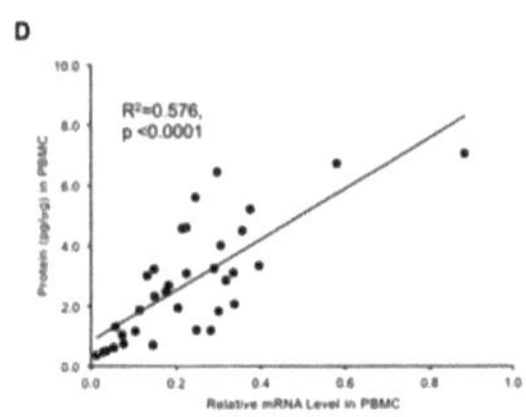

Graph name	pone.0063958.PMC3656936_figure2d_1				
X- and Y-axis relationships	quantity	inhibitors	length	hours	measured
	method	indicate	change	level	culture
	levels	frataxin	increase	compartments	correlated
	mRNA	correlation			

Fig. 9 Example results for Question 4

the relation between the X- and Y-labels (see examples in Fig. 9). Then, the users must interpret the graph relations and expressions.

The graphs used in this system were collected from a number of publications, and they are always described by captions and cited paragraphs. Sentences comprise several tokens that are dependent on one another. The fifth question is similar to Question 4, but the question investigates the relationships between two different keywords. A user may use this question to understand any implicit relations between two tokens hidden in the descriptions.

The sixth question presents information in bar graphs that feature multiple data labels. The question presents a comparison based on bar heights and legends in the bar graphs. The comparisons can be achieved with respect to one of two items: with respect to bar categories (e.g., X-label) or with respect to the legend (or data label). A user may use this question for data comparison and analysis. Figure 10 shows an example of results generated using Question 6.

3 Simulations

We conducted some simulations to evaluate our ontology-based search engine system. During these simulations, we attempted to anticipate what types of questions

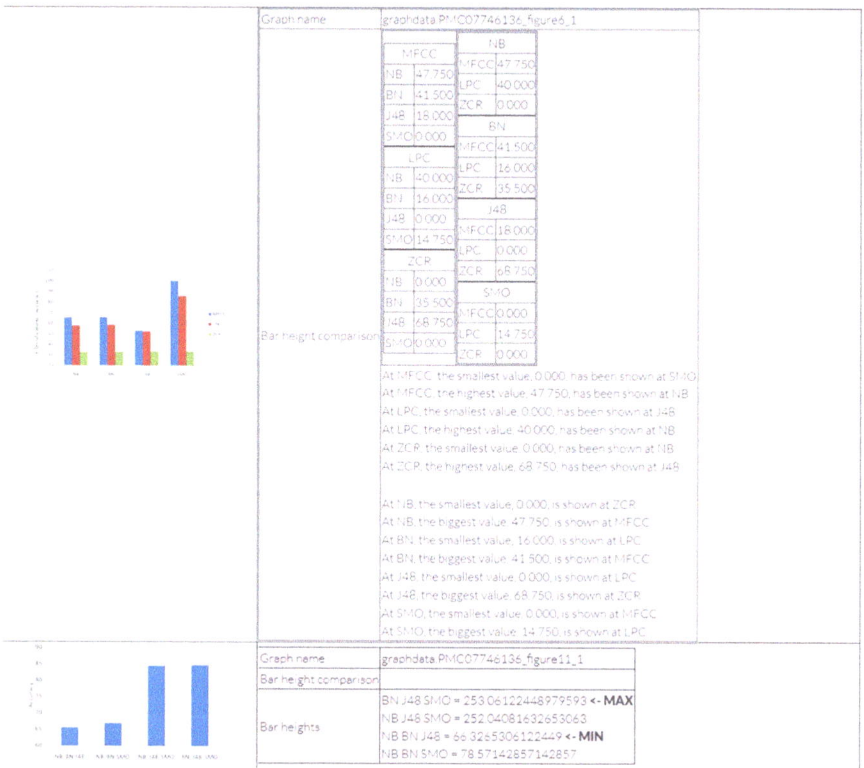

Fig. 10 Example result for Question 6

Table 1 List of issues

Issue No.	Descriptions
1	Does the system provide results relevant to other related keywords?
2	Does the system handle a variant expression of words?
3	How performance of the system is?
4	Is the extended information provided by the graphs accurate?

would be asked by users and what information they would require to fulfill their inquiries.

The objective of these simulations was to demonstrate both the capabilities of the system, which can handle several issues and its limitations. In other words, we seek to demonstrate which situations the system can or cannot handle. These were a list of possible issues that were assigned to each simulation, as shown in Table 1. To determine if the system could handle the described issues, five sets of keywords (Fig. 11) were inputted to the system and the results were presented in the results section of the system.

Fig. 11 A list of simulations, keywords, and issues

Simulation	Issue	Keywords	Questions
1	1, 3	J48	Question 1
		decision tree	Question 1
2	3, 4	intrusions, classifier	Question 6
3	3, 4	sensitivity	Question 4
4	2, 3, 4	performance, SVM	Question 5
		precision, SVM	Question 5

Although our ontology contained data relating to both biology and computer science, we selected only keywords relevant to computer science because we can evaluate the results more precisely since we have more experience in this domain.

Once the simulations were completed, we obtained many results that were relevant to the inputted keywords. Because the simulations were intentionally performed to investigate issues, we attempted to clearly identify the issues by analyzing the obtained results. All of the results were measured, and their precision, recall, and F-measures were calculated to investigate the third issue related to the system's performance. Note that F-measure is a search measurement that combines precision and recall.

We first addressed the first issue in the first simulation. For this simulation, we selected a keyword "J48" which is a decision tree algorithm and applied the keyword to Question 1. This query resulted in seven relevant graphs. When we measured the system's precision, recall, and F-measure, all three quantities were equal to one. We then queried another keyword, "decision tree" for two reasons. First, it is also a word that is related to decision tree algorithm, and second, just querying the keyword "J48" was inadequate to investigate the first issue. We obtained only one result from the system for the "decision tree" keyword. After evaluating the precision and recall, we determined that the system featured a high precision (1) but low recall (0.14) for this simulation. Additionally, a low F-measure (0.25) also resulted from this query.

For the second and third simulations, we sought to evaluate the extended information in the graphs (e.g., extracted bar heights and axis relations). We tested the system with multiple keywords "intrusions, classifier" that were applied to Question 6 in the second simulation, which returned only one relevant result. After examining the results, we determined that the system achieved very high precision, recall, and F-measure (all equal to 1). We then investigated the accuracy of the extended information, which was a comparison of the bar heights in this simulation. Unfortunately, some errors in the bar heights were reported. In the third simulation, we used keyword "sensitivity" and we obtained precision, recall, and F-measure values of 1, 0.7, and 0.82, respectively. The extra information shown in this particular simulation was the graph axis relations. We verified the extractable relations and determined that the accuracy of the relations was approximately 0.66.

Fig. 12 Simulation results for each keyword set

Keywords	Performance models		
	Precision	Recall	F-measure
J48	1.00	1.00	1.00
decision tree	1.00	0.14	0.25
intrusions, classifier	1.00	1.00	1.00
sensitivity	1.00	0.70	0.82
performance, SVM	1.00	0.89	0.94
precision, SVM	1.00	1.00	1.00
Average	1.00	0.79	0.84

For the final simulation, we attempted to address an issue involving varying word expressions using similar keywords for querying keywords such as "performance, SVM" and "precision, SVM." Note that support vector machine (SVM) is a supervised learning model utilized to classify the data using hyperplanes. We noticed that "precision" is a performance model used to evaluate a system, and the meaning of "performance" covers a definition of "precision." This simulation utilized Question 5, which investigates token relations. From the generated results, we determined several identical token relations with approximately 30% similarity for both keyword sets. The precision, recall, and F-measure values for "performance, SVM" were 1, 0.89, and 0.94, respectively. The precision, recall, and F-measure values for the keywords "precision, SVM" were all equal to one.

We evaluated the results from all simulations and obtained performance measurements representing the overall quality of the system. The average precision, recall, and F-measure values were 1, 0.79, and 0.84, respectively. Figure 12 presents a summary of the simulation results.

4 Discussion

In this study, we propose an ontology-based search engine system that utilizes a constructed ontology containing graphical and textual information from images with graphs in academic literature. The overall objective is to use the ontology to generate relevant results for users who could specify keywords, desired queries, and settings that corresponded to the users' intention.

We conducted four simulations containing specific issues as presented in Table 1 and discussed the results of the simulations to identify issues with the system.

In the first simulation, several graph images related to keyword "J48" were discovered from the ontology, whereas only a small number of results were generated from a query using the related keyword "decision tree." Even though the keywords belonged to the same category, the results from both keyword sets were different. For example, J48 is a decision tree; thus, we should obtain similar results from both queries. Note that this discussion does not focus on the meaning of the keywords. From these results, we can conclude that this system does not currently support a query from related keywords. However, this limitation should be mitigated if we propose a method to cluster keywords. With this approach, we would collect a number of keywords that were used for querying in the past and group them based on their similarities. Then, similar results will be obtained if the queried keywords belong to the same cluster. Additionally, another solution should be to integrate other ontologies with ours to bridge the relations between related keywords.

Based on the results obtained from the second simulation, only one relevant result was retrieved. We judged the one result to be relevant; therefore, the precision was very high. Additionally, we obtained a very high recall from the query because, in our data, there were no other graphs relevant to the keywords "intrusions, classifier." This situation occurred because these keywords were too specific to our dataset and the size of our ontology was small, containing only 626 graph images. Moreover, most graphs were collected from publications regarding artificial intelligence, machine learning, and classification; only a few publications were related to outlier detection or security. We can address this issue by expanding the size of the dataset to cover all study fields. However, even though the obtained graphs were relevant to the query, some errors in the extended information shown as an extra feature (in particular, bar height) arose due to the failure of the OCR. As described in the previous study [5], we used OCR to identify the numeric Y-scales from graphs to estimate their pixel proportions and to estimate the ranges of the Y-scales for bar height approximation. In this case, misrecognizing the numeric data leads to a mistaken estimation. In the past, we introduced an OCR-error correction system [11] using an ontology to suggest possible correct tokens; however, the previously proposed correction system was only applicable to tokens containing alphabetical characters. To mitigate this limitation, we must modify the system to work with numbers as well.

The third simulation used the keyword "sensitivity" for its query, and we obtained a very high precision because the result was relevant. In contrast, the recall was small because the search engine system could not offer all of the results relevant to this keyword, which has several synonyms such as "recall" and "true positive." Thus, other graphs might describe sensitivity but use its synonyms instead. As a result, the query with the keyword "sensitivity" provided only one relevant result and the system could not offer others that were related to the keyword's synonyms. Therefore, the system did not support results related to the keyword's synonyms. However, this limitation can be minimized if we use WordNet or DBpedia to investigate the synonyms of the keyword and use them all to query our ontology to expand obtainable results. In addition, after calculating the axis relations, the percentage of accurate axis relations reached as high as 66%. Hence, the system can extract precise axis relations in part.

In the fourth simulation, we attempted to investigate issues relating to variations in the word expressions. To do so, we assigned two keyword sets such as "performance, SVM" and "precision, SVM" whose partial meaning was similar. We queried the dataset using these two keywords and obtained relations between two tokens in each keyword set along with relevant graphs. Interestingly, we found that the similarity of their token relations was approximately 30%. Therefore, the system can partially handle a varying expression of words.

The ontology-based search engine system provided very high precision for most queries, including accurate extended information. Based on the presented results, the performance of this system is remarkable because the F-measure of 0.84 is considerably high.

5 Conclusions

In this study, we proposed an ontology-based search engine system that was applied to our constructed ontology with information from graphs in academic literature. The main objectives of this work were to utilize the ontology and to demonstrate the results generated from user queries. This proposed system is useful for students and researchers because the data that is used in our search engine system are graph images collected from academic literature. Thus, using the unique abilities of ontology, users will obtain new knowledge that is extracted from both graphical and textual information. The system was evaluated using four simulations that were related to four different issues. A high average F-measure was obtained; therefore, it can be deduced that the performance of the system is effective. We determined that this system can provide precise information from queries to the ontology because the ontology is highly applicable to this search engine system.

In future works, we will extend the size of the dataset to cover study areas for a wider variety of users. Moreover, we will integrate other existing ontologies with ours to minimize the related-keywords and keyword-synonyms issues.

References

1. Y. Tsuruoka, J. Tsujii, S. Ananiadou, Facta: a text search engine for finding associated biomedical concepts. Bioinformatics **24**(21), 2559–2560 (2008)
2. M. Batko, F. Falchi, C. Lucchese, D. Novak, R. Perego, F. Rabitti, J. Sedmidubsky, P. Zezula, Building a web-scale image similarity search system. Multimed. Tools Appl. **47**(3), 599–629 (2010)
3. J.-q. Ma, Content-based image retrieval with HSV color space and texture features, in *Proceedings of the 2009 International Conference on Web Information Systems and Mining* (IEEE Computer Society, Piscataway, 2009), pp. 61–63
4. S. Kanjanawattana, M. Kimura, Extraction and identification of bar graph components by automatic epsilon estimation. Int. J. Comput. Theory Eng. **9**(4), 256–261 (2017)
5. S. Kanjanawattana, M. Kimura, Extraction of graph information based on image contents and the use of ontology, in *International Conferences ITS, ICEduTech and STE 2016*, pp. 19–26 (2016)
6. S. Kanjanawattana, M. Kimura, A proposal for a method of graph ontology by automatically extracting relationships between captions and x- and y-axis titles, in *Proceedings of the 7th International Joint Conference on Knowledge Discovery, Knowledge Engineering and Knowledge Management*, pp. 231–238 (2015)

7. J. Euzenat, P. Shvaiko et al., *Ontology Matching*, vol. 18 (Springer, Berlin, 2007)
8. G. Li, S. Ji, C. Li, J. Feng, Efficient fuzzy full-text type-ahead search. VLDB J. Int. J. Very Large Data Bases **20**(4), 617–640 (2011)
9. T. Jayalakshmi, C. Chethana, A semantic search engine for indexing and retrieval of relevant text documents. Int. J. Adv. Res. Comput. Sci. **4**(5), 1–5 (2016)
10. A. Both, A.-C.N. Ngomo, R. Usbeck, D. Lukovnikov, C. Lemke, M. Speicher, A service-oriented search framework for full text, geospatial and semantic search, in *Proceedings of the 10th International Conference on Semantic Systems*, pp. 65–72 (ACM, New York, 2014)
11. S. Kanjanawattana, M. Kimura, Novel ontologies-based optical character recognition-error correction cooperating with graph component extraction. BRAIN. Broad Res. Artif. Intell. Neurosci. **7**(4), 69–83 (2017)

Improving Higher Education Quality in Jordan Using Mobile Technologies for Better Socio-economic Diversity Integration of Disadvantaged Groups Using a Mobile Multimedia/Augmented Reality Workflow

Jazz Rasool and Carl H. Smith

1 Introduction

University teachers in the country of Jordan were in need of technology enhancing approaches to learning for disabled as well as socially disadvantaged students. The mEQUITY project, an EU funded project, aimed to address engagement of students by creating multimedia and Augmented Reality (AR) content for distribution through mobile devices such as smartphones and tablets. The educational institution, Ravensbourne, based in Greenwich, London, UK, was the partner that would design the multimedia and AR content.

1.1 Establishing Minimised Workflows for Creativity

As well as designing and producing the content on behalf of Jordanian University staff, it was felt that it would be a more sustainable approach to train staff in creating multimedia and AR content by using off the shelf software coupled with tight, minimal learning curve workflows to create end-user ready content that could be completed from inception to dissemination in a matter of minutes. Having creative workflows that spanned between 5 and 10 min would allow for many iterations of content refinement to be carried out ensuring delivery and implementation of optimally engaging student learning experiences.

J. Rasool (✉) · C. H. Smith
Ravensbourne, London, UK
e-mail: j.rasool@rave.ac.uk; c.smith@rave.ac.uk

A. Reyes-Munoz et al. (eds.), *EAI International Conference on Technology, Innovation, Entrepreneurship and Education*, Lecture Notes in Electrical Engineering 532, https://doi.org/10.1007/978-3-030-02242-6_11

1.2 Production of Multimedia Content

Camtasia [2] software was used to compile existing educational digital assets such as video, images and text into videos that could be downloaded by teachers into smartphones and tablets for informing students. Typically content might need to include a video of someone using sign language as guided commentary for displayed image, animation or video accompanied by some subtitle text. The three types of assets were gathered in Camtasia software and produced into a single digital video that could be shared amongst teachers by downloading from a common, online media database that could be accessed via a mobile application loaded on a smartphone or tablet.

1.3 Production of Augmented and Virtual Reality Content

Off the shelf software from three different sources was used to experiment in producing Augmented Reality content. The mobile App, *Aurasma* [3], was originally used as a quick way of generating Augmented Reality experiences from pre-existing content. New or bespoke content could be loaded, providing asset production was prepared through an online platform. The desktop application, *Unity* [4], a 3D game engine, was also trialled to create 3D content for AR viewing using the AR plugin, *Vuforia* [5].

2 Training of Jordanian University Staff

Staff from Jordanian University partners were invited to the Ravensbourne institution to take part in a 5 day training program covering Multimedia, AR, VR and 360 content preparation, creation, production, dissemination and implementation.

- *Preparation* involved gathering of assets such as images, text, animation, sound recordings, 3D models or video. Some assets needed to be edited for quality and conciseness. Audio was prepared using free sound editing software *Audacity* [6]. Real world objects might be 3D scanned using a tool such as the *Structure IO* scanner [7]. Models could be cleaned up using 3D modelling tools such as *Meshlab* [8], *3DSMax* [9] or the online platform *Sketchfab* [10]. This was followed by exporting into file formats suitable for use in AR or VR content production software, typically OBJ or FBX file formats.
- *Creation* involved the layout, compilation and editing of finished assets that had been imported as a result of going through the preparation process. This was done in Camtasia software for video production and in Aurasma, Unity software for AR production.

- *Production* typically was centred on exporting the finished content in alignment for a particular platform such as web, desktop, mobile, tablet or an AR/VR headset.
- *Dissemination* focused on importing produced content into a specific application, platform or device. Finished videos from Camtasia were uploaded to a mEQUITY's database of assets and packaged for download to different devices. For AR/VR and 360 media, content was either uploaded as a 3D model to the Aurasma or Sketchfab website or prepared as a stand-alone mobile app. The Sketchfab platform allowed the 3D model to be viewed online or, through a smartphone, viewed in a VR headset. Stand-alone apps exported from Unity could be loaded onto a desktop, smartphone or tablet and viewed in AR or VR. The Sketchfab platform could also allow viewing content via mobile handsets, including VR views that could be experienced with a VR headset such as a Google Cardboard compatible device or Samsung Gear.
- *Implementation* was about arranging events, classes or portals for end-users or audiences to experience the created content. This would either be conducted by a teacher in a class or via online learning. Pedagogical structure was applied to the implementation including appropriate assessments and evaluation.

3 Asset Preparation

Assets for production usually come in five varieties: Video, Audio, Text, Image as well as in 3D Model form.

- *Video*: Typically footage was supplied or had to be downloaded from the Internet. Files were given in MPEG (.mpg) or 3GP format but for the sake of editing, compiling or final production in Camtasia software, were converted into MPEG4 (.mp4) format. This was done by importing an .mpg file into Adobe's *Media Encoder* software, part of its Creative Cloud (CC) suite. Footage, say of a person using sign language to describe a topic video's content, was often a key component asset in making educational videos for users with hearing issues. Often accompanying such videos were videos with educational content relating to a particular subject, such as Recycling or Alternative Energy, downloaded from the Internet. For 360 Video content, the file was imported into Unity software for it to be wrapped around a sphere, with a camera positioned in the centre to create the 360 experience.
- *Audio:* Although audio asset files were not used, often audio within a video needed cleaning up for noise, quality or volume. Camtasia software imports video as a video track and audio track in a joined or separate way. In cases of poor sound, the audio track was separated and exported into a stand-alone file. The stand-alone audio file was imported into the free software *Audacity* and its filters for noise removal and volume amplification were used to provide a clearer audio track. The old track was uncoupled from the video track in Camtasia software

and deleted. The cleaned up audio track was imported into Camtasia and placed on the vacated sound track, synchronized and aligned to the original video track.

- *Text:* Microsoft *Word* documents or ASCII-compatible Text files to be used as subtitle content for Camtasia-produced videos had their content cut and pasted into Camtasia's "Callout" text boxes and then positioned under a content video and accompanying signing video. The subtitle text began to be actively displayed at the time in the video where referred content was first shown. The duration of display of the subtitle text box was adjusted to fit the signing video or the relevant content in the topic video, whatever was being covered.
- *Image*: Images for use in Camtasia produced videos would arrive in a number of pixel formats, typically JPEG, PNG or BMP. The images may sometimes have to be cropped to just cover relevant content. Sometimes labels needed to be added to highlight part of the image's content. This was done in Adobe *Photoshop* software and saved back out in a JPEG format for importing into Camtasia. For 360 photograph or image content, the file was imported into Unity software for it to be wrapped around a sphere, with a camera positioned in the centre to create the 360 experience.
- *3D Model:* Models for viewing in Virtual Reality or Augmented Reality will typically have been exported from 3D Modelling software applications such as 3DS Max, Blender or may have been generated from 3D scanning a real world object using a handheld scanner such as Occipital's *Structure io*. For use in Unity a 3D model has to be in .OBJ or .FBX file format. Sometimes the model may need cleaning up especially if a 3D scan. Cleaning, simplification and editing of 3D scans was done by importing the model file, usually in .OBJ or .STL format into *Meshlab* software or the *Sketchfab* website. Meshlab allows excess polygons, points or mesh areas to be edited and then the model saved back out as a .OBJ file with attached Texture file in JPEG format and the integration information of Texture and Model mesh in a .MTL material file. Sketchfab also allows more user friendly finishing of 3D scanned models. The outputs from Meshlab could be taken into 3DSMax software and exported with Texture content embedded so the output was a single .FBX file. The .OBJ, .MTL and JPEG Texture files could simply be dragged into a project folder within Unity and a pre-fabricated (prefab) 3D object would be created that could be positioned in a 3D Scene. Similarly dragging a .FBX file into a Unity folder would also create a 3D prefab object.

4 Experience Creation

This involved the layout, compilation and editing of finished assets that had been imported as a result of going through the preparation process. This was done in Camtasia software for video production and in Aurasma, Unity or online Sketchfab software for AR, VR or 360 production.

4.1 Production of Multimedia Content

Creating Multimedia content typically involved taking three kinds of assets—text, content video and a signing video to produce a single MPEG4 Video as an output. The process involved the following key stages in using the Camtasia software.

1. Create new Camtasia Project.
2. Import a Content Image or Video into the Clip Library.
3. Add Content Image or Video to a track of its own.
4. Add 'Callout' text box to a track of its own for native language subtitles.
5. Add Signing Video to a track of its own.
6. Synchronise the content across tracks and align in final video space.
7. Produce the video and save in MPEG4 (.mp4) format.

It is possible with prepared assets that the entire workflow above could be completed in as little as 15 min for a content video of small duration say a few minutes. With a template Camtasia project from a video that already been created, the whole process can be completed in 5–10 min. A typical Camtasia project screen shot is shown below (Fig. 1).

4.2 Production of Augmented Reality Content

Producing Augmented Reality content was explored using two kinds of software, Aurasma and Unity.

Aurasma is by far the quickest and easiest to demonstrate Augmented Reality content.

The Aurasma app could be downloaded onto a mobile device such as a smartphone or a tablet. Using Aurasma for existing content involved the following key stages (Fig. 2).

If pre-existing content was not suitable then it is possible to go to the Aurasma website, register a user account and load up models that might be suitable.

Unity 3D desktop software is a game engine but also a 360 film, Virtual and Augmented Reality production platform. When creating Augmented Reality content an AR plugin package, such as Vuforia, can be imported to accelerate development using its AR customised components.

The process for setting up AR content to be viewed through a desktop/webcam combination or through mobile device is outlined below.

1. A user account for using content with Vuforia to be used in AR production is created on the web page https://developer.vuforia.com/user/register
2. Once logged in, the user generates a key to digitally fingerprint all AR content created through using Vuforia. The Vuforia websites Developer Portal has a *License Manager* for creating and managing keys: https://developer.vuforia.com/license-manager

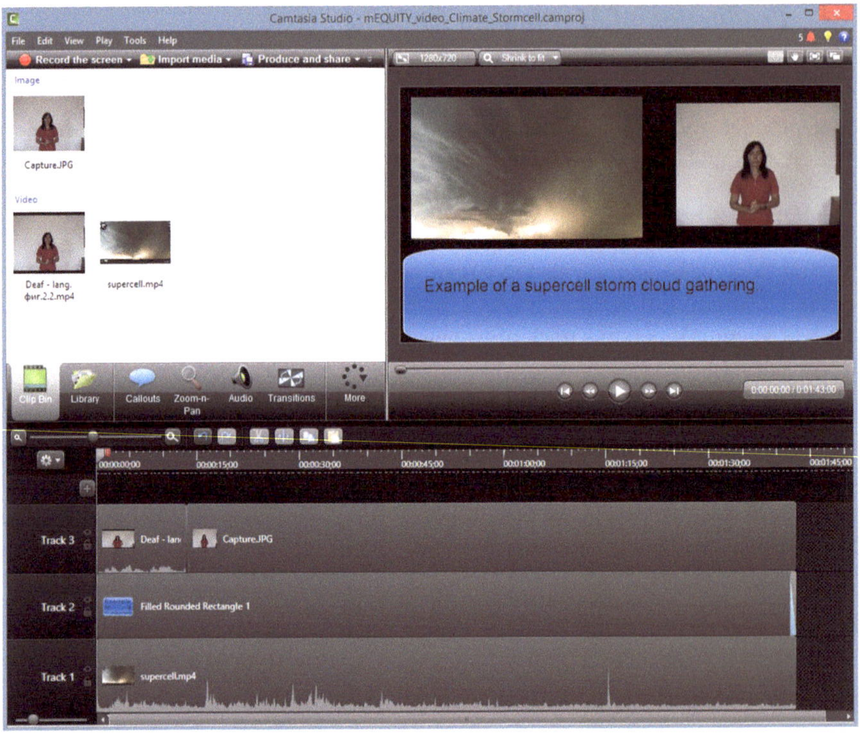

Fig. 1 Screenshot of Camtasia Software project window for a multimedia video with components of signing video, content video and subtitle Callout box

3. A database is created to house target images for AR content using the Vuforia *Target Manager*.
4. Target or trigger images are added by uploading to the database. These can be analysed to see how many unique identifiers on the image there are. The more there are the easier a trigger image can be responded to with content.
5. The database with target images is downloaded as a Unity package.
6. A Unity Project is created with the Vuforia Software Development Kit. The SDK can be used to build Android, iOS and Universal Windows applications for mobile devices and digital eyewear. Apps can be built with Android Studio, XCode, Visual Studio and Unity.
7. One of the basic AR template scenes is used to create an initial AR experience. These usually have an ARCamera to handle the associated device camera and an Image Target that will be linked to the trigger or target image.
8. The Vuforia Configuration associated with an ARCamera component has to be edited to include the license key.

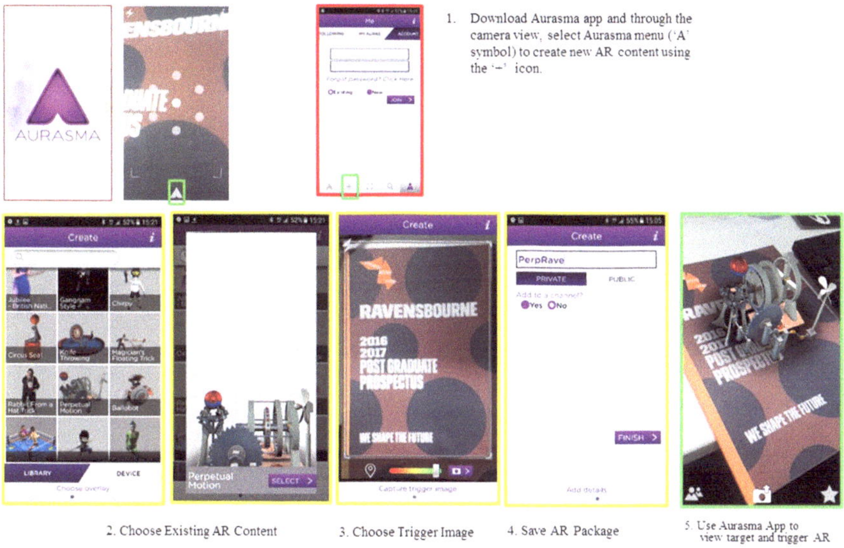

Fig. 2 Mobile device Screenshots of Aurasma AR software key stages of production

9. The database of target images is imported using the Unity Import Package menu item. After this is done the database can be selected and activated in the Vuforia configuration associated with the ARCamera component.
10. A 3D model (in .OBJ format with optional associated .JPG texture and material .MTL files) for viewing in AR is imported into the Unity project's *Assets* folder, usually into a folder specifically for models. Other content such as images or video can also be imported. The model and associated files must be dragged into the model folder together.
11. The model is then added to the Unity AR scene hierarchy and positioned on the target image on the scene. Once the scene and project are saved the application can be run through the associated camera being pointed at a printed target image, the 3D model or other content can be seen.
12. The projects build settings can be changed to build a Google Android or Apple iOS app. Providing the appropriate SDK software development kits from Google or Apple have been installed then a package that can be imported into a mobile device can be generated. For instance an *.apk* file could be generated to be imported to an Android device and run as an app (Fig. 3).

Sketchfab online software (https://sketchfab.com/) allows 3D models to be edited, shared and viewed easily. For example a scan of a real world object taken using Occipital's Structure IO scanner generates 3D files in .OBJ, .FBX and .STL formats that can be emailed from the Apple iPad device it is attached to. The files can be downloaded locally and then uploaded to the Sketchfab website providing the user has an account. The models can also be seen in Virtual Reality in a Google

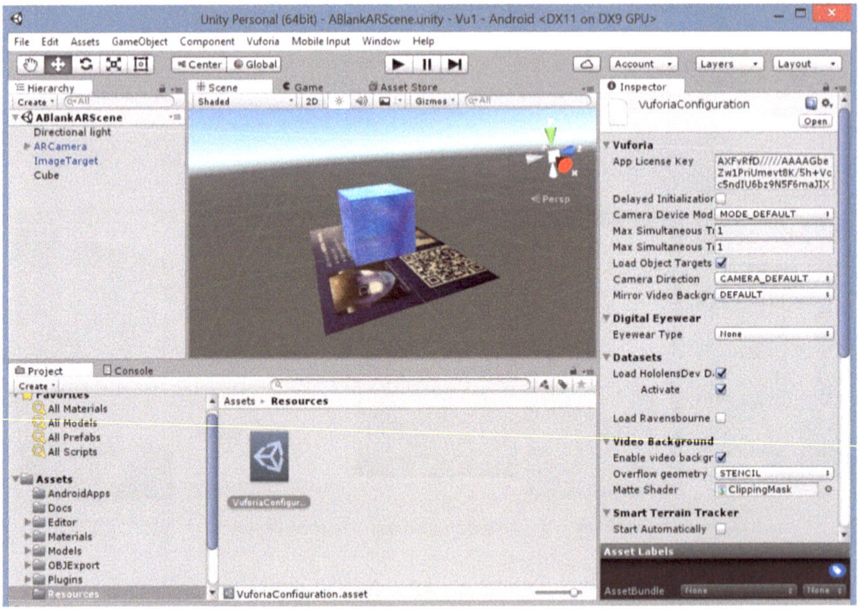

Fig. 3 Screenshot of Unity 3D software project window with Vuforia AR package installed. The top left has the scene hierarchy including ARCamera and ImageTarget components. On the right is the properties window of the ARCamera component with the Vuforia Configuration expanded, showing license key field and also Database selection options for selecting target image collections

Cardboard compatible mobile device by simply logging into the site on the mobile device and selecting the model's VR/Cardboard icon. When the mobile device is placed in a Cardboard compatible headset it can be viewed in Virtual Reality. It can take as little as 5–10 min to go through the whole process of scanning something using the Structure IO scanner, emailing it and then uploading it to the Sketchfab website to be ready for viewing in VR through a mobile device and a Cardboard compatible VR headset.

Once a user has registered and logged in at the Sketchfab website, the whole workflow for seeing something in VR in 5–10 min is as follows.

1. On an Apple iPad, email ready device Occipital's *Structure* IO app is downloaded for their Structure IO handheld scanner.
2. The Structure app is started and the real world object, to have a 3D model for it to be created, is scanned, as shown in Fig.4.
3. The full set of 3D files generated is emailed to an accessible account.
4. The set of 3D files are downloaded from the email locally onto a computer.
5. On the Sketchfab website the new model is uploaded by selecting the Upload option. The 3D files previously downloaded are selected and uploaded.

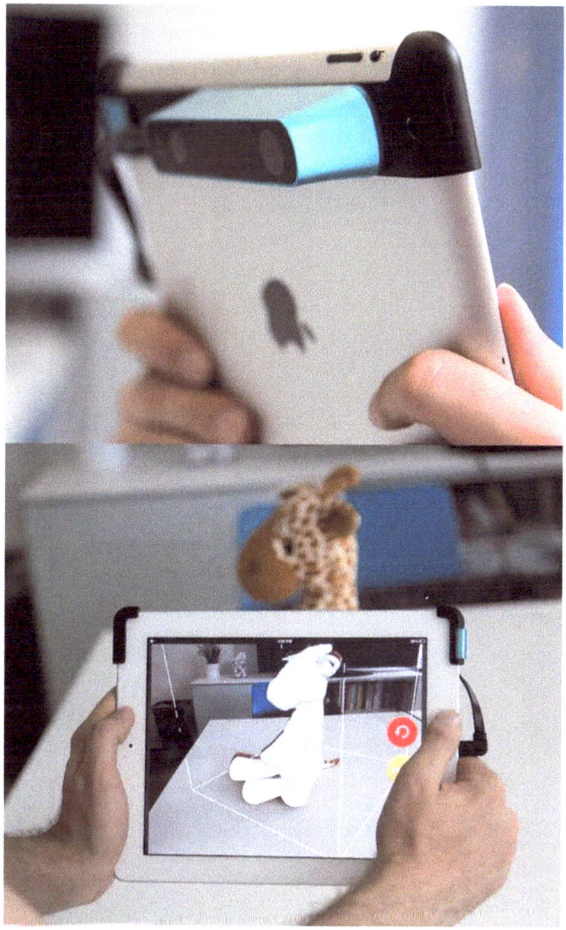

Fig. 4 A Structure IO scanner, manufactured by Occipital, is attached to an Apple iPad device. Loaded with the Structure app, a real world object can be scanned into a digital 3D format as .OBJ, .FBX or .STL files that can then be uploaded to the Sketchfab website to view in VR via a mobile device and a Google Cardboard compatible VR headset

6. Once the 3D model is uploaded, the Edit Settings option on the website can be used to change the lighting, materials or scene background as well as VR environment that will be experienced.
7. When loading and editing have been completed the model can be saved and then published for public access or simply kept private.
8. Logging into the Sketchfab website via a mobile device and then selecting the uploaded model, the VR icon on the image of the model can be pressed to generate a stereoscopic view on the mobile device (Fig 5). When the mobile device is loaded into a Google Cardboard compatible VR headset then the model can be experienced in Virtual Reality, including having the ability to change the position of view (Fig.6).

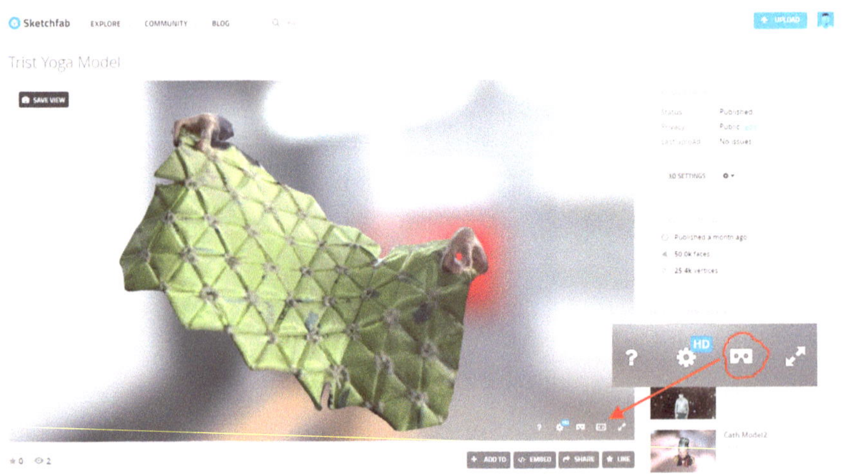

Fig. 5 Sketchfab website with an uploaded model of a 3D scanned object. When seen on a mobile device, and the glasses icon selected, it can be placed in a Google Cardboard compatible headset and viewed in VR

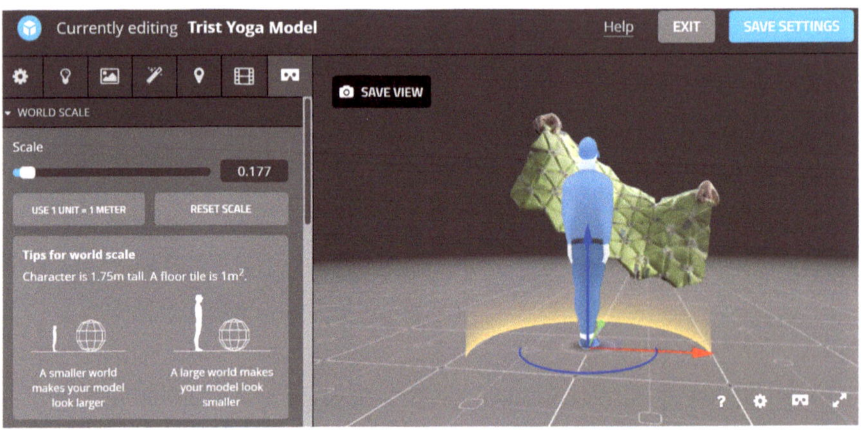

Fig. 6 Sketchfab website showing how a 3D model's qualities can be changed including Virtual Reality viewing properties

5 Dissemination of Content

In order to make content available o tutors or students it is possible to create individual experiences for each video, image or model. However having an app that connects to a server which can deliver the variety of content to a mobile or desktop device would be more sensible. The mEQUITY project has focused on creating an online database of content with an associated app that tutors can log on to and

(1) The Home screen of the Dissemination app allows a person to select their own courses through the My Courses option. This allows them to use an existing course or create a new one. When Use Course is selected existing courses can be viewed by Subject, Location (school, district, city or country) or by the Creator of the course. When Create Course is selected the user can create a new Curriculum or create new Activities for a pre-existing curriculum as well as assign users that can have access to the curriculum as a whole or specific activities.

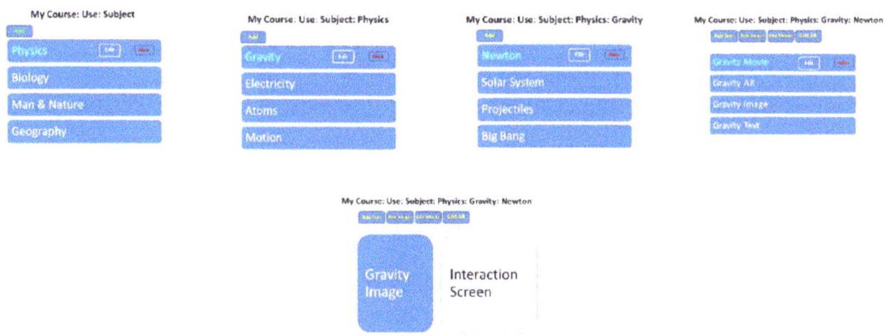

(2) When the user selects Use Course by Subject they can continue to explore by the given subject and its subtopics or use the Add option to add new content at any level. At the deepest level the user can add media including images, video, audio and 3D models in AR or VR form. The final level allows content to be experienced including through an interaction screen to provide a more active way of learning.

Fig. 7 Mock-up of app being created to disseminate Multimedia content created through a central online content server

deliver content into classes with. A refinement of this app is currently being created that will have the workflow shown in Fig. 7 below.

1. The Home screen of the Dissemination app allows a person to select their own courses through the My Courses option. This allows them to use an existing course or create a new one.

   When Use Course is selected existing courses can be viewed by Subject, Location (school, district, city or country) or by the Creator of the course. When Create Course is selected the user can create a new Curriculum or create new Activities for a pre-existing curriculum as well as assign users that can have access to the curriculum as a whole or specific activities.

2. When the user selects Use Course by Subject they can continue to explore by the given subject and its subtopics or use the Add option to add new content at any level. At the deepest level the user can add media including images, video, audio and 3D models in AR or VR form.

The final level allows content to be experienced including through an interaction screen to provide a more active way of learning.

6 Implementation of Content

Once the Multimedia, AR and VR content have been created and packaged a program is needed that will serve as an induction for tutors, staff and students at educational institutions, specifically the universities in Jordan that are partners to developing the mEQUITY deliverables.

The process to do this includes

1. Survey existing needs of learners and teachers.
2. Produce example learning content that might represent what collections of media need to be generated.
3. Train staff to act as trainers for other staff, showing how to generate content using the workflows outlined earlier. Evaluate training and improve areas that could be better packaged.
4. Run pilot trials where workflow training is rolled out to student facing staff.
5. Evaluate feedback from staff and adjust programmes for future rollout.
6. Introduce samples of multimedia or AR content into existing curriculums and expose students to learning from the content generated.
7. Evaluate feedback from students and also through assessment outputs to determine if media needs to be different in delivery or in content.
8. Reiterate the above steps at regular points in academic curriculum delivery cycles.

If the implementation follows the above cyclical process, then there is a good chance that introduction of new technology enhanced learning projects will be successful.

References

1. mEQUITY, Project Website, http://mequity.dipseil.net/
2. Camtasia, Video Editing, https://www.techsmith.com/camtasia.html
3. Aurasma, Augmented Reality Content, https://www.aurasma.com/
4. Unity 3D, Game Engine and 360, 3D, Virtual and Augmented Reality, https://unity3d.com/
5. Vuforia Plugin for Creating Augmented Reality in Unity 3D, https://vuforia.com/
6. Audacity, Sound Editing, http://www.audacityteam.org/
7. Structure IO, Occipital, Handheld 3D Scanner, https://structure.io/
8. Meshlab, 3D Model Editing, http://www.meshlab.net/
9. 3DSMax, 3D Modelling, http://www.autodesk.com/products/3ds-max/
10. Sketchfab, Online 3D Model Viewing, Sharing and Editing, https://sketchfab.com/

Part II
Technology and Innovation: Models and Processes for Managing Firm Performance

Logistic Model-Based Evaluation of Anti-counterfeiting Effectiveness for Consumer Goods: Based on Cognition of Companies and Consumers

Liu Xia, Tang Wanjin, Liu Bisong, Li Ya, Wu Qian, and Liu Tiezhong

1 Introduction

With the development of China's commodity economy, consumer goods are greatly improved both in variety and in quality. However, the huge market impact and security risks brought about by fake and forged goods are worsening. Liu Chuanzhi, the founder of Lenovo, comments (2017) that making and selling fakes is a malignant tumor in China's economy, especially in the development of industries. In response to this problem, the State Council promulgates a special guide to combat the manufacturing and sale of counterfeit and fake goods, and puts forward an objective that by 2020 the rampancy of infringement and fake goods must be checked. However, the existence of a wide range of fake and forged goods, difficulty in screening, and the lack of law enforcement personnel, among other objective conditions, make it hard for the policy market supervision department to have anti-counterfeiting enforcement, such as the fairness of law enforcement and authoritative issues caused by selective law enforcement, sports execution, centralized law enforcement and other law enforcement methods, and the rent-seeking behavior caused by the discretionary power of law enforcement officers at the basic levels. Therefore, it is of great significance to evaluate the efficiency of anti-counterfeiting enforcement concerning consumer goods and to promote the efficiency of government agencies.

Fake or shoddy products are a concept opposite to brand product. Wang Cong and Yang Deli et al. argue that imitations and fakes should be treated differently.

L. Xia (✉) · T. Wanjin · L. Bisong · L. Ya · W. Qian
Quality Management Branch of China National Institute of Standardization, Beijing, China
e-mail: liuxia1010@163.com

L. Tiezhong
School of Management and Economics, Beijing Institute of Technology, Beijing, China

A. Reyes-Munoz et al. (eds.), *EAI International Conference on Technology, Innovation, Entrepreneurship and Education*, Lecture Notes in Electrical Engineering 532, https://doi.org/10.1007/978-3-030-02242-6_12

Grossman and Shapiro were the first to give a definition of imitation and fake. Imitation goods typically have their own brand but with a lower price and try to imitate the quality and performance of brand goods, so consumers with a lower perception of brand value may regard the imitation goods as a substitute for brand goods. Unlike imitation goods, fakes are not distinguished from certified products and generally have the same price with certified ones. Foreign research on fakes tends to concentrate on individual behavior. For example, Cordell studies from the perspective of consumer psychology and concludes that consumers buying fakes is a behavior positively correlated with the expectation for product performance and is negatively correlated with legal attitudes; Luca Casola and Kemp et al. investigate the behavior of consumers who buy stolen goods or fakes in the black market, finding that purchase is more likely when the object to be infringed upon is an organization rather than an individual.

In fact, consumption anti-counterfeiting enforcement is part of government public affair. Zhao Yaxiang argues that the key of the solution to public and private disputes over maintaining the market order is to, through modification of system, and publicity and education, get ordinary consumers more involved in anti-counterfeiting in two aspects: rational driving force and behavior capacity. Since there is a legitimacy crisis in private anti-counterfeiting behavior of consumers, higher requirements are raised for the government anti-counterfeiting actions. However, the inherent mode of government public affairs will inevitably lead to ineffectiveness. Since 2011, with the implementation of the government target responsibility system, the performance management issues of government departments have begun to attract a wide range of attention. However, the connotation of government performance is very complex and rich, involving all aspects of economy, politics and society. So it is necessary to distinguish similarities and differences between enterprise performance management and government performance management. In terms of assessment criteria, the government performance evaluation falls into two kinds, one is the assessment of government activities and their results, and the other the assessment of government capacity. Currently the assessment of government management capacity is all the rage. Some scholars model the form of balanced scorecard to study the performance management issues of local governments, and propose a four-dimension evaluation model, that is, "financial capacity, community satisfaction, internal process, innovation and learning". Seung-Bum Yang and Ador R. Torneo, after studying the performance evaluation system of the Korean government, come up with government performance characteristics such as centralization, hierarchy, harmonious internal relations, and high authority.

2 Research Methods

In order to get an insight into the effectiveness of government anti-counterfeiting enforcement, from July to December in 2016, China's standardization research group created a research team to complete the research project "Research on the effectiveness of anti-counterfeiting for consumer goods". The team proposes two sets of effectiveness evaluation index systems specific to the characteristics of the evaluated object and the results of anti-counterfeiting. They gave a questionnaire to 150 consumers and 55 enterprises. Forty-five copies of valid questionnaire were received from enterprises, and 130 from consumers.

2.1 Samples of Survey on Consumers

170 copies of questionnaires were received, in which 150 were valid, a validity rate of 88%. Among the respondents, women account for 47% and men 53%, a man: woman ratio close to 1:1, indicating that the proportions of men and women are comparable. In terms of age, the survey objects were mainly the young and middle-aged people at the age of 20–50, of them the age bracket of 31–40 accounts for the largest proportion, at 38%; followed by people at the age of 20–30; people in the age bracket of 51–70 account for a small proportion of the total population, only 6%. Therefore, according to the gender ratio and age brackets involved in the samples, the questionnaire is representative. On the other hand, of all the samples, 62% are from cities of China other than Beijing, 24% from Beijing, and 14% from rural areas. Overall, more people from cities were included in the survey.

2.2 Samples of Survey on Enterprises

A total of 45 copies of questionnaires were received, in which 39 were valid, a validity rate of 87%. The survey samples cover all types of enterprises, but no enterprise type is an absolute leader, indicating that the samples are representative. Among the surveyed enterprises, enterprises from the home construction industry and the food industry account for the largest proportion, enterprises from the clothing, shoes & hats, and textile industry ranks third, and the number of enterprises which produce other types of products is relatively flat. Among the surveyed enterprises, 59% make use of anti-fake labels, while 41% do not. Those enterprises with anti-fake labels are more numerous than those without. Of the enterprises operating according to any certification, 32 have ISO9000 certification, 16 have ISO14000 certification, and 10 have ISO18000 certification. Four of them have additional certifications. In terms of registered capital, the surveyed enterprises are generally below RMB 50 million, suggesting that the enterprises surveyed are mostly small and medium-sized, so the survey results may be more representative of small and medium-sized enterprises.

2.3 Data Cleaning

This paper analyzes and cleans the data of evaluation of the effectiveness of anti-counterfeiting from the perspective of consumers and enterprises, respectively.

In Terms of the Effectiveness of Anti-Counterfeiting from the Perspective of Consumers

First, to choose and treat the variables: take as the dependent variable consumers' evaluation of the effectiveness of anti-counterfeiting. For the sake of a higher rigor, the answer "average effect" is interpreted as "no effect" (taking into account of the "moderation" mentality of many consumers, when they give an answer of "average effect" they are actually not very satisfied with the effect).

Second, through a simple data description, we can see that almost every consumer has at one time or another purchased fake or shoddy products, so we having bought, or not, fake or shoddy products is almost meaningless. So in the following modeling process, this variable is neglected.

Third, on the question of the frequency of quality inspection, people who answer "do not know" is almost as numerous as those who answer "quality inspection has never been done," so the two answers are combined into one.

Fourth, in the following four questions: 0B102—Have you ever bought any product that was fake or shoddy? 0B105—Do you know that the local quality inspection department will check such products for their fake and shoddiness? 0B109—Do you think companies have a role to play in controlling fake and shoddy products? 0B111—Do you believe that dealers have a role to play in controlling fake and shoddy products? The answer to the questionnaire should be either "yes" or "no," but more than two types of values (non-null value) have appeared in the questionnaire. After checking, we found that they are input error rather than missing values, so the values are taken as 0/1.

In Terms of the Effectiveness of Anti-counterfeiting from the Perspective of Enterprises

First, to choose and treat the variables: with a balanced consideration of "enterprises' satisfaction with relevant agencies" and "enterprises' evaluation of the role of the government in anti-counterfeiting," it can be found that there is a high degree of relevance between the two variables. So we select "enterprises' satisfaction with relevant agencies" as the final evaluation index.

Second, to choose independent variables: in the questionnaire there are a lot of questions that have nothing to do with enterprises' effectiveness evaluation of government's anti-counterfeiting effort, or questions with similar meanings. With such questions removed, a total number of 26 independent variables remain, including enterprises' emphasis degree on anti-counterfeiting, enterprises' cost of anti-counterfeiting, and benefits gained by enterprises after anti-counterfeiting. The effectiveness of anti-counterfeiting will be evaluated in these three aspects, and the relationship between them and enterprises' satisfaction with the anti-counterfeiting effort of the government will be established.

2.4 Data Analysis

Reliability Test

The results show that Cronbach's alpha value reflects the internal consistency of the questionnaire. When Cronbach's Alpha value exceeds 0.7, and Cronbach's Alpha values of the variables are about 0.7, the reliability of the questionnaire is better. Cronbach's Alpha values provide an indicator of the reliability of the questionnaire, since the retest reliability test cannot be performed (Table 1).

The above table shows that the variables "cultural, educational and sports products, electrical appliances, electronic information products" represent whether consumers buy cultural, educational and sports products, electrical appliances, or electronic information products regularly; correspondingly, the variables "fakes of cultural, educational and sports products, fakes of electrical appliances, fakes of electronic information products" represent whether consumers have purchased fake and shoddy cultural, educational and sports, electrical appliances, or electronic information products. We see that Cronbach's Alpha value is 0.72, and the variables of the Cronbach's Alpha values are about 0.7 or above, so the original questionnaire gives a good reliability.

At the hypothesis stage of the study, consumers' perception of the efforts of quality inspection of the government departments, and the types, frequency, quantity, and adverse effects of purchased fake and shoddy products are factors that affect consumers' satisfaction with anti-counterfeiting enforcement.

Validity Test

Factor analysis and KMO statistics are used in analyzing the structural validity of the questionnaire (Table 2).

It can be seen that the KMO value is greater than 0.7, indicating the original questionnaire is structurally valid. The questionnaire measures the structure of the hypothesis in the design of the questionnaire. The factor rotation load generated by this factor analysis serves also as the basis for further analysis of the index system.

In summary, botht the reliability and the validity of the questionnaire are good, but there are still two possible improvements: question formulation with more powerful distinction degree should be used on the two questions of the degree of consumer satisfaction and consumers' perception of the frequency of quality check;

Table 1 Reliability statistics of the questionnaire

Items	Deleted scale mean	Deleted scale variance	Normalized total relevance	Cronbach's alpha value
Whether they know something about the anti-counterfeit law	12.76	23.371	0.135	0.722
Whether they know something about quality check	14.71	24.300	0.045	0.724
Frequency of quality check	11.91	26.851	−0.281	0.847
Culture, education and sports	14.64	22.702	0.410	0.701
Electrical appliances	14.71	22.222	0.495	0.694
Electronic information products	14.70	23.494	0.217	0.713
Children's products	14.78	22.504	0.423	0.699
Furniture	14.93	21.409	0.691	0.680
Clothing and shoes	14.66	23.398	0.245	0.711
Chemicals and hygiene products	14.78	21.863	0.566	0.689
Transportation	15.11	23.473	0.289	0.709
Food	14.73	23.434	0.226	0.712
Daily necessities	14.67	22.427	0.462	0.697
Fakes of cultural and educational products	14.88	22.197	0.496	0.694
Fakes of electrical appliances	14.90	22.138	0.512	0.693
Fakes of electronic information products	14.91	22.257	0.487	0.695
Fakes of children's products	14.98	22.476	0.460	0.697
Furniture fakes	15.01	21.586	0.692	0.682
Fakes of clothing shoes and hats	14.88	23.150	0.286	0.708
Fakes of chemicals and hygiene products	15.00	22.437	0.480	0.696
Fakes of means of transportation	15.26	24.176	0.239	0.714
Fakes of food	15.09	23.773	0.203	0.714
Fakes of daily necessities	14.84	22.116	0.509	0.693

Note: In general, Cronbach's Alpha value = 0.719, 23 items

Table 2 KMO and Bartlett test

Kaiser-Meyer-Olkin with enough samples		
Bartlett's Spherical Test	Measurement	0.792
	Chi-squared approximant	1156.724
	df	253
	Sig.	0.000

adding the specific frequency and quantity of consumers' purchases of fake and shoddy products better reflects the reality than just asking them whether they have ever bought a certain kind of fake and shoddy product.

3 Modeling Analysis

3.1 Model Design

This is a scoring model based on logistic regression and is often used in quantitatively evaluating default risk and risk factors that involve clients of banks (supporting document). The scoring model consists of a number of feature scores, each corresponding to a question on the questionnaire or an indicator based on a question there (e.g., whether a consumer has ever purchased a fake and shoddy electrical appliance, the indicator that shows "the proportion of each category of fake and shoddy products among the products often bought by customers" according to the purchase situation of the ten categories of products, and the benefits gained by the enterprise after the anti-counterfeiting actions). Each feature includes a number of possible attributes, corresponding to a range of possible answers to each question. In the scoring model, the relationship between each feature and consumer/enterprise satisfaction is firstly determined, and then an appropriate score weight is assigned to the feature in view of the correlation. The higher the score weight, the greater the effect of the feature on the final satisfaction. The score of a consumer/enterprise is a simple weighted sum of its individual feature scores.

The satisfaction degree of the consumer/enterprise is taken as the dependent variable, and the characteristic of the consumer/enterprise as an independent variable, to explore the relationship between the satisfaction degree of a consumer/enterprise and the relevant index. Since this is a two-category dependent variable, so the logistic regression model is used to calculate the weight of the features, assuming that the values of independent variables affect the probability of the overall satisfaction of consumers.

The basic method based on the logistic scoring model is to establish a logistic regression model between the dependent variable and multiple independent variables. According to the established model:

$$\widehat{p} = \frac{\exp\left(\beta_0 + \beta_1 x_1 + \cdots + \beta_t x_t\right)}{1 + \exp\left(\beta_0 + \beta_1 x_1 + \cdots + \beta_t x_t\right)} \tag{1}$$

Assess whether the consumer is satisfied with the probability of anti-counterfeit. After normalization using mark $= 100 * \widehat{p}$, the model is used to evaluate the comprehensive score of consumer satisfaction with anti-counterfeiting actions.

The above model can be transformed by Logit transformation as follows:

$$p' = \beta_0 + \beta_1 x_1 + \cdots + \beta_k x_k \tag{2}$$

$$p' = \ln\left(\frac{p}{1-p}\right) \tag{3}$$

Therefore, the fractional weight of the consumer's individual feature can be estimated by the following expression:

$$\frac{\beta_{\mathrm{i}} x_{\mathrm{i}}}{\beta_0 + \beta_1 x_1 + \cdots + \beta_t x_t} \tag{4}$$

The steps involved in the modeling of logistic are: to use the standard of VIF variance with an expansion factor >10 to cycle out the multiple collinearity variables; to use stepwise regression (AIC optimal criterion) to establish the optimal model in (Eq. 5); and to carry out significance test and interpretation.

When the logistic regression equation becomes known, a method of assessing individual consumers/enterprise satisfaction with anti-counterfeiting is obtained. In order to estimate consumer/enterprise's overall satisfaction with anti-counterfeiting, the interval estimates have been made on the performance of consumer/enterprise overall in terms of various features. And then the estimated value is introduced into the score model to have an overall assessment of consumer/enterprise on anti-counterfeiting work.

According to the theory of sample estimation, the interval estimation method of the overall performance of consumers/enterprises in each feature is as follows:

$$\widehat{x} = \frac{1}{N} \sum x_i \tag{5}$$

$$\mathrm{var}\left(\widehat{x}\right) = \frac{1-f}{N} \sum \left(x_i - \widehat{x}\right)^2 \tag{6}$$

$$\mathrm{sd}\left(\widehat{x}\right) = \sqrt{\mathrm{var}\left(\widehat{x}\right)} \tag{7}$$

$\widehat{x} \pm 1.96\mathrm{sd}(x)$ is the final result of interval estimation

3.2 Model Establishment

In accordance with the modeling steps, what comes first is logistic regression on the satisfaction of consumers, and you will get the following model:

$$\widehat{p} = \frac{\exp\left(- 0.03175 - 0.037x_1 - 0.101x_2 - 0.107x_3 - 0.135x_4\right)}{1 + \exp\left(- 0.03175 - 0.037x_1 - 0.101x_2 - 0.107x_3 - 0.135x_4\right)} \quad (8)$$

where, x_1, x_2, x_3 and x_4 represent consumers' awareness of the frequency of quality inspection, whether they have bought fake and shoddy appliances, whether they have bought fake and shoddy food, and the proportion of the category of fake and shoddy products among the products often bought by consumers respectively. The model is significant and reaches the minimum AIC value.

Therefore, four variables will be put into the model. They are electrical appliance fake, food fake, (consumer awareness) of the frequency of quality check, and the proportion of the category of fake and shoddy products among the products often bought by consumers respectively; since variables from 1 to 4 are ordered in frequency from "frequent" to "almost no," when the coefficient is negative, we can conclude that the higher the frequency of quality check is, the higher the degree of satisfaction will be. It can be seen from the model coefficients of each independent variable that in the mind of customers the less the fake appliance and the less the fake food, the higher the frequency of the government department's quality inspection is and the lower the proportion of the category of fake and shoddy products among the products often bought by consumers is. In this way, the overall satisfaction of consumers (the probability of consumer satisfaction) will be higher.

As can be seen from the model results, food, electrical appliances, and cognitive quality check effort of the government departments are important factors affecting consumer satisfaction. On the one hand, this is in line with our impression that consumer satisfaction with anti-counterfeiting work is related to intensity of government's actions; since food safety is integral to our daily life, customers will be certainly concerned with anti-counterfeiting of food fake and shoddy products; appliances typically involve a large sum of money, so fake and shoddy appliances may bring greater losses to consumers. On the other hand, it also provides us with a viable reference to find out the main influence factors of consumer satisfaction and to carry out more in-depth anti-counterfeiting work on these factors.

Second, in conducting logistic regression on the company's satisfaction, we get the following model:

$$\widehat{p} = \frac{\exp\left(1.2724 + 0.4603x_1 + 0.1644x_2 - 0.6444x_3\right)}{1 + \exp\left(1.2724 + 0.4603x_1 + 0.1644x_2 - 0.6444x_3\right)} \quad (9)$$

where x_1, x_2, and x_3 represent whether the company has a special anti-counterfeiting mechanism (which reflects the enterprise's effort at cooperation with the government in anti-counterfeiting work), the company's own cost for anti-counterfeiting work (which reflects enterprise's own input in anti-counterfeiting), and benefits gained by enterprises after anti-counterfeiting (which reflects the benefits brought to the enterprise after anti-counterfeiting) respectively. The model is significant and reaches the minimum AIC value.

Therefore, three variables are put into the model. The company's own cost of anti-counterfeiting and the benefits gained after anti-counterfeiting work decrease as the variable value rises; given the value of each variable and the model coefficient of independent variables, it can be learned that enterprises which have anti-counterfeiting mechanism (attach more importance on anti-counterfeiting work) incur less cost of anti-counterfeiting, gain more benefits after anti-counterfeiting work, and achieve higher overall enterprise satisfaction (the probability of enterprise satisfaction).

From the model results, we can see that, for enterprises, special anti-fake mechanism, company's own cost of anti-counterfeiting work, and benefits after anti-counterfeiting are important factors affecting enterprise satisfaction. On the one hand, this is in line with our understanding that the company's satisfaction with anti-counterfeiting work is related to the company's readiness to cooperate with the government in anti-counterfeiting work. Enterprises which cooperate more actively are more satisfied with government anti-counterfeiting work; the enterprise will be more satisfied if government anti-counterfeiting work helps reduce enterprise cost of anti-counterfeiting; if government anti-counterfeiting work can bring greater benefits to enterprises, they will be more pleased with the work. On the other hand, it also provides us with a reliable reference to find out the main influence factors of consumer satisfaction and to carry out more in-depth anti-counterfeiting work on these factors.

The interval estimation on the overall performance of enterprises in the three features is then made after the main component described in the modeling steps is integrated, and the results are as follows: First, the proportion of the establishment of the enterprise's anti-counterfeiting mechanism is estimated to be $x_1 = 0.216$, and the interval is estimated at (0.149, 0.283); the enterprise's own anti-counterfeiting cost is estimated to be $x_2 = 2.10$, and the interval is estimated at (1.96, 2.24), so the degree is "moderately high." Second, the consumer awareness of the frequency of quality check is estimated to be: $x_3 = 2.97$, and the interval is estimated at (2.88, 3.06), so the degree is "there is no obvious income."

According to the logistic regression model and the scoring model, which are presented at the model review stage, we have $\widehat{p} = 0.451$. Therefore, the score for the overall performance of enterprises on anti-counterfeiting can be taken as $0.451 \times 100 = 45.1$, that is, enterprises are not satisfied with the anti-counterfeiting work (if enterprises are considered as a whole, there is a 45.1% probability that they are satisfied with the anti-counterfeiting work). From the values of various features, we can conclude that the main reason why enterprises are not satisfied is that in the light of the relationship between the current anti-counterfeiting efforts of the government and its cooperation with enterprises, the enterprise's own cost of anti-counterfeiting is high, and the benefits gained by them after anti-counterfeiting actions is not significant.

3.3 Analysis of Results

After the logistic regression and scoring models have been established, we can draw the final conclusion that the overall satisfaction probability of consumers on anti-counterfeiting work is 0.459, or 45.9%, when consumers are considered in a general case; the overall satisfaction probability of enterprises on anti-counterfeiting work is 0.451, or 45.1%. This shows that anti-counterfeiting work should be enhanced to "make consumers and enterprises more satisfied." On the other hand, the analysis results show that the factors most affecting the overall satisfaction of customers with anti-counterfeiting work are food safety, shoddy appliances (large), quality inspection frequency of anti-counterfeiting work, and the proportion of fake and shoddy products among the products bought by the consumers; the factors that significantly affect the overall satisfaction of enterprises with anti-counterfeiting work are cooperation between enterprises and the government, anti-counterfeiting cost of enterprises, and benefits gained by enterprises after anti-counterfeiting actions.

Therefore, in the process of pursuing anti-counterfeiting enforcement, in order to improve the overall satisfaction of consumers, it is very important to lower the proportion of fake and shoddy products in the market, strengthen food safety, investigate and treat fake appliances, and publicize the effect of regular quality inspection. In particular, from the above factors, we can see that the proportion of consumers having been cheated by fake and shoddy electrical appliances is about 0.4 (40%). Therefore, anti-counterfeiting actions against fake and shoddy electrical appliances should be strengthened. In addition, after the communication and cooperation with enterprises are further strengthened and the related fake and shoddy products are recorded, law enforcement efforts made on important products should be increased to reduce the cost of enterprises and bring benefits to enterprises. In this way, the overall enterprise satisfaction can be enhanced. At present, enterprises which have established a sound anti-counterfeiting mechanism only account for 21%, strongly demonstrating that there are many deficiencies in the communication channels of the Anti-counterfeiting Bureau.

4 Management Recommendations

This study suggests that in order to carry out anti-counterfeiting work more efficiently, the government should focus on key industries and regional special products that people care about, and further improve and perfect the responsibility system in which local governments take the overall responsibility, enterprises take the entity responsibility, and supervisory departments take their own share of responsibilities. Besides, the government should form a work mode managed jointly by the government, enterprises, intermediaries, and consumers, and a large data platform should be created for anti-counterfeiting enforcement in order to promote the sharing and co-governance of quality society and promote economic structure optimization and overall upgrade of product quality level. Specific policy recommendations are as follows:

First, more publicity and input should be given to anti-counterfeiting work. From the questionnaire for enterprises and consumers, it may be concluded that most enterprises and consumers, while calling for a higher frequency of anti-counterfeiting actions, are not quite aware of the government efforts and input in anti-counterfeiting. It can be seen that the government should enhance the publicity of anti-counterfeiting actions and increase the input of law enforcement personnel, funds and equipment according to the characteristics of the region and the products, so as to improve the efficiency of work and scientific normality.

Second, the mechanism that combines anti-counterfeiting work and government performance evaluation should be established. When the appraisal on the government performance of the local government at all levels is carried out, anti-counterfeiting enforcement should be included in the appraisal as an important indicator of evaluation, and a sound government performance evaluation management system and program should be established. Evaluation can be performed at three levels: self-evaluation, mass assessment, and assessment by an assessment team. The evaluation results will be directly added to the year-end performance appraisal results of the local government. Measures should be taken so that anti-counterfeiting enforcement can be emphasized and implemented.

Third, a large data platform for anti-counterfeiting should be constructed. Relevant information and resources from government regulators, local governments, consumer associations, industry associations, enterprises and other anti-counterfeiting products should be integrated. According to the product category, region, time, and other factors, relevant information should be integrated and analyzed to construct a large data platform capable of query, complaints, and comments, and other functions. Moreover, through the platform, anti-counterfeiting data sharing can also be enabled to improve the efficiency of government regulation and reduce the cost of enterprises and consumers in safeguarding their rights and interests.

Fourth, the mechanism of enterprise-government joint anti-counterfeiting enforcement should be strengthened. Focusing on promoting self-declaration of corporate standards and supervision system, enterprises should renovate the law enforcement inspection methods of product quality. In centering on the corporate self-declaration to fulfill the commitment, enterprises should combine the regular and irregular spot inspection of the government with the law enforcement inspection. The inspection results will be uniformly released to the society by the General Administration. Enterprises can also take a variety of anti-counterfeiting solutions, such as anti-fake labels and telephone inquiries, to safeguard their rights.

Fifth, the anti-counterfeiting mechanism that combines the government and the third party should be improved. Forces in various fields should be extensively united to participate in the government decision-making in anti-counterfeiting work. The work and duties shall be divided reasonably among the government and the Consumers Association, the industry associations related to consumer goods, NGOs, and other third-party intermediaries, so as to form a bridge linking the communication and cooperation of the government and third-party intermediaries and to create a social atmosphere in which anti-counterfeiting work is taken up by the whole society.

Sixth, the long-term mechanism of joint law enforcement by ministries and commissions should be established. The new mechanism by which the Ministry of Public Security, the Ministry of Commerce, Industry and Commerce, the Ministry of Industry and other ministries and commissions jointly carry out law enforcement should be established to achieve a cross-ministry work mode of "information sharing, early intervention, joint action, complementary advantages." It is recommended that AQSIQ take the lead in combating counterfeiting of consumer goods, convening inter-ministerial joint meetings, and discussing the cooperation mechanism of law enforcement in a timely manner. At the same time, the ministries should reach consensus on centralized publicity and unified action of "anti-counterfeiting," promote the effective convergence of administrative law enforcement and criminal justice, and raise the deterrence and influence of anti-counterfeiting work for consumer goods.

Acknowledgments This paper has been funded by the national key research and development project "Research on key technical standards for quality and safety control of consumer goods" (2016YFF02022600), and the central basic research business project "Research on method for consumer product safety risk assessment based on the simulation technology—taking children's products as an example" (552015Y-3990), and the project "Development of general chemical safety standard for 10 consumer products" (201510203).

References

1. M. Hao, Research on anti-counterfeit enforcement governance of product quality. J. Guangxi Univers. **6** (2016)
2. M. Chen, S.X. Tong, X. Youquan, Z. Boyi, Quality Supervision Bureau, Qingpu District, Shanghai: Practice and thinking on network anti-counterfeit enforcement. China Qual. Tech. Super. **10** (2015)
3. Y. Tengfei, H. Maqin, Thinking on the issues of "Large Numbers of Fakes" and "Hard Anti-counterfeit" from the perspective of economics. Contemp. Eco. **4** (2013)
4. Z. Zhu, Differences between domestic anti-counterfeit and international anti-counterfeit. Qual. Standard. **12** (2016)
5. Z. Huijun, Exploration on the issues of professional anti-counterfeit. Qual. Stand. **3** (2017)
6. S. Jingjing, Combating against the "Latent Rules" of Fake Products – "Sword of Quality Control" Typical Case of Anti-counterfeit Special Action announced by AQSIQ in 2016, China Quality Technical Supervision 2016/12 (2016)

7. B. Qianwen, Z. Chunyan, Static comparison and dynamic analysis of the hazards of fake and forged products. J. Shandong Inst. Bus. Technol. **3** (2016)
8. B. Zheng, Explanation on the difficulties of governing the Fake and Shoddy Phenomenon in China Based on Game Theory. Modern Econ. Inf. **11** (2016)
9. Q. Jinhua, L. Kai, On the current professional anti-counterfeit phenomenon and response. China Bus. Admin. Res. **7** (2014)
10. Q. Ruoying, Current situation and countermeasures analysis of rural fake and Shoddy goods market. Enterprise Technol. Dev. **9** (2016)
11. Z. Zhu, Differences between domestic and foreign anti-counterfeit. Qual. Standard. **12** (2016)
12. L. Jinchun, Why do fake and shoddy goods appear endlessly - Economic thinking of anti-counterfeit measures. Modern Econ. Inf. **4** (2014)
13. Z.A. Collier, S. Walters, D. DiMase, et al., A semi-quantitative risk assessment standard for counterfeit electronics detection. SAE Int. J. Aerosp. **7**(1), 171–181 (2014)
14. S.H. Ang, P.S. Cheng, et al., Spot the difference: Consumer responses towards counterfeits. J. Consum. Mark. **18**(3), 219–235 (2001)
15. V.V. Cordell, R.L. Kieschnick, et al., Counterfeit purchase intentions: role of lawfulness attitudes and product traits as determinants. J. Bus. Res. **35**, 41–53 (1996)
16. Tan, Understanding consumer ethical decision making with respect to purchase of pirated software. J. Consum. Mark. **19**, 96–111 (2002)
17. S. John, Cracking down on counterfeit parts. Assembly **59**(7), 69–77 (2016)
18. A. Chia, L.S. Mee, The effects of issue characteristics on the recognition of moral issues. J. Bus. Ethics **27**(3), 255–269 (2000)
19. J. Kujala, A multidimensional approach to Finnish managers' moral decision-making. J. Bus. Ethics **34**, 231–254 (2001)
20. T.W. Dunfee, N. Craig Smith, W.T. Ross Jr., Social contracts and marketing ethics. J. Mark. **63**(3), 14–32 (1999)
21. D. DiMase, Z.A. Collier, J. Carlson, R.B. Traceability and risk analysis strategies for addressing counterfeit electronics in supply chains for complex systems. Risk Anal. **36**(10), 1834–1843 (2016). https://doi.org/10.1111/risa.12536
22. D. Jin, Y. Qing, Y. Lihua, *A Study of Moral Decisions on Consumer Purchasing Counterfeit Product Behavior*. 2011 International Conference on Computer Science and Service System, CSSS 2011. pp. 2937–2942

Idea Validation for an Acceleration Program Connected with Equity-Based Crowdfunding

Rastislav Petráš, Jaromíra Vaňová, Martina Horváthová, and Miloš Mrva

1 Introduction

In recent economic environment, start-ups play important role in local and global economy. Focus on flexibility, shortening of life-cycle of the companies and pressure on cost reduction and accelerating time to market provides market gap for start-ups. Large companies have significant advantage in economies of scale or in better access to financial sources, on the other hand, they have limited flexibility and longer reaction time for change.

Number of institutions and entities is already aware of the importance and vulnerability of start-ups and tries to support them. For instance, we can mention incubators, accelerators, business angels and other institutions and individual investors and supporters.

Start-ups generally face many problems, especially in the early stages of its existence. One of the most common ones is lack of capital needed for the start of the business. Start-ups are trying to fill gaps in financing with external financial sources. However, the ability of start-ups to receive external funding is usually weak.

In this context, equity-based crowdfunding starts to be rapidly growing tool of external funding. Equity-based crowdfunding allows start-ups to raise capital through online platforms that have an impact on a large group—crowds of investors. The second tool, which is mentioned in the context of support of start-ups is the

R. Petráš (✉) · J. Vaňová · M. Horváthová
Institute of Industrial Engineering and Management, Faculty of Materials Science and Technology in Trnava, Slovak University of Technology in Bratislava, Trnava, Slovakia
e-mail: rastislav.petras@stuba.sk; jaromira.vanova@stuba.sk; martina.horvathova@stuba.sk

M. Mrva
Fakulta Managementu Univerzity Komenského v Bratislave, Bratislava, Slovakia
e-mail: milos.mrva@fm.uniba.sk

A. Reyes-Munoz et al. (eds.), *EAI International Conference on Technology, Innovation, Entrepreneurship and Education*, Lecture Notes in Electrical Engineering 532, https://doi.org/10.1007/978-3-030-02242-6_13

acceleration program, which also include the venture capital instruments, as they often provide seed capital to their participants. In addition, other elements of these programs are mentoring and networking which further increases the chances of the participants to succeed. We want to underline these facts as these features make accelerators more than just a source of finances.

Both instruments equity-based crowdfunding and acceleration programs are new tools that arose a few years ago but their use is growing rapidly every year. Our paper examines the possibility of linking these two instruments to achieve the reinforcing of the efficiencies that achieve these tools separately without any interconnection. In this paper, using the lean startup methodology we validate potential interest of start-ups in such program which combines the equity-based crowdfunding and acceleration program.

2 Background

Start-Up and Its Definition To define what a start-up is we shall cite Silicon Valley "Godfather" Steve Blank who describes it as a company, a partnership or temporary organization designed to search for a repeatable and scalable business model [1]. We could also make a reference to the work of Eric Ries, the author of the lean startup methodology who sees a start-up as a human institution designed to create a new product or service under conditions of extreme uncertainty [2]. As noted in Business dictionary, we can besides this say that a start-up is early stage in the life cycle of an enterprise where the entrepreneur moves from the idea stage to securing financing, laying down the basis structure of the business, and initiating operations or trading [3].

Crowdfunding With crowdfunding, an entrepreneur raises external financing from a large audience (the crowd), in which each individual provides a very small amount, instead of soliciting a small group of sophisticated investors [4].

The categorization of the four main types of crowdfunding (donation-based, reward-based, lending, and equity) is based on what, if anything, investors receive for their contributions.

Equity-based crowdfunding is a form of financing in which entrepreneurs make an open call for funding on the Internet, hoping to attract a large group of investors. The open call and the investments take place on an online platform (e.g., Crowdcube) that provides the means for the transactions (the legal groundwork, preselection, the ability to process financial transactions, etc.). In recent years, the equity branch of crowdfunding has become an increasingly important financing alternative for companies and especially for start-ups, and volume has doubled every year since 2009 [5].

Bradford [6] explains equity crowdfunding as a model in which funders receive an interest in the form of equity or equity-like arrangements in the ventures they fund. Belleflamme et al. [7] point out that the central difference between equity crowdfunding and traditional capital-raising is the funding process itself: Entrepreneurs make an open call for funding on a crowdfunding platform, and investors make their decisions based on the information provided therein. Moreover, the crowdfunding platform facilitates the transaction by providing a standardized investment contract and settling the payments.

We believe that this form of financing opens up new opportunities in the capital market, since it can allow (depending on individual country law) nonprofessional investors from the public to get a share of ownership in starting innovative company in exchange for capital. Ordinary people would be able to buy a small share of innovative enterprises far before their IPO. Of course, with different terms. The biggest limitation is low liquidity. The sale of this equity is far less liquid in comparison to shares of companies traded on the stock exchange. Moreover, the amount that can be raised by using the equity-based crowdfunding is significantly lower when we compare it to the IPO. Unlike shares, for equity-based crowdfunding is regulated the upper limit of amount which can be raised and this amount depends on the law of a particular country. The highest limit is at the moment in Italy—five million euros within 12 months. Bradford [6] nevertheless believes that equity-based crowdfunding opens up new opportunities for people to invest and source of finance for businesses.

Accelerators Accelerator with its acceleration program is the second tool for financing early stage start-ups which we study within our research and which we want to link to equity-based crowdfunding in order to create new tool or concept of acceleration program which will increase the benefits of equity-based crowdfunding.

Accelerators are accepted as a *source* of finance as they often provide to their participants seed capital. However, acceleration programs give participants more than just capital. They last normally between 3 and 4 months and provide besides capital also other services. We shall refer to the work of Miller and Bound [8] who list in their study called "Startup factories" the following four main functions of acceleration programs.

1. Funding—money that accelerators give to participants are a valuable attribute for people who apply to participate in the acceleration program. The amount of money offered varies from program to program. However, the benefit of this funding was rarely rated by participants as the most important element of the program. The main advantage of funding identified by the participants was that it allowed them to focus on their business during the program full-time, without having to rely on another source of income.
2. Mentoring—Accelerators give founders of new businesses a chance to meet people from industry, get feedback on their products and learn tips and hints how to deal with obstacles they have to face. Accelerators such as Seedcamp, TechStars, and Springboard cooperate with external mentors, while Y Combinator invite external speakers for their group events and individual mentoring is carried out by managers of this accelerator. Quality and relevance of such advices have been

confirmed by a large number of entrepreneurs who believe that the guidance on this extent and quality would be difficult to get without participating in the accelerator.

3. Introductions to investors—Participants of accelerators are introduced to investors. Such introduction is usually difficult to achieve for a small business in this extent. Since accelerators represent for investors a source of new investment opportunities, which have gone through the selection process of the accelerator, investors often want to get to know participants of respected acceleration programs and attend their demo days where companies graduating from the program present their products in front of sometimes hundreds of investors. In fact, most accelerators have built its reputation on the basis of data on what proportion of their graduates proved shortly after their program to get a new investor.
4. Building credibility—Entrepreneurs identified that the fact that they participated in the program increased their credibility and value in front of investors, customers or even the attention of journalists. It helps them to increase the trust in their product and trust in the company if they can say that they have been selected by an accelerator as a "promising business," because accelerators are known for being very selective. The value of this acknowledgment, however, is associated with how credible and respected is a particular accelerator.

Demo day is important aspect of the accelerators. This is the culmination of the whole program. Participants have an opportunity to present their businesses and achievements to business angels and venture capitalists to get their attention and prepare the ground for further communication and negotiations with investors. These events provide the participating teams access to large groups of investors. Moreover, demo day is an opportunity to present their products or services to the public as the media coverage of these events is usually significant.

3 Integration of Typical Acceleration Program with Equity-Based Crowdfunding Campaign

We already described acceleration programs and equity-based crowdfunding. The importance and usage of both tools are growing thanks to their capability to support start-ups at the early stage of their life cycle. However, we believe that suitable combination and synthesis of these two tools could bring bigger value to its users than these tools provide separately without each other. You can see basic outlines how this could combination could look like on following Fig. 1.

Left part of this figure refers to an example of structure of a regular acceleration program. Participating start-ups are selected during selection process before the program. Afterwards start-ups deal with topics such as fine-tuning their product, working on their marketing strategy and similar. The entire program normally culminates with Demo Day where companies present their projects to investors

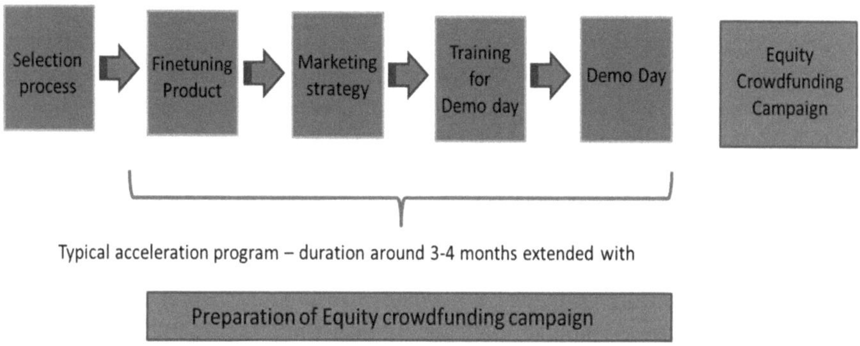

Fig. 1 Outlines of combination of acceleration program and equity-based crowdfunding (drawn by authors)

and further communication is held after the program with investors which showed interest in a company after the Demo Day. We suggest extending such regular program with equity-based crowdfunding which would start at Demo Day so projects would not just passively present their product to investors. In addition, they would also actively start raising money from investors within the program. This could be done theoretically without any integration of crowdfunding campaign and acceleration program too. Campaign can just be a next step after acceleration program. However, we believe that deeper integration of these two tools would positively affect final outcomes. Many crowdfunding campaigns fail as many entrepreneurs underestimate knowledge and efforts required for a crowdfunding campaign, and a campaign fails because it is not prepared properly. Within such combination of acceleration program and equity-based crowdfunding, preparation of the campaign would be part of the acceleration program so entrepreneurs would work meanwhile on their campaign too with the assistance of mentors who are seasoned crowdfunding experts. In our opinion, the combination of acceleration program and equity-based crowdfunding would increase chances of entrepreneurs to succeed because of the following advantages created by merging these two tools:

1. Start-ups raising money through equity-based crowdfunding will have bigger chance to achieve their target amount of capital they want to raise after participation in acceleration program which will help them to prepare their campaign properly. Structured way of preparation of crowdfunding campaign would be a part of the program.
2. Entrepreneurs coming to the crowdfunding platform from an accelerator will have bigger chance to successfully raise the money because investors will trust more such companies who graduate form an accelerator as they know that these accelerators help them to understand how they need to manage their business to succeed. They help them to adopt state of the art methodologies and leverage experience of other start-ups and mentors to properly deal with crucial aspects of running a start-up company and avoid common mistakes of new companies.

3. Entrepreneurs coming to the crowdfunding platform from an accelerator will have bigger chance to successfully raise the money because investors will trust more such companies as they know that these companies are not average companies since they already passed selection process of the accelerator. The fact that a company already passed a tough selection process before the acceleration program could increase their credibility.
4. Entrepreneurs coming to the crowdfunding platform from an accelerator will have bigger chance to successfully raise the money as they would be normally only presented to investors within regular acceleration program. Thanks to equity-based crowdfunding they will get an exposure into theoretically unlimited number of investors who can immediately invest into their company in shorter time than normally.
5. Entrepreneurs from regions with weak start-up ecosystem can get this way important support in their region without any need for moving to other place as crowdfunding is purely web based service and acceleration program can be organized online too. This can enable cross border investments as well as communication between mentors and start-ups from different regions/countries.

4 Lean Startup and Its Principles

We suggested above a concept combining equity-based crowdfunding and acceleration program what can be understood also as an idea for a new product, where the program is a product and start-ups are customers. We decided to use in this paper lean startup methodology to assess the potential of such product even without any need to run any experimental program with those features.

Lean startup is a methodology for start-ups which help them to develop the right product which its customers are willing to buy efficiently and as soon as possible using ongoing interaction with its customers. Eric Ries's seminal work laid the foundations for this methodology. His ideas on developing a new product were set out in his book The Lean Startup. A lot of accelerators apply this methodology while working with their participants and we want to use it too to prove that start-ups would be interested in participation in the accelerator which would put together typical acceleration program with equity-based crowdfunding as we suggest.

Too many start-ups begin with an idea for a product that they think people want. They spend months, sometimes years, perfecting that product without ever showing the product, even in a very rudimentary form, to the prospective customer. When they fail to reach broad uptake from customers, it is often because they have never spoken to prospective customers and examined whether or not the product was interesting.

A base hypothesis of Lean Startup methodology is the build–measure–learn feedback loop. The first step is figuring out the problem that needs to be solved and then developing a minimum viable product (MVP) to begin the process of learning as quickly as possible. Once the minimum viable product is established, a start-up can work on tuning the engine. This will involve measurement and learning and must include actionable metrics that can demonstrate cause and effect question [2].

Specific form of the minimum viable product is “Fake door” method. Start-ups can quickly see whether customers will engage with a new feature by launching just the first part of it. Instead of laboriously building the whole feature, you just launch the first button. When start-ups observe a large number of visitors clicking the button to access, they know that they are onto something and can build the rest of the feature. We decided to use this method to assess interest of start-ups in the new concept of acceleration program combining typical acceleration program with equity-based crowdfunding campaign to raise funds for graduates of this acceleration program.

5 Methodology

The main goal of this paper is to validate whether start-ups could be interested in acceleration programs connected with equity-based crowdfunding and in addition partially feasibility of this model from the perspective of costs needed for the acquisition of new customers (start-ups applying for the program) compared to the potential income. Based on the facts and assumptions described above we can say that there are several reasons why we could expect positive outcomes of such extended program. These outcomes might be motivation for early stage start-ups to apply for such program to increase their chances to launch successfully their product and raise funds for further growth of the company. However, we similarly described basics of lean startup methodology which teaches us to test such assumptions and validate idea with customers before the development of any product. That is how we decided to proceed also with our idea of combining acceleration programs with equity-based crowdfunding campaign. Validation can be used for different areas and assumptions, for the purpose of this paper, we decided to validate following questions or assumptions.

5.1 Validation of the Problem (Need)

Firstly, we wanted to validate that there is an existing problem among start-ups which the product, in this case combination of accelerator program and equity-based crowdfunding, can solve. It means it is important to prove that there is a need on the market for such product. In this case it means that it is important to show that there is a problem which can be solved by the solution which we suggest. We decided to validate this assumption indirectly. We surveyed the external sources to prove that there is a problem which accelerators and equity-based crowdfunding portals aim to solve and which start-ups have to deal with.

5.2 Validation of the Solution

This validation aims to prove or reject our assumptions that start-ups would be interested in such a service like combination of acceleration program and equity-based crowdfunding. We mentioned already in the background lean startup methodology and one of its important fundamentals is to check before the development and realization of an idea that there is a demand among its customer segment for such solution. For this validation, we decided to use the method of offering a fake product called "fake door" which we described above. The main idea of this technique is to offer our solution to our customers already in the beginning when this solution does not exist at all. The reason of doing this is that this method reveals more accurate data than questionnaires or focus groups since sometimes people answer in questionnaire in one way, but at the end in real practice their customer behavior is different. Whereas thanks to offering them fake product (fake door method), we could study their real behavior as they were not aware of the fact that such program which we offered them does not even exist.

5.3 Validation of the Business Model

We offered within this experiment a program which would not charge them any fixed payment for the participation in the program but 6% of the amount which they would raise during the crowdfunding campaign within the program. We have to admit that it is complicated to validate this model at this time as it is difficult to estimate total costs of realization of this program. We decided to realize this validation at least partially through the comparison of acquisition costs or in other words estimated marketing costs necessary to get enough applicants for one selected participant with potential income per one participant. We used data from already existing onsite accelerators to figure out how many applicants they need on average to find one suitable start-up selected for the participation in the program. Afterwards we multiplied this number by the amount of marketing costs which we needed in our experiment to get one applicant. Potential income from one customer was calculated as a 6% of the amount which is normally raised by graduates from the accelerators shortly after the program and by start-ups raising money through equity-based crowdfunding.

This part of the experiment cannot completely confirm feasibility of this business model as we do not have relevant sample about other costs but we can still reject this model if acquisition costs are higher than potential income or very close to the income what would not leave much space for other costs arising from the realization of the program.

5.4 Procedure of the Experiment

We decided to use for this experiment combination of Facebook ads and landing page to promote start of our fake program, offer start-ups an opportunity to apply for coming program and follow if there is any positive feedback and demand for this program in the form of applicants. We needed to prepare for the realization of our experiment:

Landing Page We created landing page promoting fictitious company/accelerator program which combines acceleration program with equity-based crowdfunding. We were very careful with the content and the way how we promoted this program as we had to emphasize and to pay the biggest attention within our value propositions on the fact that start-ups can participate in the program which is unique by its combination of acceleration program and equity-based crowdfunding. We had to avoid promotion of the program through presenting its mentors, previous successes and so on as we had to follow reactions of people referring to the fact that we offered them a new concept of acceleration program. That is why all the promotion and the descriptions had to be focused on the presentation of this concept. Including other information would make it difficult to distinguish whether reactions of people refer to the new concept we suggest or something else. For instance people could be interested in this program, but for its famous mentor, not the program itself. Therefore, we made website simple and focused on combination of equity-based crowdfunding and acceleration program.

Top of the page included just following basic short description of our fake product: "Accelerate with EBAcceleration! EBAcceleration is an acceleration program for early stage start-ups which helps them during three months not only to fine-tune the product and meet investors on demo day, but also raise money through equity-based crowdfunding from investors from different countries."

Below we provided visitors of this website more detailed information about the program, explanation why is this program different from other acceleration programs, definition of equity-based crowdfunding and we furthermore explained how this program works (especially connection of equity-based crowdfunding and acceleration program). Moreover, we listed in six bullets key features and benefits of the program which were mostly focused on features of the concept which we suggest.

The call for action of this landing page was call for applications for upcoming program. We promoted within this website an opportunity to apply for the coming program which should start few weeks after the time of this experiment. Website included two apply buttons which did not direct "applicants" to any application, just to another link where they figured out that this program is only part of our experiment.

In order to increase authenticity of the website, we added on the bottom of this landing page contact form which visitors could use for asking their questions regarding to the program.

Landing page is still possible to visit here http://ebacceleration.wix.com/apply.

Facebook Campaign Landing page itself was not sufficient for this experiment. We had to promote this landing page and get exposure for this simple website. In order to achieve this we created a small Facebook campaign with the budget of 40 €. We used for this experiment "external website ad" located on the upper right part of Facebook website. Title of this ad was "Accelerate your start-up" with following description "New concept of program connecting typical acceleration program and equity crowdfunding" Of course, this ad included also link to our landing page so the flow was as following. People saw a Facebook ad promoting our acceleration program, then they could click on the link and continue to landing page with description of this fictitious program where they could apply for the program.

For this experiment we decided to address people from six countries. We decided to create two groups of countries:

1. Countries from western Europe with well developed economies and new legislative for equity-based crowdfunding which legalizes usage of equity-based crowdfunding in these countries and eventually creates good environment for its development with maximum amount possible to raise through equity-based crowdfunding over 1 mil. €—France, Spain, Italy.
2. Countries from central and eastern Europe which do not support equity-based crowdfunding through their legislation and maximum limit for equity-based crowdfunding campaign under 1 mil. €—Lithuania, Poland, Romania.

We created for each country small crowdfunding campaign separately which aimed to get 50 clicks on the ad, what means 50 visitors for landing page from each country. This campaign was executed between 5 August 2014 and 11 August 2014.

Target groups for this ad in each country consisted of people between 18 and 50 years, both men and women, and who have between their interests at least one of following items or keywords: Business incubator, Crowdfunding, Design thinking, Entrepreneur, Lean startup, Product design, Seed accelerator, Silicon Valley, Startup company, Startup Weekend, TechCrunch, or Wayra.

Tracked Variables and Tools for Tracking of the Statistics In terms of validating the interest of customer it is not enough only to get applicants (or not). For the validation of the idea we had to also track more detailed statistics about behavior of people which we addressed through our Facebook ads and later via landing page. Below we list the variables we decided to track.

1. Click through rate of our Facebook ads—click through rate represents the number of clicks divided by number of impressions per ad—is an effective metric for measuring user response. The higher the click through rate is, the greater likelihood a customer will sign up to follow a brand, or purchase something on the ad's other end [9]. Click through rate basically refers to the ratio of how many times people see the ad, and how many times people click on this particular ad. Higher click through rate means higher interest of customers in offered product.

2. Conversation rate—conversion rate refers to the ratio of the people who became our customers (applied for the program) and people who visited our landing page. In other words, this variable says how many people, out of all people visiting our landing page, decided to apply for our program after reading information of our program.
3. Acquisition costs of one applicant—total budget of the campaign divided by total number of applicants. This number will be used for the calculation of the estimated acquisition costs of one participant of the program what will be compared to potential income for the purpose of validating the business model as we discussed above.

To simplify these information about tracking of our statistics, we can say that we followed mostly how often people clicked on our Facebook ad and how often people decided to apply for the program after they visited our landing page promoting the program.

Evaluate Results of the Experiment We will compare click through rates and conversation rates achieved within our campaign with the results of other campaigns to figure out whether the interest of our target group was within campaign similar, higher or lower than the average. We calculate estimated acquisition costs of a customer based on the acquisition costs of one applicant and the number of applicants needed in the selected accelerator for one selected participant, (in this case actually user – selected participant of the program). Afterwards we compare these costs to potential income.

6 Results

6.1 Validation of the Problem (Need)

We believe that in the case of this product is validation of the problem quite simple and does not require any field experiments to prove it. The main purpose of equity-based crowdfunding is to facilitate start-ups the access to finance what acceleration programs aim to solve too. Moreover, acceleration programs help start-ups to get ready for the market and make the management of their participants as efficient as possible. It is widespread fact that start-ups often have to face problems with financing of their activities and problems regarding to the management of the company as they often fail to develop their product properly in respect to the market needs, they fail with the presentation of the product, building the team while the company is growing and so on.

The list of the most serious problems of SMEs is treated in the report of European Commission. The report includes results of a survey in which entrepreneurs from SME responded to the question, what are the major issues they have to deal with. We want to use these results to indirectly prove that there is a need in the market for

Table 1 Summary of Facebook campaign for imaginary program

Country	Reach	Freq.	Impressions	Clicks	CTR	Costs (EUR)	CPM (EUR)	CPC (EUR)
Romania	11,866	4009	47,574	50	0.105	3.94	0.083	0.079
Poland	9599	3848	36,937	50	0.135	5.03	0.136	0.101
Lithuania	651	8712	49,234	50	0.102	3.9	0.079	0.078
Italy	17,306	2049	35,465	50	0.141	6.82	0.192	0.136
France	11,363	3337	37,922	50	0.132	9.56	0.252	0.191
Spain	10,816	2962	32,038	50	0.156	8.91	0.278	0.178
RO+PO+LT	27116	4932	133,745	150	0.114	12.87	0.099	0.086
IT+ES+FR	39485	2670	105,425	150	0.143	25.29	0.241	0.169
Total	66601	3591	239,170	300	0.128	38.16	0.170	0.127

such financial tools. Thanks to this we can prove the first important assumption—existing need for our solution. As documented in this survey, access to finance and lack of managerial skills are the second and the fourth most common problems that SMEs must face so there are apparently problems which accelerators and equity-based crowdfunding portals want to solve [10]. Moreover, it is very important that entrepreneurs named these problems as it shows that they are aware of these obstacles what means that they probably eventually look for solutions for these problems.

6.2 *Validation of the Solution*

As we discussed in the section devoted to the methodology, we followed interest of target group into program which combines acceleration program with equity-based crowdfunding. We created Facebook campaign for imaginary product with such features and followed their reactions on this product firstly through click through rate on external ads on Facebook. More than 66,000 people from our target groups in six countries saw our Facebook ad of our fake acceleration program with the average frequency of 3.591 impressions for one person what makes all together almost 240,000 impressions of our ad. This number of impressions was necessary to get 50 clicks from each country. You can see the exact numbers of clicks, impressions and other stats for each country, group of countries and all countries together in the Table 1. Click through rate (CTR) of the campaign which indicates interest of viewers of the ad in the product reached the level 0.128% what could look like small number. However, the average value for this kind of advertisement on Facebook (External Website Ad) is only 0.02%.

Andrew McDermott, expert on Facebook advertising, dealt with the topic of ideal CTRs in the detailed benchmark of click through rates on Facebook where he describes success of Facebook campaign based on its click through rate based on the normal distribution of click through rates of campaigns on Facebook. According to his benchmark, average click through rates varies between 0.04% and 0.05%. Our campaign for fake acceleration program combined with equity-based crowdfunding achieved very good results according to this distribution as it belongs with its 0.128% into the top category of optimal click through rate, based on the benchmark of McDermott [11]:

Optimal CTR: 0.11–0.16%
Above Avg CTR: 0.07–0.09%
Average CTR: 0.04–0.05%
Below Avg CTR: 0.02–0.03%
Poor CTR: 0.01%

We can see a difference in the results. In western countries with legislation more favorable towards equity-based crowdfunding (Italy, Spain, France) click through rate reached even higher level of 0.143%. On the other hand, results were lower in the countries of Central and Eastern Europe (Poland, Romania, Lithuania) with legislation facilitating equity-based crowdfunding less, though still above average with 0.114%.

CTRs of our campaign indicate high interest of people who saw our ad of the acceleration program combined with equity-based crowdfunding. Nevertheless, this number could be potentially even higher because we targeted within this campaign people who are interested in start-ups (targeting according to their interests) but it does not mean that all of them are involved in a start-up. That is why we assume that this ad had part of its impressions showed to people for which this ad is not relevant. However, we still had to work with this limitation as it is difficult to separate within Facebook campaign people who are only interested in start-ups.

The second statistic we tracked was conversion rate. As we said above 0.128% of people who saw our ad on Facebook clicked on that and continued to our landing with more detailed description of the program and the option to apply for the coming program. All together 300 people visited this landing page and conversation rate refers to the number of people who applied for the program out of all visitors. We got during this short campaign five applicants which gives us the conversion rate 1.67%.

According to the conversation rates distribution prepared by Kim [12], this value is below average but still not among the worst results. However, we can again see difference between western countries and countries from Central and Eastern Europe. At the same time the conversion rate reached in western countries the level of 2.67% what is already above average of campaigns analyzed by Kim [12].

To conclude, the results of conversion rates are slightly worse than results of click through rates of Facebook ads though still at the level where we can say that there is existing interest of potential customers into this product/program. Especially in Western Europe countries where the conversation rate exceeded conversation rate of average campaign.

Still we believe that real program with features we wanted to simulate would achieve higher results as promotion of this fake program has several limitations. We have identified several factors which in our opinion decreased conversation rate on our landing page:

1. We did not promote on our landing page top class mentors, testimonials from participants of previous programs or value of various features (which are normally part of acceleration programs and would be probably part of promotion of such program) since we wanted to follow how people react on the idea of combining acceleration program and equity-based crowdfunding so we focused the entire landing page on the promotion of this concept. We assume that for example promotion of respected mentors could cause that we would get applicants who would apply because of their interest in the cooperation with those mentors, not because of the interest in program with features we suggest what would decrease value of data got from this experiment. We believe that provider of such program would better promote also other features what would increase the interest as well as the conversion rate.
2. As far as the quality of the landing page is concerned, authors of this paper are not marketing experts and we did not cooperate with any during the preparation of our landing page. We believe that more complex pages for real programs, prepared by marketing experts will reach higher numbers thanks to the higher quality of the landing page inviting people to apply for the program.
3. As a fake product, program which we promoted does not have any online presence. We assume that people considering the option of applying for an acceleration program do not gather information only from the website of its provider but search also for other information and references on the internet. We believe that part of people who visited our landing page tried also to find some other information from other sources what decreased the number of applications and conversation rate as they did not find any other information.

6.3 Validation of the Business Model

You can find insights related to this validation in the Table 2. Total costs of the campaign were 38.16 € and we reached 5 applicants what makes 7.63 € per one applicant. We used for the comparison of estimated acquisition costs per one participant and potential income from one participant historic data of Tech Stars accelerator. This world known accelerator accepts for their program only 1% of applicants what means that if we wanted to run such selective program as Tech Stars is, we would need 100 applicants to get one participant. As we mentioned acquisition costs per one applicant within our campaign were 7.63 € so the program would need according to these results 763 € for marketing online campaign to get one participant. Of course, we have to admit that the reliability of this result is very limited as it is based on five applicants what is not a representative sample and we are going to repeat this experiment in bigger scale in the future.

Table 2 Conversion rates and comparison of acquisition costs with potential income

Countries	Website visitors	Applicants	Conversion rate (%)	Costs (€)	Costs per applicant (€)	Number of applicants per participant	Acquisition costs per participant (€)
RO,PO,LT	150	1	0.67	12.87	12.87	100	1287
IT,ES,FR	150	4	2.67	25.29	6.32	100	632
Total	300	5	1.67	38.16	7.63	100	763

Business model we tested is based on the commission from the amount of money raised through the program (6%). We wanted to calculate this potential income in two ways. The first one would use as a baseline average amount of money raised by graduates from acceleration programs and the second one would use as a baseline amount of money normally raised through equity-based crowdfunding. We again drew the data from Tech Stars to get information referring to the accelerators and report on equity-based crowdfunding market to figure out average amount of money raised by start-ups by using this source of finance. It turned out that both amounts are almost identical and reach around 1.5 mil. €. Therefore, if we expect that graduates of the program we suggest would raise on average 1.5 mil. € as graduates of the acceleration program and as start-ups raising money through equity-based crowdfunding at the same time, then the estimated potential income as a 6% of this amount could reach the amount 90,000 €.

Estimated acquisition costs per one participant 763 € look like too high, it is still less than 1% of estimated income from one participant. For this reason we believe that this business model based on commission looks like viable model at least from the perspective of marketing costs as it leaves more than 99% of the income for the costs covering realization of the program and profit. Naturally this is not yet the solid evidence for the business model validation as this simplified experiment takes into consideration only marketing costs regardless the rest of the cost structure.

7 Conclusion

We provide in this paper a brief description for the areas of acceleration programs and equity-based crowdfunding as a baseline for the introduction of a new model of a program combining acceleration programs with equity-based crowdfunding. We also briefly presented principles of the lean startup methodology which we used in this paper to validate the idea of the new kind of a program which we suggest. In sum, our results proved that there was a demand for such program among people we approached within our experiment. At the same time, we understand it will be necessary to continue in the research of this topic to provide more proofs and get more insights about the potential customers and the ideal structure of such a program. We want to continue in our research with similar experiments on larger scale. Moreover, we would like to extend this validation by validations with potential investors of equity-based crowdfunding to examine and validate their interest in such concept as it would not be feasible only without start-ups as its customers, but neither without investors.

Acknowledgements This work was supported by research project VEGA 1/0218/16 on "The Model of the implementation of controlling as a management tool within medium enterprises in the engineering and electronics industries."

References

1. Steve Blank, http://steveblank.com/2012/03/05/search-versus-execute/
2. E. Ries, *The Lean Startup*, 1st edn. (Crown Business, New York, 2011)
3. Webfinance Inc, http://www.businessdictionary.com/definition/start-up.html#ixzz3AH7uUoFZ
4. C. De Buyers et al., http://eurocrowd.winball2.de/wp-content/blogs.dir/sites/85/2013/06/framework_eu_crowdfunding.pdf
5. G.K.C. Ahlers, D.J. Cumming, G. Christina, S. Denis, Signaling in equity crowdfunding (2012). Available at SSRN: https://ssrn.com/abstract=2161587
6. S.C. Bradford, Crowdfunding and the federal securities laws, Columbia. Bus. Law Rev. **2012**(1) (2012)
7. P. Belleflamme, T. Lambert, A. Schwienbacher, Crowdfunding: Tapping the right crowd. J. Bus. Ventur. **29**(5), 585–609 (2013). CORE Discussion Paper No. 2011/32. Available at SSRN: https://ssrn.com/abstract=1578175
8. P. Miller, K. Bound, The Startup Factories; The rise of accelerator programmes to support new technology ventures. (2011)
9. S. Cooper, http://www.businessinsider.com/facebook-ads-still-lag-paid-search-ctr-2013-6
10. European Commission, SBA for Europe, report on the results of the open consultation. (2008)
11. A. McDermott, https://www.quora.com/What-is-a-good-CTR-for-a-Facebook-ad
12. L. Kim, http://www.wordstream.com/blog/ws/2014/03/17/what-is-a-good-conversion-rate

Science Fiction Prototypes Illustrating Future See-Through Digital Structures in Mobile Work Machines

Tiina Kymäläinen, Olli Suominen, Susanna Aromaa, and Vladimir Goriachev

1 Introduction

Full visibility of the terrain and targets is an important challenge in the operation of mobile work machines. The machine operator must see the work targets at all times, while the machine and its parts are moving. Furthermore, the operator must be aware of any pedestrians in the proximity of or passing by the machine. However, providing full visibility is not always possible due to occlusions caused by for example boom motion or obstructions caused by the machine structure. The lack of visibility slows down the use of the machine, endangers others working and moving in the area, and physically stresses the operators, due to constantly avoiding and peering around obstacles. From the operator's standpoint, having as many of the structures as transparent as possible, would provide a more ergonomic, safe and productive work environment.

The see-through structures, comprised of new visualization methods and Augmented Reality (AR) solutions, may provide several possibilities for enhancing mobile machines with virtual elements and thus improve the operators' user experience (UX) and usability. Currently, such technologies for augmenting the driver's field of view in traffic have mainly been studied by the car industry, but not excessively in the mobile machine context. Consequently, the research introduced here presents such opportunities for the work machine context, essentially through

T. Kymäläinen (✉) · S. Aromaa · V. Goriachev
Human Factors, Virtual- and Augmented Reality, VTT Technical Research Centre of Finland Ltd., Tampere, Finland
e-mail: tiina.kymalainen@vtt.fi; susanna.aromaa@vtt.fi; vladimir.goriachev@vtt.fi

O. Suominen
Department of Signal Processing, Tampere University of Technology, Tampere, Finland
e-mail: olli.j.suominen@tut.fi

A. Reyes-Munoz et al. (eds.), *EAI International Conference on Technology, Innovation, Entrepreneurship and Education*, Lecture Notes in Electrical Engineering 532, https://doi.org/10.1007/978-3-030-02242-6_14

Science Fiction Prototypes (SFPs) that illustrate the diminishing of occluding objects and enhancing the work machine operators' field of view.

In recent years, design has increasingly been applied as an instrument to challenge and speculate about future technological and societal developments, rather than merely as a method for developing durable products [1]. Science Fiction Prototyping has been identified as a practical method for this line of work in technology research [2, 3]. Principally, the SFPs created by the method have been presented as stories grounded in current science and engineering research that are written for the purpose of acting as prototypes for people to explore a wide variety of futures [2]. In general, the earlier technology-driven SFPs have conversed the discussed research relating to Augmented/Virtual/Mixed Realities (AR/VR/MR), such as those demonstrated, for example, in [4, 5]. In this paper the aim is yet to designate potential technologies more thoroughly, engage many other supporting technologies to the SFPs and, conclusively, evaluate the prototypes. For supporting the earlier research, the firm emphasis in this paper is on the two final steps of the SFP method, "human inflection point" and "exploring the implications, solution or lessons learned" as validated, for example, in [6, 7]. Furthermore, for advancing the method, this paper introduces Vision Concepts [1, 8], which may well be seen to complement the SFP method in general. Hopefully this will extend the more visual approach to the creative prototyping process that has been previously initiated briefly, for example, by [9, 10].

Consequently, the remainder of this paper is structured as follows: first, there is a brief description of the context-relevant earlier works, followed by a detailed account of the process, methods, and data sets of the research; accompanied by a brief discussion of the field study investigations and their results. Subsequently, the main contribution is the SFPs complemented with the Vision Concepts. As the reflecting outcome of the process, the paper concludes with a brief evaluation of the creative prototypes and a short-term consideration of the potential technologies.

1.1 Background of the Context and Related Work

In general, the research presented in this paper refers to such terms as Augmented Reality (AR) and diminished reality. Augmented reality is defined as an interactive real-time system that combines real and virtual elements in 3D [11], whereas diminished reality is generally considered to be a system where objects are removed from the real environment and, according to Herling & Broll it is a sub-area of Mixed Reality (MR) [12].

An important aspect, in this context, is the management of occlusion, that is, the way that occluding and occluded objects are treated and visualized. Aside from simple video analysis-based approaches, for example by [13, 14], the more sophisticated methods are based on extracting objects from 3D reconstruction, feature detection and segmentation [15]. In cases where foreground occluding objects are contextually unimportant, the application has used diminished reality solutions to deal with occlusions [12, 16]. In those cases, the occluding object is completely removed and the whole virtual object is shown in its place. On average, the methods using background information collect and process knowledge of the 3D structures of the scene, which they then take into account. These methods typically achieve good visual quality, but require more processing capacity. Another proposed solution for occlusion management is "X-ray vision," where the real environment and the augmentation are blended in a manner that creates an illusion of seeing through or inside a real object [17].

In general, the industrial vision-based systems for displaying occluded and invisible objects, have mainly been investigated by the car industry, with Land Rover's Transparent Hood[1] system being a notable example, as well as the Urban Windshield concept by Jaguar Land Rover. Within the mobile work machine industry, there have been lesser attempts at reducing the amount of occluding structures through the physical design of the cabins and vehicles. One example in the tractor industry, is the Driver Extended Eyes by Deutz-Fahr[2], which presents the occluding areas in a separate monitor. An interesting issue in the car industry concepts has related to how the cameras are placed (topology), and this has been specifically studied in solutions targeting a bird's-eye view synthesis (e.g., Daimler's 360° camera[3]). The systems usually embed multiple sensors (RGB cameras, time-of-flight depth sensors and infrared sensors) for multi-modal sensing and subsequent data fusion. Another trend is the consideration of AR applications and wearable devices, such as head-mounted displays (HDM), hand-held devices or AR- and smart glasses. Such devices have been introduced, for example, with BMW's Mini Augmented Vision concept[4] and devices, such as Microsoft's HoloLens, and Innovega's iOptik[5] cybernetic contact lenses.

Although the automotive industry has been working on similar technologies as those of interest here, it should be noted that their application areas are not as complex as that of the mobile machines, with respect to robustness, accuracy, prolonged use, etc. In this context, it is also critical that the provided solutions do not conflict with other requirements, such as structural integrity, safety, and operational capacity. Moreover, as most of the described technologies in the car industry are still in their infancy or in concept form, timely research in this context is motivated.

2 Process, Methods, and Datasets

The research presented in this paper places the work machine operator as the center of attention. The conducted research investigates primarily the operator's core tasks and the working environment, and secondarily, the supporting technologies. To study the role of human activity in the mobile work machine context, the project has

[1]https://www.youtube.com/watch?v=L7j1daOk72c

[2]https://www.youtube.com/watch?v=CqCLS1reawM

[3]http://techcenter.mercedes-benz.com/en/360_degree_camera/detail.html

[4]https://www.youtube.com/watch?v=j7TYxmsUBuA

[5]http://innovega-inc.com/index.php

employed the practice-theory approach to human factors [18], which emphasizes the role of user experience [19], encourages the active participation of users [20] and demonstrations through effective concept design [6, 21, 22], which in this case was carried out with the means of the Science Fiction Prototyping [2, 23] and Visual Concepts [8].

The research process included three main tasks. The first one was to become familiar with the application areas through visits to different machine vendors and work sites, at four different locations. The second task was to create four SFPs, including Visual Concepts, which were the design outcome of the field study investigations. The third task was to evaluate the creative prototypes and assess their short-term technical feasibility.

2.1 Field Studies

The objective of the site visits was to identify the visibility, human factor and user experience needs of current machine cabs, in addition to operators' expectations of new see-through and augmentation opportunities. The investigations included four locations in Finland: tractor machine vendor's facilities at Suolahti, a work site arranged by a harvester machine vendor in Luopioinen, a machine vendor's test mine in Tampere and an operational mine in Orivesi, and a vendor's test area for cargo handling solutions in Tampere.

To evaluate the visibility needs, safety aspects and UX expectations, the expert machine operators were interviewed, in situ, at the abovementioned machine vendors' work or test sites, during January–February, 2016. The participants were selected from among the test drivers or customers of the project's participating companies. The evaluations included seven operators (all male) with an average age of 36.9 years. The participants had work experience ranging from 3 to 23 years; they operated a machine from a couple of hours to 50 h/week and all were very interested in new technologies. The data collection methods were observations, semi-structured interviews, a questionnaire, video recordings, and photographing on location [18, 24, 25]. The procedure included audio recording, signing a consent form, collecting operator demographics, filling in the questionnaire and core task investigations concurrent with the operator carrying out the work task.

2.2 Concept Evaluation

The Science Fiction Prototypes were based on the aforementioned user research and illustrated the anticipated future see-through and augmented information concepts. To evaluate the UX of the prototypes, there was a user research setup established with experts in the work machine industry. In general, the participants' backgrounds were related to industrial design, machine development and management as well as test driving. The web questionnaire was active for a 3-week period in October–November 2016. In all, ten (10) experts (all male) participated in the Web survey; their average age was 41.3 years and their work experience ranged from 6 to 30 years. Again, the participants were selected from among the employees or the customers of the participating companies: three operated tractors, five harvesters, one in the mining industry, and two in the lifting business. All were interested or very interested in new technologies. The participants were introduced to the visual prototypes via a presentation[6] that was embedded in an online questionnaire. The online survey consisted of both closed and open-ended questions, and the participants assessed important features of the UX concepts (using a 5-point Likert scale: 5 = very important and 1 = unimportant) specifically created for this project. The interview and online survey data were further transcribed and qualitatively analyzed.

3 Results

3.1 Field Studies

In essence, the aim of the field studies was to define the core tasks and activities of the machine operators, visibility issues relating to occlusion caused by the cab/machine parts, technologies that are currently used to help the situation and expectations relating to the forthcoming see-through structures and augmented information.

Tractor

The first field visit was carried out with a test driver from Valtra Tractor (T4) in the machine vendor's outdoor facilities. In general, the core tasks, with the tractor, are related to multiple chores such as farming, front loader work, road maintenance, transport, lumbering, special working solutions (related for example to airports and military operations). Specific challenges concerning the machine cab were related to a constantly changing environment such as one involving location (field, farm, road, forest) as well as the season/weather. For example, in wintertime, ice and snow caused unexpected occlusions and low sunlight was described as blinding the driver.

Consequently, the visibility challenges related to the design of the machine, which was seen to be a compromise between different features (e.g., engine frame/visibility, different tools in front of the machine (the large size of the tools),

[6]The full presentation can be found on: https://indd.adobe.com/view/bb24016d-3eca-48ee-b646-894571cc831b

the need for backing up with the trailer and lifting loads, for example, to the truck, snow when snowplowing (where it is important to see what has already been done)). In summary, the most notable visibility issues for the machine were related to:

- Tractor tools in front of the machine, especially the front loader
- The engine cover being too high/big
- The driver being unable to see the target object
- The driver being unable to see the corners of the machine
- The driver being unable to see people around the machine
- General visibility too close and below.

Harvester

The second field visit was carried out with a forest machine operator using John Deere harvester (1270 E) in a privately owned forest. The core task was to cut down trees (motivated by delivering a profit to the forest owner). Operator characteristics were described as: capable of working alone and skillful in minor maintenance operations. Environmental challenges were related to location and weather changes and the fact that the forest has diverse conditions during the year, that is, the best season for work was stated to be when the ground was frozen (encompassing a few weeks in autumn/spring) in Finland.

The most critical visibility issue was related to the boom in front of the harvester, with other challenges being associated with the engine frame, snow in the trees, darkness in wintertime, the fact that the operator should be able to see the whole tree (top) and rocks in the ground while driving. Consequently, to summarize, the most notable machine visibility issues were related to:

- The boom being in front of the machine
- The boom operator being unable to see the target object
- The cab and engine frame blocking the view
- The operator being unable to see the corners of the machine
- Snow in trees, darkness in wintertime
- The inability to see to see the whole tree (top).

Mining Drill

The field visits were continued by studying several mining machines in two locations: a test mine (with a test operator) and an operational mine (with a driller). During the visits, the research group was introduced to several drills; ultimately, the Sandvik Jumbo (DD530-S60C) was selected as presenting most of the vital occlusion problems, during its use with core tasks, related to tunneling (e.g., drilling holes). Unlike the previous two environments, the mine was described as being a social environment, in which the communication with coworkers and other stakeholders was carried out during the tasks by walkie-talkie. The particular challenges in the mine environment related to darkness, dustiness and muddiness.

The main visibility challenges of the machine were related to drilling upfront, booms, manipulation of booms, darkness, the machine frame, pedestrians, and driving. To summarize, the most notable visibility issues for the machine were related to:

- Machine frame blocks the view
- Drills/booms, in front, block the view
- Engine cover is too high/big
- Unable to see all parts of the machine
- Difficult to estimate the space needed by tools
- Unable to see people around the machine
- Difficult to see around curves.

Cargo Handling

For defining the visibility issues relating to the cargo handling, the machine vendor offered their outdoor testing area for the investigations. At the location, several machines were introduced (e.g., an RTG, reach stacker, terminal tractor, and forklift), yet during the visit, there were no operators working in the environment. The main core tasks were stated to relate to the moving of containers (from ship to shore, stack to truck) and the work environments comprised ports, terminals, distribution centers and heavy industry. The work settings were described as social environments. The physical environmental challenges were seen as being diverse, with conditions being highly dependent on the location and weather, including snow, ice, wind and sand.

The main visibility challenges were related to the cab frame, spreaders, containers, driving tasks, other vehicles, and weather conditions (including, for example, fog, rain, dark, sunlight, and wind). The most notable visibility issues were related to:

- The cab frame or spreader blocking the view
- The load (container) blocking the view
- The operator being unable to see the target object
- The operator is unable to see other vehicles (especially behind them)
- Difficulty understanding depth via the camera view
- Too small a camera display.

Visibility Issues and Augmented Information

In the questionnaire, the expert interviewees were requested to estimate (in percent) how well they could currently see the target object outside the cab. The average of the first interview group was 72.86% (responses varied between 55 and 85%). The participants clearly indicated that the occlusion problems were related to the cab structures and machine parts. Furthermore, the experts stated that they were

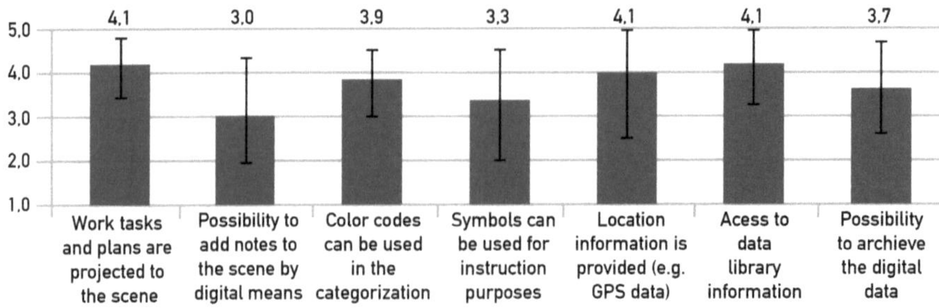

Fig. 1 Augmented information for the field of view, as defined by users

not always able to trust the information that they received with current assisting instruments (monitors and mirrors): 57% stated that they were satisfied with them, 43% were not.

As the aim in this research was to consider making parts of the machine cab transparent by digital means, it was suggested that simultaneously some instructional data and augmented information could be presented to the operator. Consequently, the participants were required to assess what kinds of augmented information they preferred in order to help them with the tasks pertaining to the operation of the machine. These issues were traced from a literature study comprising, for example, [13, 15, 16, 26–29]. Figure 1 presents the results.

3.2 *Science Fiction Prototypes*

Distinguishing the visibility problems and demands for augmentation led to the creation of Science Fiction Prototypes with Visual Concepts that provided further means for continuing discussion with the expert machine operators. In essence, they illustrate tractor work for snow plowing, harvester work in a frozen forest, mining machine operation in narrow tunnels and future cargo handling solutions. The focus was on future see-through solutions and augmented information cases and it is expected that some of their features will be exploitable in the very near future, some after 5–20 years.

Snow Plowing (SFP 1)

In the first SFP, the target machine was a tractor and the case focused on front loader work during a snow plowing task (see Visual Concepts in Figure 2). The user experience issues related to control, safety and common sense; issues that the expert tractor drivers described as being important operator characteristics. The opportunities for the see-through structures, in this case, were based on the remark that the driver was unable to see the corners of the machine and the people around the machine (see Figure 2, left). Furthermore, the prototypes were inspired

Fig. 2 Left: With transparent tractor pillars, it is easier to observe people in proximity of the machine. Right: Static information about the plowing area is projected to the driver

by remarks from the surveyed tractor drivers who stated that winter conditions for snow plowing were a particularly challenging task.

Figure 2 (right) presents information on how the plowing area is projected to the driver. Overall, the concept was inspired by an earlier study by Schall et al. [17], titled "Simulated looking through solid objects." The research, relating to civil engineering, demonstrated an AR application that displayed underground gas pipes and power lines targeting maintenance and network planning. Regarding user studies, the concept was inspired by a remark from a driver who proposed that the digital information should provide some static structures of a city (e.g., information on the pavements, location of obstacles, and manhole covers). In general, augmented information should support the moving of the machine and provide additional visual cues that encourage fast decision-making. Here, technological challenges are related to accessing and maintaining up-to-date maps of the environment that provide information of the surrounding infrastructure, accurate positioning of the machine in relation to the environment and the robust registration of the AR information on top of the real-world view.

Cutting Trees in a Winter Forest (SFP 2)

In the second case, the target machine was a harvester, and the SFP focused on the task of cutting trees in a winter forest. The concept was based on user clarification, according to which the best season for work was when the ground is frozen; although in wintertime, snow causes various other challenges, as it falls from the tree tops and covers the operating scene. The Visual Concept presenting the see-through structures (in Fig. 3, left) focused on the critical visibility issues with the harvester: the boom in front of the machine that restricted the capability to see the target object. In this SFP, the means to make the boom transparent is referred to as "optical camouflage" technologies presented, for example, by Kiyokawa et al. [29].

The augmented information in Fig. 3 (center and right), focused on the problem that the operator cannot see the whole tree top when making assessments, and during wintertime the forest is dark and there is snow in the trees. The Visual Concept

Fig. 3 Left: semi-transparent harvester boom; Center: areal drones inspecting the forest conditions; Right: color-coded guides for indicating the conditions of trees

(in the center of Fig. 3) presents how aerial quadcopters scan the logging area, for example, with point-and-shoot cameras and long-range sensors, before the machine operator begins the work task. In such a case, the augmented information allows the operator creating a full real-time map of the work area, and the condition of each tree within the area (in Fig. 3, right). Furthermore, the operator may use the information to plan the categorization tasks and bucking measures. Also, the digital scene may include other information such as work area information, notes, and symbols.

As a reflection, the technology required for building such an environment map may be based on a variety of different sensors, as the reconstruction task is not time-sensitive. This falls under the concepts of photogrammetry or Simultaneous Localization and Mapping (SLAM), where a geometrical model of the environment is built, piece by piece, by a sensor moving, in the space, over time. It can be done by cameras on the machine, itself, while it moves through the forest, or, for example, by a drone independently mapping out the trees beforehand. Most likely, a visual map is sufficient for getting the optimal information from the scene, so some scene analysis has to be performed, either through conventional machine vision to extract straightforward measurements (tree thickness, straightness, and height) or through more sophisticated machine learning (the identification of trees, the planning of which trees to leave, and what length to cut logs into).

Drilling Holes in the Mine (SFP 3)

In the third SFP, the target machine was a driller, and the case concentrated on mining tasks that related to manipulating booms and driving in the dark. The opportunities for the see-through structures focused on the machine frame and drills/booms that blocked the view (see Fig. 4 at left), an engine cover that is too

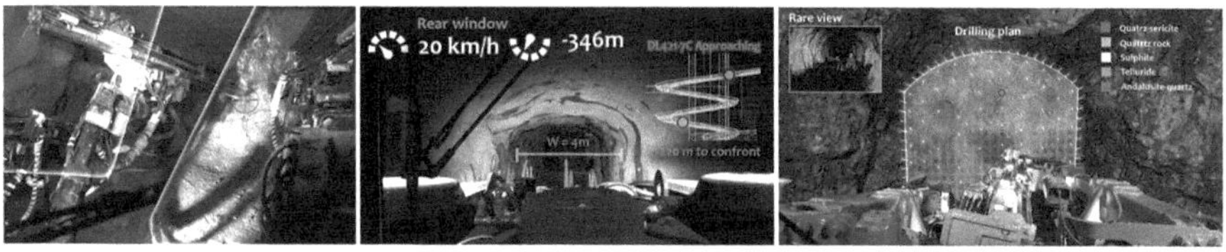

Fig. 4 Left: pillar is made transparent for revealing the drill; Center: augmented information when driving in a tunnel; Right: the drilling plan and colorful analysis of the stone material

high, the difficulty of the operator to estimate the space the tools need and the difficulty of seeing around curves. The human factor requirements related to the challenging lighting conditions, the need to improve the safety of the people in the proximity of the machine and the need to support the moving of the machine.

The SFP was inspired by the interviewed mine machine operators, who explained that while driving in narrow tunnels, the booms obscured the view and it was difficult to observe other traffic in the tunnel. In Fig. 4 (center), the augmented information provides statistics about the machine, tunnel, and distances. Because the mine is a social environment, the interviewees mentioned that they would prefer to communicate with other coworkers by visual means. Also, the machine operators mentioned that it would be convenient to have the drilling plan placed exactly on top of the drilling area. As a fictional idea, the operators imagined that it would be convenient to have an analysis of the stone material of the drilling area, as it influences the drilling and "pull out" maneuvers of the drill; these suggestions are illustrated in Fig. 4 (right). From a technical point of view, as the mining environment is relatively static, it allows for time-multiplexed reconstruction. A specific need is to have reliable communication, both on the inside and the outside of the mine, to relay the maps and the drilling plan information. Since coordinating the actions inside the mine due to the tight spaces is a significant issue, tracking the position of vehicles and personnel and synchronizing the information between all actors, is of key importance.

Lifting Cargo in the Sky (SFP 4)

In the fourth SFP, the target machine handles cargo and the concentrates on a task of moving containers. The opportunities for the see-through structures, in this case, were related to the cab frame that blocks the view from the target object. In this situation, the operation position is usually not ergonomic and current camera displays are too small and restrict depth perception from the view.

In the future, the cargo handling terminals, harbors and port yards could, themselves, be seen as a complex intelligent machine that is constructed of sensors, cameras and remotely controlled intelligent machines (see Fig. 5). In the last SFP, the future cargo handling terminals are operated from a remote-control center that visualizes the sensor and camera data, and provides intelligent operating systems

Fig. 5 Left: Sensor-embedded port yard that is remotely controlled

Table 1 Preferred Augmented Information

Topic	Percentage (%)
Information about the work area is presented in the operator's field of view	70
Areal drone scan the working area and the information is projected to the field of view	60
Location of other staff is presented in the field of view	20
Target work (e.g., the drilling plan) is projected to the field of view	10
Instructions are projected to the field of view	10

for remote operators. Operators can control intelligent machines for example by using new interaction techniques such as speech and gesture control, as described, for example, in [7].

3.3 *Results of the SFP Evaluation*

When analyzing the user expectations of the SFPs, the Web respondents' interest in each of the four scenarios correlated with their machine's operation background. In general, all of respondents found the Visual Concepts to be plausible, as well as desirable. The transparency level of the concept figures was sufficient according to 60% of the respondents, while 40% expected the pillars and tools to be even more transparent. Again, 60% expected to have a color outline marking the digital transparency area. As to color preferences, 30% preferred orange/yellow; 20% blue/turquoise, 10% green and 40% another color or no color. As a further commentary, all respondents reported that the main distance they needed to be able to see from the cabin was between 5–50 m (on average 23.5 m). All of the respondents reported needing the see-through information of the scene when the machine was moving.

Table 1 demonstrates which of the augmented information topics the expert respondents found to be useful, based on the SFP concepts.

3.4 Consideration of the Technologies

In general, the main challenge in the building of see-through solutions is to dynamically capture changing visual information of outdoor scene elements, especially those that are outside the operator's field of view; then to convert it to a representation suitable for fast and error-free processing, and finally, to visualize it to the operator. With technology that is available today, creating such a visualization system is generally possible by simply stacking enough sensors and hardware. However, the main challenge for the development is to consider what kind of design is realistically feasible and practical for actual use. To achieve a robust and practical implementation, a compromise must be reached between visual quality, operating speed and the amount of additional hardware required to run it. Another challenge is the fact that the information should be collected in a robust way that is invariant to lighting and environmental changes (e.g., presence of dust or haze). The implementation of see-through structures can be presented as two interconnected components, namely, environment sensing and in-cab visualization, with some amount of processing steps in between, needed to map sensor data into various displays.

In environment reconstruction, there are several different sensors available that have their suitable properties. A key aspect in processing is to be able to deal with incomplete and noisy data. While similar topics have been extensively studied for example in compression and transmission research under the umbrella of error concealment, such approaches are inherently problematic considering the operator may have to make safety critical decisions, based on the information provided by the system. Therefore, whatever interpolation or other filling methods used to deal with missing scene information must be applied carefully, in order to avoid providing false information.

Due to the complex structure of work machines and their cabins, visualization must be partially done in areas not covered by opaque cabin elements, but window surfaces. This prompts the use of transparent visualization media. In automotive and aviation applications, heads-up displays have been investigated, where the intent is to show for example selected numerical information to the operator. The existing solutions typically use either direct or virtual image formation techniques for producing the images. In this context, the disadvantage in virtual image systems is that all the components require significant amounts of space, are expensive to produce and provide only small viewing areas and angles. Direct image formation appears to be more pertinent, as it forms the image, in one way or another, on the glass itself, and can be implemented by either projecting images onto an appropriate semi-transparent medium (glass embedded with refractive elements, glass beads etc.), or by placing light emitting components into/onto the glass itself (transparent LCD, OLED, TASEL).

4 Conclusions

New see-through applications and advanced augmentation solutions are expected to influence strongly on future industrial work tasks. It is obvious that clear visibility of the work target means remarkable improvements in the accuracy and productivity of mobile work machines. In addition, the novel see-through technologies are expected to increase the safety and user experience of the machine operation.

This paper presents Science Fiction Prototypes illustrating see-through and augmented information concepts that do not merely focus on the machine structures and technical solutions, but also emphasize the targeted operator practices and preferences. This approach is seen to benefit the industry in the long run, as supporting technologies are currently developing fast. Future work in the area will continue by building a physical Mixed Reality demonstrator for a tractor by employing some of the abovementioned technologies. Taking into account the results of the study, the digital information in this demo will be delivered straight into the windshield of the operator.

As a further note, the concept cases illustrated subtly how, in the longer term, the presented technologies are evidently moving towards remote operations, which may well build upon the digital augmentation technologies presented in the Science Fiction Prototypes. It should be noted, that the case study focused on four explicit machine operation contexts, but, the core ideas may be further employed with other similar intelligent machine working environments.

Acknowledgements The work was carried out under the SeeWork project with the VTT Technical Research Centre of Finland, Tampere University of Technology and FIMA (Forum for Intelligent Machines) partners. We gratefully acknowledge the financial support from TEKES (the Finnish Funding Agency for innovation).

References

1. R.M. Sarmiento, G. Pasman, P.J. Stappers, Vision concepts within the landscape of design research. Proceedings of DRS 2016, DRS International Conference Series, **4**, 1659–1676 (2016)
2. B.D. Johnson, *Science Fiction Prototyping: Designing the Future with Science Fiction*, Synthesis Lectures (Morgan & Claypool, San Rafael, CA, 2011)
3. T. Kymäläinen, *Science fiction prototypes as design outcome of research* (Aalto University Press, Helsinki, Finland, 2015)
4. D. Hales, in *The Decadence of Mimetic Science: Against Nature 2.0*. Workshop Proceedings of the 9th International Conference on Intelligent Environments (IOS Press, 2011)
5. G. Clarke, M. Lear, in *We All Wear Dark Glasses Now*. Workshop Proceedings of the 9th International Conference on Intelligent Environments (IOS Press, 2010) pp. 242–250
6. T. Kymäläinen, P. Perälä, J. Hakulinen, T. Heimonen, J. James, J. Perä, in *Evaluating a Future Remote Control Environment with an Experience-Driven Science Fiction Prototype*. International Conference on Intelligent Environments (IE 2015), pp. 81–88
7. T. Kymäläinen, E. Kaasinen, M. Aikala, J. Hakulinen, T. Heimonen, H. Paunonen, J. Ruotsalainen, L. Lehtikunnas, P. Mannonen, in *Evaluating Future Automation Work in Process Plants With an Experience-Driven Science Fiction Prototype*. International Conference on Intelligent Environments (IE 2016), pp. 54–61

8. T. Keinonen, R. Takala, *Product Concept Design: A Review of the Conceptual Design of Products in Industry* (Springer, Berlin, Germany, 2010)
9. J.O. Schwarz, F. Liebl, Cultural products and their implications for business models: Why science fiction needs socio-cultural fiction. Futures **50**, 1–25 (2013)
10. The tomorrow project anthology – Cautions, dreams and curiosities, Open access electronic article (2013), http://csi.asu.edu/wp-content/uploads/2017/07/Conversations_About_the_Future_Anthology.pdf
11. R. Azuma, R. Behringer, S. Feiner, S. Julier, B. Macintyre, Recent advances in augmented reality. IEEE Compu. Graph. **2011**, 1–27 (2001)
12. J. Herling, W. Broll, *Advanced Self-Contained Object Removal for Realizing Real-Time Diminished Reality in Unconstrained Environments*. 9th IEEE International Symposium on Mixed and Augmented Reality (ISMAR, 2010), pp. 207–212
13. J. Chen, X. Granier, N. Lin, Q. Peng, *On-Line Visualization of Underground Structures Using Context Features*. Proceedings of 17th ACM Symposium on Virtual Reality Software and Technology (2010), pp. 167–170
14. B. V. Lu, T. Kakuta, R. Kawakami, T. Oishi, K. Ikeuchi, *Foreground and Shadow Occlusion Handling for Outdoor Augmented Reality*. 9th IEEE International Symposium on Mixed and Augmented Reality (ISMAR 2010), pp. 109–118
15. N. Kawai, M. Yamasaki, T. Sato, N. Yokoya, *AR Marker Hiding Based on Image Inpainting and Reflection of Illumination Changes*. 2012 IEEE International Symposium on Mixed and Augmented Reality (ISMAR), pp. 293–294
16. J. Herling, W. Broll, *PixMix: A Real-Time Approach to High-Quality Diminished Reality*. 2012 IEEE International Symposium on Mixed and Augmented Reality (ISMAR), pp. 141–150
17. G. Schall, E. Mendez, D. Schmalstieg, *Virtual Redlining for Civil Engineering in Real Environments*. 2008 IEEE International Symposium on Mixed and Augmented Reality (ISMAR), pp. 95–98
18. L. Norros, *Acting Under Uncertainty - The Core Task Analysis in Ecological Study of Work* (VTT, Espoo, Finland, 2004), p. 241
19. E. Kaasinen, V. Roto, J. Hakulinen, T. Heimonen, J.P. Jokinen, H. Karvonen, T. Keskinen, H. Koskinen, Y. Lu, P. Saariluoma, H. Tokkonen, M. Turunen, Defining user experience goals to guide the design of industrial systems. Behav. Inform. Technol. **34**(10), 1–16 (2015)
20. T. Kymäläinen, S. Siltanen, Co-Designing Novel Interior Design Service That Utilises Augmented Reality – A Case Study (2012), pp. 1–12
21. S. Aromaa, S.-P. Leino, J. Viitaniemi, *Virtual Prototyping in Human-Machine Interaction Design* (VTT Technology, Espoo, 2014)
22. K. Helin, J. Karjalainen, T. Kuula, N. Philippon, in *Virtual/Mixed/Augmented Reality Laboratory Research for the Study of Augmented Human and Human-Machine Systems*. 12th International Conference on Intelligent Environments (IE 2016), pp. 163–166
23. S. Ylirisku, J. Buur, *Designing With Video* (Springer, Berlin, Germany, 2007)
24. M. Kuniavsky, *Observing the User Experience* (Morgan Kaufmann, Burlington, MA, 2003), pp. 419–437
25. A.P. Rios, Exploring Mixed Reality in Distributed Collaborative Learning Environments, Doctoral dissertation (University of Essex, 2016)
26. B. Avery, C. Sandor, B. H. Thomas, in *Improving Spatial Perception for Augmented Reality X-Ray Vision*. Virtual Reality Conference (2009), pp. 79–82
27. S. Feiner, B. MacIntyre, T. Höllerer, A. Webster, A touring machine: Prototyping 3D mobile augmented reality systems for exploring the urban environment. Pers. Ubiquit. Comput. **1**(4), 208–217 (1997)
28. M. Inami, N. Kawakami, S. Tachi, Optical Camouflage Using Retro-Reflective Projection Technology, in *Proceedings of the 2nd IEEE/ACM International Symposium on Mixed and Augmented Reality*, 2003, p. 34
29. K. Kiyokawa, An optical see-through display for mutual occlusion with a real-time stereovision system. Comput. Graph. **25**(5), 765–779 (2001)

Developing the Business Process Management Performance of an Information System Using the Delphi Study Technique

Hisham AbouGrad, Jon Warwick, and Amare Desta

1 Introduction

Developing high quality decision-making processes are major elements of an organisation's business process systems success. Accordingly, appropriate decision-making business practice techniques are required to ensure that business process systems perform to high standards. The Delphi study technique is well-known for exploring organisational issues and helping to develop a business decision-making process. The Delphi study technique was first developed as a communication control technique for forecasting based on a panel of experts [1, 2]. Nowadays, the Delphi study technique is also used as a method for resolving important business process management (BPM) issues, make predictions, and develop decision-making indicators for decision-makers who look to improve their workflow systems policy, procedures, or quality assurance [3].

Business process management is a business management approach that concentrates on studying business process systems with the aim of improving business performance based on improved effectiveness and efficiency levels. BPM systems have mainly looked to improve the efficiency of workflow systems by enhancing the role of information systems. Thus, BPM systems have evolved to be a new business management approach for forecasting business technical requirements to optimise workflow systems. BPM approaches are used to find out about business process (workflow) systems effectiveness and efficiency using business practices and studies to successfully achieve the business objectives [4].

Information systems (IS) are the main component of business development and innovation that aim to improve business profit and return levels [5]. Information

H. AbouGrad (✉) · J. Warwick · A. Desta
School of Business, London South Bank University, London, UK
e-mail: abougrah@lsbu.ac.uk; warwick@lsbu.ac.uk; destaa@lsbu.ac.uk

A. Reyes-Munoz et al. (eds.), *EAI International Conference on Technology, Innovation, Entrepreneurship and Education*, Lecture Notes in Electrical Engineering 532, https://doi.org/10.1007/978-3-030-02242-6_15

systems have three key elements, which are employees (people), responsibilities (tasks), and technology [6]. The business process operates through an IS model that contains the BPM framework for the organisation's business operations and activities.

The Delphi method can be used to enhance the validity of a BPM framework as it produces results with high validity when compared to other data collection procedures. In fact, the Delphi method has resulted in a higher number of insights compared to other decision-making methods [7]. Hence, to develop information systems performance, an iterative study method is required to ensure the validity of BPM framework operations and activities. The Delphi method can be used to conceptualise and analyse the elements of an organisation's business process performance [8]. Thus, the Delphi method rounds can develop a decision-making instrument, which can be used as a model to improve a BPM framework for developing business information systems performance.

This paper looks at different Delphi and BPM research studies to confirm the usability of the Delphi method for developing business information systems by evaluating and improving business process systems performance. The paper presents a literature review where business process management and the Delphi study technique have been discussed in relation to business performance and information systems, and then the Delphi method details have been presented including Delphi's rounds (Table 1). Thereafter, the application of the Delphi method is discussed in relation to several published Delphi studies and supported by the empirical evidence reported in these studies. The paper then concludes by discussing Delphi's use in evaluating and developing business information systems performance.

Table 1 The Delphi method rounds

Round	Input of the codification panel	Output of the expert panel
1	Brainstorming • Identify indicators/factors • Initial list of criteria	• Initial criterion: – Rate its importance – Make open comments • For the criteria: – Rate overall importance – Make open comments • Propose missing criteria
2	Narrowing down • Consolidate criteria	• Per criterion: – Rate its importance – Give open comments • For all criteria: – Rate overall importance – Give open comments
3	Weighing • Determine final criteria • Request weightings	• For all criteria: – Rate overall importance – Give open comments • Weigh criteria and options

2 Literature Review

2.1 *Business Process Management Evaluation*

BPM is a complete business management approach that focuses on the organisation's processes without looking at the organisation's functions [8]. BPM has been developed when investment in enterprise information systems have been made that aim to transfer non-information technology business processes to systemic information technology (IT) business processes [9].

The Delphi method has been used as a business study technique to produce empirical evidence supporting decision-making in several BPM situations [9]. An example is in studying BPM job descriptions, categories, technical skills, and systems experience and ability to find the ideal profiles and types of BPM professionals. Another example is in developing BPM project aims, objectives, phases, terms and conditions, and contractors' portfolio by BPM experts based on specific industries and locations. In fact, BPM is growing as a process-oriented management research field within the information systems area of research [4]. Also, BPM is recognised as an interdisciplinary approach to organisational business processes analysis, design, implementation, and improvement, as well as supporting the sustainability of information systems [9]. Thus, to evaluate the BPM mechanism for an organisation's information system effectiveness and efficiency, a methodology such as Delphi is essential to enable an evaluation process.

The Delphi method can be used to formulate key indicators to evaluate the organisation's BPM through factor analysis criteria such as variance inflation factor (VIF) and critical success factors (CSF) where an organisation can identify the BPM success and ensure sustainable implementation of the organisational BPM systems [4, 10].

2.2 *The Delphi Method and Business Process Management*

The Delphi method has been used to identify BPM success values by developing controlled decision-making criteria that aim to find consensus among a panel regarding BPM systems. The Delphi method is operationalised by setting sets of questions to a group of selected experts. The Delphi study questions are designed based on the use of questionnaires or interviews as data collection procedures. The Delphi study data collection procedures are used to improve the effectiveness and efficiency levels of an organisation BPM system [4].

A global Delphi study has recommended that the Delphi and BPM methods can be used to design and develop an organisation's BPM system components [4]. This study has recognised the use of BPM values in achieving BPM success. Hence, BPM values can be used along with the Delphi method to evaluate BPM systems performance.

In fact, BPM approaches aim to improve an organisation's operational effectiveness and efficiency using BPM systems core values rather than an organisation's functional components [9]. For example, an improvement could be achieved by developing an organisation's product quality and customer services compliance.

2.3 The Business Information System Performance

Business information system performance is based on an organisation's business processes. Accordingly, when an organisation has established a BPM workflow system, the organisation decision-makers who are responsible for making decisions need an accurate business performance evaluation process to improve the business processes and information system performance, as well as ensure sustainable growth [11]. An organisation's decision-makers want a method that can help them understand BPM success values, so they can decide which business processes need improvement and what exactly should be done to boost business information systems performance.

Information systems are used to understand an organisation's information flow and solve business issues using BPM framework [12, 13]. Information systems and BPM frameworks together build the organisation's workflow information system (WIS).

In fact, WIS combines information systems software and hardware applications to achieve employees' daily tasks through automated business processes [13].

The Delphi method is a business practice methodology that uses expert feedback about business process information systems, so that decision-makers can develop and improve their organisation's business performance. This requires an appropriate WIS to ensure that the organisation has the right systematic business actions, and the Delphi method can enhance the understanding of the key efficient and effective business process factors to achieve the best business performance. In fact, WIS is a complete systematic solution that manages workflow technologies for information systems development and management [13]. Systematically, WIS allows the information system to achieve the required business performance. WIS needs business process management as a means of understanding business process efficiency and effectiveness by focusing on BPM values within the workflow of an information system implementation [4, 13].

2.4 The Delphi Method and Information Systems

To ensure the efficiency and effectiveness of an information system, the Delphi method as an iterative methodology has been shown to enhance the validity of the collected feedback data [7]. Results from applications of the Delphi method have included highly successful and innovative decision-making. In fact, Delphi as a decision support method results in a higher quality and quantity of decisions in comparison to other methods [7]. In general, reported business practice studies on information systems using the Delphi method are growing, and more BPM, WIS and business performance studies are emerging that suggest using Delphi to optimise business practices.

Business process management and workflow information systems join to form an organisation's systematic and automatic business process procedures. These deliver the services and products with the required levels of quality to ensure organisational business sustainability. Information systems are the engine of the organisation's complete workflow system, which is based on the organisation's BPM full lifecycle. An organisation's BPM is designed to understand business process efficiency and effectiveness using BPM organisational business values, which are used within the workflow process of an information system implementation [2, 4, 7]. On the other hand, the Delphi study technique can utilise BPM values as comprehensive key indicators to evaluate workflow information systems performance. Also, the Delphi method can discover business performance key issues and indicators by weighing decision criteria using its multistage decision-making steps to reach the right decisions (Table 1). We now discuss the Delphi method in more detail.

3 Methodology

3.1 The Delphi Method Context

The Delphi method was initially developed as a study technique for a project that utilised feedback from a panel of experts. The first project was named "Project DELPHI" at The RAND Corporation [1]. The Delphi study technique as a business practice method can be achieved by collecting expert opinions and identifying consensus positions using a process of controlled and structured feedback within different staged rounds fully utilising the experts' business experience [9]. There are several ways to use Delphi, for example some research studies depend on the experts for finding issues and key indicators while others make use of a literature review to formulate a set of indicators prior to the Delphi rounds [2, 4, 7].

The Delphi method has three basic rounds where each Delphi study round has a different purpose and aims [2]. The basic Delphi study rounds are Brainstorming, Narrowing down and Weighing. Table 1 summarises the conventional Delphi method rounds and their inputs and outputs [7].

Brainstorming

Delphi's Round 1 is Brainstorming, which is used to identify or select key indicators or factors and set up an initial list of criteria for decision-making. Brainstorming is usually implemented as an initial study for collecting a slice of the study data to propose the initial list of criteria and important key indicators. The Brainstorming round identifies the initial decision criteria, which will be the subject of Narrowing down in the second Delphi round [2, 7].

Narrowing Down

Delphi's Round 2 is Narrowing down, which narrows down the initial list of criteria in order to validate the factors or key indicators, and then rate or rank them based on their importance or feasibility. The narrowing down round validates the results from Delphi's first round (Brainstorming) and searches for a complete ratio, rating, or ranking to recognise the key indicators by measuring their importance or feasibility, with a degree of consensus obtained. This Delphi round produces a list of factors or key indicators to use in the Weighing, Delphi's third round [2, 7]. Moreover, the narrowing down confirms the key indicators of the business process system, which are used to evaluate an information system. This evaluation process identifies the business information systems key variables from the BPM values, and then rates the degree of consensus of each key variable. The narrowing down round results in both the recognised business information systems key variables and the rate of the consensus of each key variable, which will be used in Delphi's Weighting round [4, 8].

Weighing

Delphi's Round 3 is Weighing, which is used to weigh the rated indicators to reach consensus or verify the decision and present the Delphi study results. The weighing round is used to conduct the final evaluation to reach the Delphi study final answers and findings, so then the Delphi study results can be published. Thus, the study will determine the final criteria; an example is developing a weighting system for a workflow information system model [2]. The weighing round seeks the most reliable and valid business process key indicator variables of BPM values to obtain the best business performance. These key indicator variables of BPM values are an information system running principles that determine business process system structure and behaviour in the workflow information system [4, 8].

3.2 Delphi Validity

The Delphi method has been used to identify, select, conceptualise, and validate business process factors in many business studies (see for example [2]). Also, the Delphi method has been used to examine business process systems validity in the weighing final round through a quantitative evaluation of the validity of the workflow information system model. Thus, we believe that the Delphi method as an expert analysis methodology is appropriate as a basis for an organisational study to evaluate business process systems. The Delphi method results in quantitative measurement instruments that can be used to define specific business process system operations using BPM values multi-dimensional construction [4]. This gives empirical insights on the descriptive and predictive rule of the new business workflow information system model.

In fact, the Delphi method has a key methodological role that ensures the validity of Delphi results as experts can be asked to validate their results in the context of the final business process management study findings [2, 8]. The Delphi method rounds are procedures to ensure the decision quality and Delphi acts as a quality control and assurance strategy for enhancing the quality of an organisation's business decisions.

3.3 Delphi Toward the Right Business Process Management Performance

The Delphi method is used to produce appropriate decision factors for evaluating business workflow information systems. In parallel, BPM is an approach to improving process quality; for example, total quality management (TQM) aims to provide high quality products and services. Indeed, as we have argued BPM increases organisational efficiency and effectiveness through improvement and innovation using the Delphi method as a development concept that produces a set of indicators [4]. Thus, BPM values and Delphi study technique can be used to ensure the quality of the business process performance using Delphi's decision-making rounds and BPM key business quality factors, indicators, or values.

Information systems are used to improve business performance and individual's excellence by formulating BPM values as a BPM solution to develop the business environment [4, 8, 13]. Workflow is a BPM technology, which is used to deliver business performance requirements in order to achieve the business objectives. Workflow and information systems formulate workflow information systems, which are important for the organisation business development. WIS is a way to improve business performance, automate BPM, cut business costs, and manage BPM time. In fact, WIS is a complete systematic application to solve business process issues, as its key characteristic is to automate the business processes by linking staff tasks to IT systems [13].

Practically, WIS needs BPM values to understand an organisation's business processes and systems mechanism in order to automate them successfully. Thus, WIS strategic techniques are used to evaluate business information systems performance as WIS has an influence on running the business environment [4, 8]. The combination of WIS, BPM values and Delphi method enables an organisation to identify the right business decisions and have the advantage of identifying the performance indicators using BPM values and the Delphi method leading to the right business performance levels.

4 Application and Discussion

There have been a number of studies published which have discussed the empirical results obtained by the use of Delphi method. These Delphi studies focus on finding consensus and determining the effectiveness of the criteria for measuring business activities and rules [3]. In addition, the Delphi method has generated effects on different business performance components, which impact on business processes efficiency and effectiveness. In this section, we discuss some elements of published Delphi studies and their results with the effects of Delphi and BPM values on business performance.

4.1 The Delphi Method and Business Process Management Values

Business process management values are the core values of the key indicators, which are recognised through a Delphi study to measure an organisation's workflow systems [4, 8]. The recognised key indicators are the workflow information system driving factors; these factors are an instrument to quantitatively measure the workflow systems achievement of BPM objectives and of optimising business performance. Results from the Schmiedel et al. [8] global Delphi study have revealed key BPM core values. This study has identified four BPM core values (Table 2) which are Customer orientation (C), Excellence (E), Responsibility (R) and Teamwork (T). The BPM core CERT values have sub-indicators, which are used as key indicators to measure the workflow system. For example, the BPM value 'Excellence' has the following key indicators: Continuous improvement, Innovation, Leanness, and Quality; these indicators as described in Table 2 could

Table 2 The Delphi study identified BPM values [8]

BPM core values	Key indicators (variables)	Description
Customer orientation	Customer orientation	The business driver and target of the workflow system
Excellence	Continuous improvement	The process to satisfy business and customer needs
	Innovation	The business processes creativity and upgrade
	Leanness	The business processes efficiency
	Quality	The ideal and right business processes performance
Responsibility	Responsibility	The workflow commitment, engagement, and mission
Teamwork	Cross-functional orientation	The workflow system non-functional elements

be evaluated to produce the workflow system excellence rate (percentage). If the excellence rate is found to be low, then the organisation can use the excellence value to improve the workflow performance by focusing on the business excellence key indicators (variables). Therefore, workflow systems characteristics can be examined to evaluate the business processes in order to have more efficient and effective business performance. Also, the Delphi method and BPM values can be used to have a systematic BPM construct as a concept in the first stage of the decision-making process until completion of the concept test stage, and then the concept can be implemented as a complete BPM system for the organisation.

4.2 Results From Delphi Method

The application of the Delphi method has resulted in publication of important and valuable research studies and results, which lead to successful business decisions, and the Delphi study final stage "weighing round" confirms the appropriate systematic structured instrument to develop workflow systems performance [2, 3, 8].

Schmiedel et al. [4, 8] have completed a research study to identify BPM values and used them as validation of an instrument to measure an organisation's BPM system. This makes possible improved organisational process efficiency and effectiveness levels using Delphi's framework for decision-making improvement and innovation. Thus, the Delphi study has identified CERT values as BPM core values to achieve successful BPM, and allowed the construction of an instrumental procedure to measure and validate workflow information systems.

The Quyên [2] Delphi study has developed reliable and comprehensive key indicators to evaluate an organisation's governance practices. Five potential key indicators have been defined as dimensions of the organisation's practices. These five dimensions are (1) Management and Direction; (2) Participation; (3) Accountability; (4) Autonomy and Transparency; and (5) Management and Direction. The five dimensions identify resources at different levels within the organisation which are used for management purpose to realise the organisation's business tasks and duties. Thus, the Delphi method was very useful in identifying and measuring the five dimensions and in producing the initial key indicators of the weighting system.

A multi-criteria decision-making (MCDM) research study has been developed by Pangsri [10] using the Delphi method to make an MCDM tool for supporting experienced decision-makers in making several different and complex evaluations for construction projects. The research study used the Delphi method to evaluate alternatives and applied 'the order preference by similarity to ideal solution' (TOPSIS) weight technique to find the final decision-making criteria (Table 4). This resulted in the ranking of the projects. Thus, the use of the Delphi method along with other methods using the Delphi rounds can result in powerful decision tools for use in complex situations. Each of these published applications of the Delphi method is discussed in the following sections.

The Empirical Evidence of Delphi Studies

The Delphi method has generated many empirical evidence in application to BPM, workflows, and information systems. Generally, empirical studies in this area of research should have a methodology which can be used to validate decision-making. In fact, the Delphi method in combination with BPM values can provide this as it confirms requirements and makes decisions based on Delphi's multiple round process (Table 1) using BPM values from the organisation's workflow systems. The Schmiedel et al. [4] Delphi study has implemented an evaluation of a measurement instrument designed to identify and evaluate four BPM key dimensional values, which can be used to ensure BPM objectives success (Table 3). This Delphi study has six stages to ensure the implementation of the BPM construct (BPMC) of CERT values, and how the CERT values concept is suitable for supporting decision-making by measuring BPM.

The Schmiedel et al. [4] Delphi study has two core strategic entities, which are the development core and validation core. The development core includes the study first stage for the items creation to identify the key indicators (dimensions) using a literature review, questionnaires, and interviews; the second stage for the substrata identification to identify the important items in order to construct categories; the third stage for the key items selection which come from item pools in stage two and ranking the items based on experts' responses; the fourth stage for indexing (key items revision) by conducting an index-card-sorting test, which is used to ask experts to identify key indicators from stage three and calculate the ratio of each indicator based on their feedback.

Overall, the identified key indicators from the development core entity are appropriately acceptable to be implemented in the validation core entity which includes: the fifth stage for the BPMC (instrument) preparation, which refines the identified indicators by developing a pre-test and a pilot test studies, and then a valid translation of the measurement BPMC (The pre-test study is used to evaluate and expand the understanding of the BPMC in such iterative process, while the pilot test

Table 3 The Schmiedel et al. [4] research study process

Stage	BPMC development core	Stage	BPMC validation core
1	Items creation Literature review Questionnaires Interviews	5	BPMC (Instrument) preparation • Pre-test • Pilot test • BPMC validity and reliability (Instrument translation)
2	Substrata identification • Construct categories		
3	Key items selection • Ranking the indicators	6	BPMC (Instrument) Application • Factor analysis techniques (field survey)
4	Indexing (key items revision) • Index-card-sorting test		

study has been conducted to evaluate and confirm the BPMC validity and reliability using an exploratory factor analysis.); the sixth stage for examining the application of the validated and reliable BPMC using confirmatory factor analysis techniques based on a field survey (questionnaire) to ensure the practicality of the CERT values instrument.

The Quyên [2] Delphi study collected results from four levels. The first level described an indicator set and constructs where ninety-one indicators were recognised. The second level was importance levels used to rank the importance of each indicator and the study found that all indicators should be included as 92.3% were rated very high (very important, or extremely important) and 7.7% rated relatively important. The third level examined consensus levels where three groups (A, B, C) were set to have the coefficient of quartile variation (CQV) along with interquartile range (IQR) and median levels to evaluate the convergence level within the experts' opinions as well as the indicators importance level. CQV was used to evaluate the consensus level within the ratings, and the CQV formula is $CQV = (Q3 - Q1)/(Q3 + Q1)$. The fourth level is the weighting system, which resulted from the indicators importance scores. The indicators mean scores are used to calculate the final indicator importance levels (indicator weights). An indicator weight is the ratio of an indicator importance score to the sum of all indicators importance scores within a factor. A factor weight is the ratio of the sum of all indicators importance scores within a factor to the sum of all indicators importance scores within a dimension. A dimension has a set of factors. In fact, the Quyên [2] study has recorded a high level of agreement in experts' opinions on the indicators importance of 70% indicating high reliability of the set. Also, the weighting results have shown significant differences in the indicator weights within 23% of the factor groups and in the dimensions' factor weights of 80%. This indicates that the weighting is significant and confirms the results reliability.

The Delphi Method Validity of Business Process Study Results

The Delphi method has been chosen to develop business process studies because of Delphi's iterative measurement, which ensures the validity of the results [7]. In fact, Delphi's concentration on experts' responses has been found to be an appropriate approach for validating BPM decisions, key indicators, and important factors in several business studies [2, 10]. Hence, BPM and workflow systems studies need the Delphi method validity to ensure the flow of decisions throughout the business process.

The Pangsri [10] research study has used three decision-making methods (Fig. 1) as multi-criteria decision-making (MCDM) methodology for project management to successfully achieve specific objectives. The MCDM methodology consists of the Delphi method, the analytic hierarchy process (AHP) and technique for order preference by similarity to ideal solution (TOPSIS).

The analytic hierarchy process is a process for determining the weights of criteria to be used in complex decision situations. The AHP process consists of the

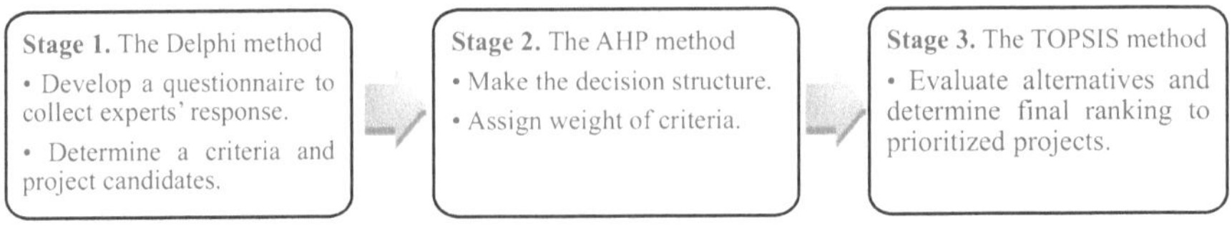

Fig. 1 MCDM research study methodology

following five steps: First, define the problem and develop the goal of the decision; Second, identify the decision structure (solution) based on the goal and develop the objectives; Third, develop a set of comparison matrices; for example, consider n elements to be compared C_1 to C_n then denote the relative *significance weight of* C_i relating to C_j by a_{ij} and then form a square matrix $A = (a_{ij})$ of order n as $a_{ij} = 1/a_{ji}$ where $i \neq j$, and $a_{ii} = 1$, all i; Fourth, the significance weight elements can be used to weigh the priorities for every element, then reweight the elements by adding its weighed values to have the overall global significance; Fifth, calculate the consistency index (CI) using formula λ_{max}–$n/(n–1)$. The MCDM process will apply the AHP process in the assignment of weight of the criteria (Fig. 1).

The technique for order preference by similarity to ideal solution (TOPSIS) is a multi-criteria process for identifying solutions based on set of predetermined elements. The TOPSIS identifies a positive solution as criteria to maximise the benefits and/or minimises the cost. In contrast, a negative solution minimises the benefits and/or maximises the cost. TOPSIS process has five steps which are:

Step 1: Inputs the x_{ij} for matrix of priorities as $i = 1 \ldots m$ alternatives and $j = 1 \ldots n$ criteria by $r_{ij} = \frac{x_{ij}}{\sqrt{\sum_{i=1}^{m}\sum_{j=1}^{n} x_{ij}^2}}$

Step 2: Use the weight normalised matrix

$$v_{ij} = w_i r_{ij}, \text{ where } i = 1 \ldots m \text{ and } j = 1 \ldots n$$

Step 3: Process the normalised decision matrix where the positive ideal solution A^+ is identified by choosing the largest normalised and weighted score for each criterion, and the negative ideal solution A^- is identified by choosing the least normalised and weighted score of each criterion by the calculate of these formulas

$$A^+ = \left\{v_1^+, .., v_n^+\right\}, \text{ where } v_j^+ = \left\{\max\left(v_{ij}\right) \in J \min\left(v_{ij}\right) if\ j \in J'\right\}$$

$$A^- = \left\{v_1^-, .., v_n^-\right\}, \text{ where } v_j^* = \left\{\min\left(v_{ij}\right) \in J \max\left(v_{ij}\right) if\ j \in J'\right\}$$

Table 4 The rank of the projects using TOPSIS

Rank no.	Project ID	The project weight
1	Project 5	0.747
2	Project 7	0.746
3	Project 3	0.614
4	Project 2	0.441
5	Project 4	0.386
6	Project 1	0.358
7	Project 6	0.264

Step 4: Measure the positive and negative measures for each alternative using these formulas

$$S_i^+ = \sqrt{\sum_{j=1}^{n} \left(v_j^+ - v_{ij}\right)^2},$$

$$S_i^- = \sqrt{\sum_{j=1}^{n} \left(v_j^- - v_{ij}\right)^2}, i = 1 \ldots m$$

Step 5: Obtain the relative closeness to the ideal solution (final weighting) by

$$C_i^+ = \frac{S_i^-}{\left(S_i^- + S_i^+\right)}, 0 < C_i^+ < 1, i = 1 \ldots m$$

The implementation of the Pangsri [10] MCDM methodology has ranked the studied projects using TOPSIS at the final weighting to order the projects based on their importance and indexed values (Table 4).

4.3 Delphi Results' Quality and Business Process Performance Development

The quality of Delphi study results is high compared to other BPM study methods for evaluating and developing business process decisions and workflows [2, 4, 8]. The Delphi method's main advantage is in the reliability of the study results and the validation of the Delphi study findings as the experts' panel ensures this [4, 8]. In fact, several business studies have used Delphi's framework to develop key indicators and factors as Delphi study results have been confirmed to be appropriate in constructing complete weighted and validated indicators for evaluating business practices in such a systematic process [2]. Also, the Delphi method has been recognised as a structured communication exercise that allows the emergence of agreement in dealing with complex business issues in systematic way. The Delphi study method has the capability to identify the critical success factors (CSFs) necessary for achieving business objectives at the right level of performance [10].

The development of the business process workflow system and performance need an effective business study method to convert results to objectives in order to achieve the right business tasks at the right time with the right levels of quality and performance [14]. The results from Delphi's framework have demonstrated such effectiveness as Delphi studies have achieved the business objectives and enabled improvement in BPM systems for different organisations. Also, the Delphi method CSFs capability makes high levels of quality achievable, so business process performance can be developed to ensure the quality of business products and services throughout the workflow systems. The Delphi method‘s ability to identify and validate key indicators as a multi-dimensional construct for examining BPM can be used to explore empirically the relationship between BPM, workflow systems and business performance [4, 15]. Hence, the Delphi method and BPM values can play an important role in developing business process and workflow systems performance to achieve business objectives.

5 Conclusion

An information system needs a decision-making development framework to improve its business processes and workflow systems performance and the Delphi method can develop a workflow‘s decision-making framework. This paper has discussed three Delphi studies with different approaches to the use of the Delphi study technique to enhance business decision-making. Schmiedel et al. [4] has used the Delphi method to construct and validate business process management values, which can be used to ensure business workflow systems success. Quyên [2] has used the Delphi method to develop key indicators, importance and consensus levels, and a weighting system for measuring business practices to facilitate business processes. The Pangsri [10] multi-criteria decision-making study has developed a framework where the Delphi study technique has an important role in initiating the MCDM framework process and enabling the application of other decision-making techniques.

Together, the Delphi method and BPM values can be used to improve workflow framework and information systems performances by measuring BPM performance and identify the key indicators for enhancing business workflow information systems. Delphi study analysis and results have demonstrated several advantages where BPM performance success can be reached by applying the Delphi method. These advantages include the ability to:

- Identify key important business process indicators and/or factors.
- Recognise business performance using ranked keys.

- Validate the BPM keys using Delphi's weighing system.
- Achieve business process performance success by implementing Delphi's BPMC.

In practice, the Delphi method has enhanced our ability to evaluate BPMC models and an organisation's workflow system in order to choose the right workflow mechanism that supports the organisation's services and productivity [2, 7].

To conclude, Delphi and BPM research studies have confirmed that Delphi's framework can be used to build a decision-making structure to develop an information system by improving the BPM and workflow system performance.

The research studies within the BPM framework, workflow decision-making, and workflow information systems development are still limited and there are many further opportunities and applications that could be explored. Hence, future use of the Delphi method to develop BPMC models and evaluate WIS performance is recommended as, to date, the Delphi method has demonstrated its ability to enhance business decision-making processes and systems.

References

1. N. Dalkey, O. Helmer, An experimental application of the Delphi method to the use of experts. Manag. Sci. **9**(3), 458–467 (1963)
2. Đ. Quyên, Developing university governance indicators and their weighting system using a modified Delphi method. Procedia Soc. Behav. Sci. **141**, 828–833 (2014)
3. J. Nworie, Using the Delphi technique in educational technology research. TechTrends **55**(5), 24–30 (2011)
4. T. Schmiedel, J. vom Brocke, J. Recker, Development and validation of an instrument to measure organizational cultures' support of Business Process Management. Inf. Manag. **51**(1), 43–56 (2014)
5. J. Becker, J. vom Brocke, M. Heddier, S. Seidel, In search of information systems (grand) challenges. Bus. Inf. Sys. Eng. **57**(6), 377–390 (2015)
6. O. Thomas, J.A. vom Brocke, Value-driven approach to the design of service-oriented information systems–making use of conceptual models. IseB **8**(1), 67–97 (2010)
7. A. Looy, M. Backer, G. Poels, *Towards a Decision Tool for Choosing a Business Process Maturity Model*, DESRIST 2012. LNCS, vol 7286 (Springer, Berlin Heidelberg, 2012), pp. 78–87
8. T. Schmiedel, J. vom Brocke, J. Recker, Which cultural values matter to business process management? Results from a global Delphi study. Bus. Process. Manag. J. **19**(2), 292–317 (2013)
9. O. Müller, T. Schmiedel, E. Gorbacheva, J. vom Brocke, Toward a typology of business process management professionals: identifying patterns of competence through latent semantic analysis. Enterprise Inf. Sys. **10**(1), 50–80 (2016)
10. P. Pangsri, Application of the multi criteria decision making methods for project selection. Univers. J. Manage. **3**(1), 15–20 (2015)
11. C. Sonnenberg, J. vom Brocke, The missing link between BPM and accounting. Bus. Process. Manag. J. **20**(2), 213–246 (2014)
12. J. vom Brocke, A. Simons, A. Cleven, Towards a business process-oriented approach to enterprise content management: the ECM-blueprinting framework. Inf. Sys. e-Bus. Manage. J. **9**(4), 475–496 (2011)

13. J. García, J. Vanderdonckt, C. Lemaige, J. Calleros, in *How to Describe Workflow Information Systems to Support Business Process*. 10th IEEE Conference on E-Commerce Technology and the Fifth IEEE Conference on Enterprise Computing, E-Commerce and E-Services (IEEE Computer Society, Washington DC, 2008), pp. 404–411
14. A. Draghici, A. Popescua, L. Gogan, A proposed model for monitoring organizational performance. Procedia Soc. Behav. Sci. **124**, 544–551 (2014)
15. M. Kohlbacher, H. Reijers, The effects of process-oriented organizational design on firm performance. Bus. Process. Manag. J. **19**(2), 245–262 (2013)

A Low-Cost Portable Health Platform for Monitoring of Human Physiological Signals

Keiran Brown, Emanuele Lindo Secco, and Atulya Kumar Nagar

1 Introduction

Low-cost and wearable devices for the monitoring of physiological and environmental parameters have become more approachable due to the miniaturization of the electronics, the increasing performance of rechargeable battery and the development of wireless communication protocol. Such devices have a variety of application from the monitoring of health condition of elderly users to the life monitoring of soldier and emergency operators [1, 2].

In this context some devices have been developed in the scientific community. Some notable developments are the monitoring camera [3], which monitors health conditions using visual information; it uses small changes in color of the skin, or slight changes in light reflection to infer parameters such as heart rate or body temperature. Another development is the eHealthGuard [4], which performs many individual elements where each one measures a specific aspect of a person's health and communicates the same using Bluetooth radios. Another similar development is the Fitbit [5]. This device is now a consumer product. It has a focus on physical activity of the person wearing it; it goes on the wrist and is able to measure various aspects of health. It uses an accelerometer to monitor walking and running activities as well as sleep activity to determine how well the wearer is sleeping and what part of the sleep cycle they are in. The device uses an application to display variousstatistics about health and progress towards goals set by the user. It can also

K. Brown (✉) · E. L. Secco · A. K. Nagar
Robotics Laboratory, Department of Mathematics and Computer Science, Liverpool Hope University, Liverpool, UK
e-mail: 15000194@hope.ac.uk; seccoe@hope.ac.uk; nagara@hope.ac.uk

A. Reyes-Munoz et al. (eds.), *EAI International Conference on Technology, Innovation, Entrepreneurship and Education*, Lecture Notes in Electrical Engineering 532, https://doi.org/10.1007/978-3-030-02242-6_16

measure heart rate, altitude, ambient light, and orientation relative to gravity and a digital compass. These sensors change with different versions (further details are available on [6]).

Pantelopoulos and Bourbakis presented a health monitoring system in 2010: this system performs the same task as the one of this work but using less expensive hardware and offloading tasks to existing hardware [7]. This paper outlines a design for a system that takes the raw signals from the various sensors and performs the electronic processes that convert the analogue signals into digital signals. Then the processing that would typically be done by the E-Health board is done by a wearer's smartphone. Smartphones are an already common technology throughout the general population and even if the wearer does not have one, providing one would cost much less than an E-Health board.

Another similar development is presented in [8]: it is very similar to the aforementioned one, but it actually uses the E-Health board to perform the signal processing from the sensors themselves. The signals are still sent to a smart phone; however, this seems to be the best solution for most health monitoring solutions as this is a computer that can gather data and send them to the cloud where they may be stored on servers [8].

Both of these solutions use XBee communication modules (IEEE 802.15.4) to transfer their data wirelessly to the reading device. This will be due to the low cost, low-power consumption and ease of use of XBee modules.

Finally, another development using again, independent sensors is a proposed system for ubiquitous health monitoring using a wireless body sensor network [9]. This connects to the internet to communicate various health related parameters. There is an emphasis on making the sensors very small to minimize effect on everyday life for the wearer. Although, noted, are the challenges faced with accuracy and precision of such sensors when performing miniaturization.

In this paper we propose an integrated low-cost platform for the low-cost and wearable monitoring of physiological parameters: the platform is based on a commercial hardware solution combined with a wireless communication protocol and a customized visualization software (Fig. 1). This development is important as it could be a solution for monitoring health parameters of patients both inside and outside of treatment facilities enabling health issues to be realized earlier than if only checkups were utilized.

The paper is organized in four main parts concerning the hardware and software components (Sect. 0), their functional integration (Sect. 0) and some preliminary results (Sect. 0). Further two parts report the final discussion and conclusion (Sects. 0 and 0, respectively).

2 Materials: Hardware and Software

This section is broken down into the description of the hardware (HW) and software (SW) components of the proposed architecture. A further part details the implemented graphical user interface (GUI) for the project, allowing for its function directly.

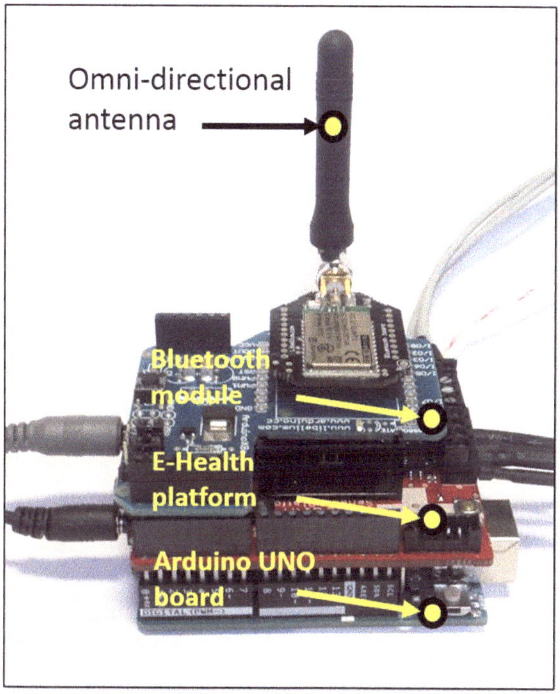

Fig. 1 Overall hardware design

2.1 *Hardware*

All the hardware components which have been used for this architecture are detailed in this paragraph.

Arduino Uno Board

Arduino is a low-cost open source platform. The board task is to perform data collection and transformation of data for serial transfer to a computer capable of extracting and displaying individual data points. It does this using a microcontroller which is situated on a small printed circuit board that has many components enabling the microcontroller to function as, for example, a linear voltage regulator that fixes the incoming voltage to 5 V DC to ensure no damage of the components.

E-Health Board

The E-Health board is a commercial device, which can read, sample and measure all the signals of a set of physiological sensors and then send these information to the Arduino Uno board for their collection [10]. The board, in the proposed setup, costs less than 500 Euro and allows the monitoring of multiple physiological parameter. Current portable device for monitoring just one parameter, that is, the ECG signal for example, can cost the same amount of money [11]. The board contains a processor that is designed to read incoming Analogue and Digital signals from the various medical sensors. This board is used as an easy way of gathering the information from all of these sensors and outputting their respective signals to be made use of. In a future development, this may be used as a custom integrated solution with lower cost and improve performance.

E-Health Transducers

The E-Health sensors are able to measure various stimuli determining the state of various parameters of the human body. Each sensor is unique, some of them contain some local processing features for collecting information and digitizing it, some are simply passive transducers that modulate the electrical signals in proportion to the outside stimulus.

(a) *Temperature Sensor*
This sensor measures the temperature of the body using a resistive element, namely a thermistor, which has a resistance that is proportional to the temperature. The thermistor consists of a material in which the electrical resistance grows with temperature, this limits the current flowing through it, although if the current is kept constant the changing voltage, which rises with resistance, can be measured to determine the temperature.

(b) *Pulse Oximeter*
This sensor uses two LEDs, one infrared and one red to detect the oxygenation of hemoglobin in the bloodstream to determine the oxygenation of the blood. It also detects the heart rate by using the expansion and contraction rates of blood vessels in the fingertip.

(c) *Electrocardiogram (ECG) and Electromyography (EMG) Electrodes*
Both ECG and EMG signals require the same physical interface, which measures the potential differences between three superficial electrodes to be applied on the subject's skin. The ECG detects this information across the heart muscles to determine the change in electrical signals across the heart. This determines the health of heart muscles, as the higher signal strengths tends to mean healthier muscles. The same sensor can be used elsewhere on the body in order to detect the muscle activity, namely the EMG: This measures the electrical signals across other muscles such as in the arm, but infers information about the health of motor neurons controlling those muscles, which can become weak because of some subject physical impairments.

(d) *Glucometer*
This sensor uses a small strip that collects a small amount of blood to measure blood sugar levels. This strip is inserted into a glucometer, which has electrodes that—when they encounter the glucose in the blood—they react,

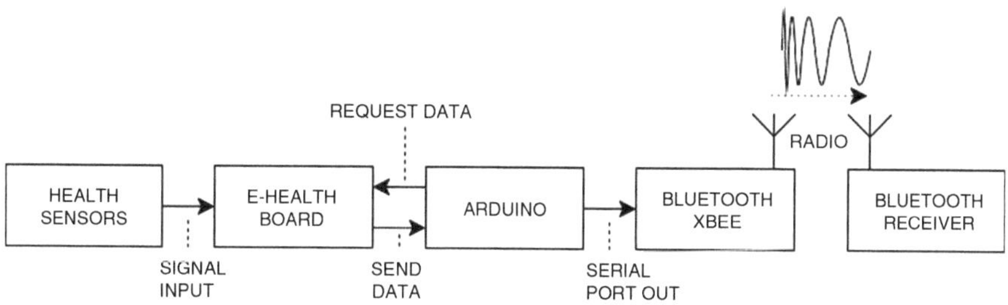

Fig. 2 Overall design of framework (J Graph, 2017)

producing a small electric current [12]. This current can be measured: The more intense the current, the more the glucose concentration present in the subjects' blood sample.

(e) *Galvanic Skin Response*

This sensor is a constant current sensor which is able to measure the resistance between two points by producing a voltage performing a constant current source. The resistance can then be calculated using this reading. This sensor shows the conductivity of skin, which rises as a person sweats due to stress.

(f) *Breathing Rate Sensor*

This further device uses a thermocouple to measure the change in the temperature of the air around two probes which have to be inserted on the nose of the subject. This only detects the warmer air from outward breaths. The higher the outward breathing rate, the higher the temperature reaches at the probes and thus, a higher reading is detected.

(g) *Bluetooth & X-Bee Wireless Communication Board*

This board is mounted on top of the E-Health shield (Fig. 1) and it collects the serial data from the Arduino Uno, buffer them temporarily, and then—after encrypting them—it sends those data via a modulated radio signal to the a Bluetooth dongle receiver (i.e., the IEEE 802.11 communication protocol).

(h) *Power Supply*

The power supply package is set up in order to contain eight AA batteries providing a 12 V DC: It is soldered to a DC jack for supplying such power to the Arduino board, which will cut down the voltage to 5 V DC by means of the aforementioned linear voltage regulator.

Figure 2 shows how the hardware interfaces together. Each arrow represents a process between two different parts of the system. The general flow of information is from the left to the right since the sensor information is modulated into the radio signal which is received by the Bluetooth dongle.

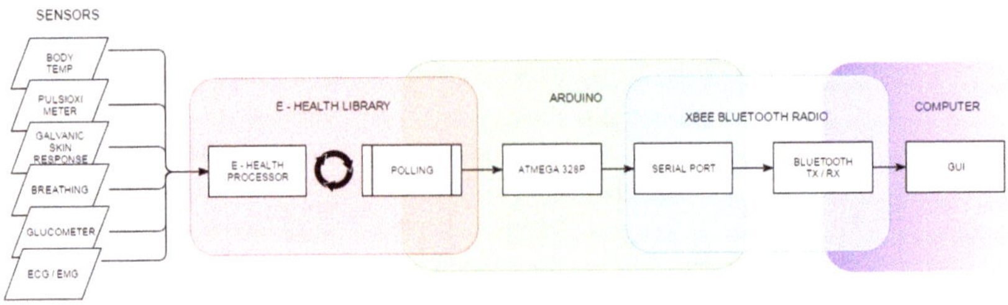

Fig. 3 Flowchart showing software processes

2.2 *Software*

All the Software components are detailed in this paragraph. In particular, Fig. 3 shows the processes involved in the Software. The figure also displays how the different parts are linked together through the Software: for instance, an E-Health Library allows linking the E-Health hardware to the Arduino board by giving access to the sensors' reading; this process is controlled through the polling of each sensor indirectly by the Arduino microcontroller (i.e., the ATmega 328P—[13]).

Arduino Code

This comes in the form of a piece of C code that is uploaded to the Arduino board: it allows controlling how the board micro-controller collects and processes the sensors' data. This does not require any outside software, only ASCII is used to interface different modules together (Arduino—[14]).

Java Code

A Java software-based program is used on a remote computer integrating the Bluetooth dongle, which display the information in a more readable fashion: the GUI is developed by using the processing development environment (PDE). This leads to a flexible application that can be executed on any device as long as it has Java installed, which is a commonplace software [15, 16].

Bluetooth Wireless Communication Module

The Bluetooth module is controlled by using the serial port of the Bluetooth module itself and sending ASCII commands to set up the password, ask for acknowledgements, and so on. This gives flexibility to the module as well.

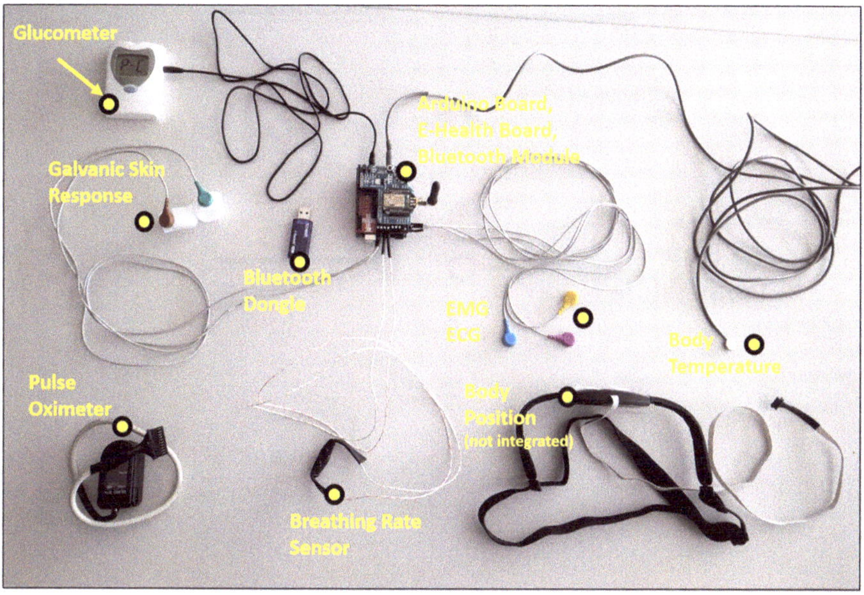

Fig. 4 Overall components and set of sensors

Integration

Figures 1 and 4 show all the components of the integrated device, namely (1) the platform with the Bluetooth X-Bee board, which slots into the top of the E-Health board and (2) the integrated boards, respectively.

3 Functional Integration

This is the functional approach of the developed system: the aforementioned sensors convert the energy from the external and physiological stimulus to electrical signals by using various methods. These signals are appropriately gathered by the E-Health board. Then, the Arduino board interrogates the E-Health board for each of the latest taken readings by using the E-Health Board Library (Fig. 3). Once all sensorial information has been obtained, data in its raw form are concatenated together into a single long string with semicolon markers in order to delimit each information. The string is then sent down to the Arduino serial port, which connects to the Bluetooth module buffering these data and sending them via the modulated radio waves to the receiver who has been initially paired and authorized (Fig. 2). The Arduino is sending information down a Serial Port and data are packaged with a marker—the word "ADRUINO" is used at the beginning as a one-way handshake—enabling the software at the receiver to correctly identify the source. The receiver can then split

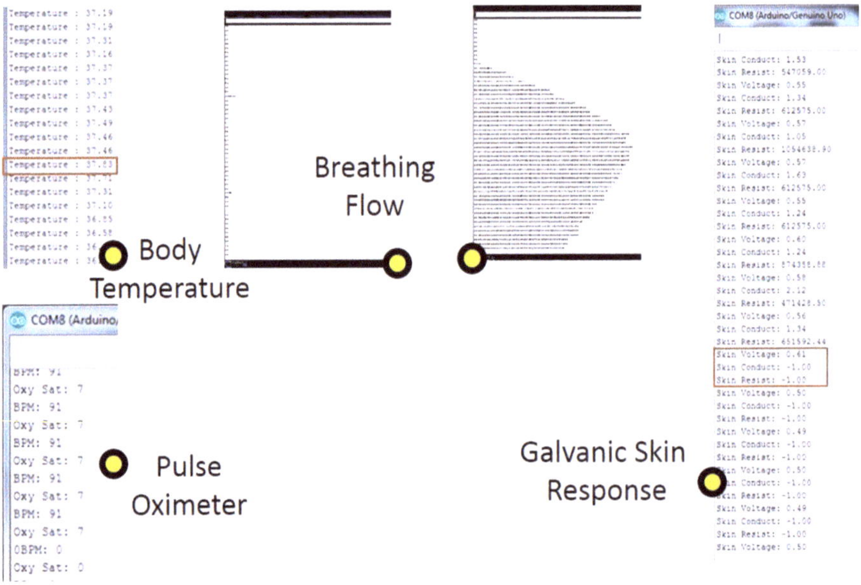

Fig. 5 Raw output of the sensor set

the string of data, remove the semicolons, and read each data as a floating point number which refers to a specific sensor, according to his position. Finally the data are then displayed on the screen of the computer via the developed GUI (Fig. 5).

4 Testing

This section reports the data that has been gathered by the device. These readings can be displayed in their raw form, before being used within the GUI. These results are gathered by using the sensors on a wearer. No gold standard equipment or certified medical equipment were used during these preliminary testing. Readings were taken to verify that sensors are properly functioning and that their signals are robust vs noise and movement artifacts.

In the future when hardware will be modified, tests will be performed to verify that these sensors function well compared to verified medical equipment [2]. This would properly validate their readings and ensure accuracy [1, 17, 18].

4.1 Airflow

Figure 5 displays idle and breathing sensor readings, respectively as well as the other sensor output. As it can be observed from the graphs, a small amount of noise can be noticed, although it is quite small compared to the actual reading, which is clear.

According to the output, the measured airflow is an absolute data and not a rate of the airflow (i.e., the subject's breathing rate): this latter one can be determined by using a differential of the absolute reading. Similar information can be displayed for the other sensor.

4.2 Temperature

The raw output of the temperature sensor is also shown in Fig. 5: the figure reports a set of temperature readings, where the idle reading is around 35 °C, since the typical sensor range is tailored to measure the human body temperature. The highlighted temperature reading in Fig. 5 shows when the sensor was taken away from the body of the subject and left open to the air again, at which point the temperature began to fall again. Each reading was performed every half a second, to give an indication of the time frame for readings of this sensor.

4.3 Pulse Oximeter

This sensor returns two outputs: the heart rate and the oxygenation of the blood. Figure 5 shows these readings on the bottom left panel, the last reading shows that when the sensor is removed from the body, then the sensor is returning a proper zero value. Oxygen saturation readings only show a single integer; here it reports a value of *7* which infers *97%* blood Oxygen saturation. The readings were quite consistent and show little error.

4.4 Galvanic Skin Response (GSR)

The readings for galvanic skin response are shown in Fig. 5, left panel. The highlighted reading shows when the transducer was removed from the body as well: this sensor produced a small voltage which is being produced to attempt to retain the constant current source, whilst other readings are far out of range and so a value of -1.00 is displayed as a default.

Table 1 Average, standard deviation, and—if applicable—the mean rate of change of sensor readings

Sensor	Number of readings	Average	Standard deviation	Mean rate of change
Body temperature (warming) [°C]	25	36.81	0.736	0.194 °C s^{-1}
Body temperature (cooling) [°C]	19	35.698	1.230	−1.230 °C s^{-1}
Skin conductance [S]	8	1.436	0.3079	–
Skin resistance [Ω]	7	689,175	188,458	–
Skin conductance voltage [V]	7	0.569	0.016	–
Heart rate [bpm]	6	75.833	33.914	–
Blood oxygenation [% O_2]	6	80.833	36.150	–

°C stands for Celsius degrees, S stands for skin conductance, Ω stands for Ohm, V stands for voltage, bpm stands for beats per minute, and % O_2 stands for the percentage of oxygen in blood

4.5 Glucometer

The readings for the Glucometer were taken on a separated module and then the E-Health board was performing the reading to be passed on to the Arduino board. The Glucometer sensor also has a built-in memory, such as the taken sensor readings may be read as long as they are stored.

In order to report an overall data acquisition, Table 1 shows the average of each sensor reading. It also shows the standard deviation of each reading, according to a preliminary set of trials, which have been performed.

Figure 6 shows the Java application, which was developed in order to practically visualize the values and the time pattern of the values of each sensors in real-time. This interface was developed in order to make the data more acceptable from end-users with no expertise vs. physiological and medical parameters.

The software is combined with a one-way handshake in order to manage the wireless communication. All data are displayed graphically for ease of use. This provides a level of abstraction for any user, as they may use all components and wirelessly transfer data to a computer without knowledge of the intermediate components.

According to the organization and implementation of this GUI, the end-user only needs to know how to apply the sensors and how to read the data.

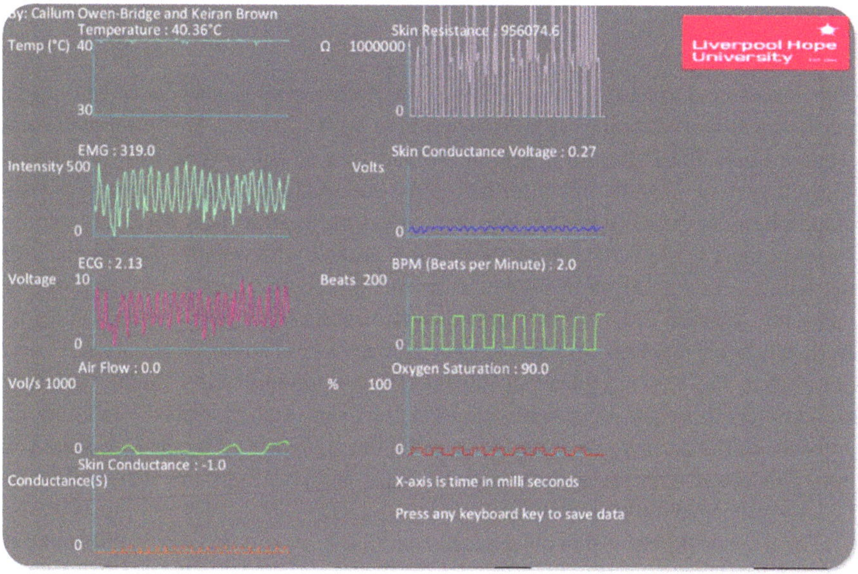

Fig. 6 The friendly and customized graphical user interface (GUI) software

5 Discussion

This platform provides some reliable operation for a relatively low cost compared to the typical medical equipment seen in hospitals. It is an example of how miniaturization can be achieved with current technology. If medically certified, this would provide a platform for easily measuring many details of a person's health which was before only possible at a hospital, with much more expensive equipment.

As for the hardware reliability, vulnerabilities only lie in the sensors themselves. The E-Health board is very accurate and able to take very fast readings and buffer them. The Arduino Uno is very capable at gathering and sending out data. A one-way handshake has been implemented using the Arduino with Bluetooth Security on top, allowing for a receiving machine to listen for the Arduino. Due to the nature of this one-way handshake, the Arduino can be disconnected and reconnected without having the restart the receiving application.

In order to consolidate these results and optimize the software and hardware, further experiments should be performed to test and validate the system in a real scenario. These test should also verified the performance of the system vs. the movement artifacts, which typically occur on wearable application for daily life activity of the subject.

6 Conclusion

The development process for this device involved working out how to get various platforms to work together as well as working with software. This prompted some work on how the E-Health board works and how sensors work with that board. This helped with gathering information by the Arduino, which must not make assumptions or errors in readings so the correct polling rate was necessary and the Arduino must wait for the sensors if readings are not yet available. Interfacing with the wireless communication module (i.e., the Bluetooth communication protocol) using the Arduino involved inserting a level of abstraction between the platforms. As it has been reported in Sect. 0, the Arduino was sending information down a Serial Port and was not aware of the Bluetooth communication shield: therefore, the data has to be packaged with a marker (i.e., one-way handshake) enabling the software at the receiver to correctly identify the source. The wireless module was only concerned with taking any received bytes and packing them into packets for Bluetooth encryption/transmission.

The proposed platform is quite small and all the sensors are removable and will not affect operation if not present (should the user only want to use a subset of the sensors). Such type of systems can be used for different purpose, due to their probability [1, 17, 18].

The completed platform is able to measure every sensor with proper accuracy. This is done at reasonable rate, considering that all readings can be taken within 2 s. This can vary depending on some sensors, for instance, the Pulse Oximeter does some independent processing and storage of the readings and so, if there are more readings stored that usual, the Arduino can take a little longer to detect, which is the last updated reading.

7 Future Developments

This project could be modified to reduce the cost and improve accuracy, and possibly, increase its usefulness in patient monitoring and diagnosis. To reduce costs, the E-Health board could be replaced with a board containing amplifiers, filters, and analogue to digital converters (ADC) necessary to read the signals using the Arduino platform, this would reduce cost and processing could be done with either a smartphone or a small processor. This may eliminate the need for an Arduino; also depending on the processor used, it may be able to read inputs from sensors. A better processor of the information from the sensors may also improve the accuracy of the device and enable it to take in information at a faster rate, for instance, by having more control over the pulse oximeter hardware the polling rate would be increased as the system would not have to wait for it to finish reading as it has to now.

Another possible development is the integration of this monitor with advisory software or artificial intelligence [19]. By having a suite of health parameters about a person, intelligent software may be able to advise the wearer on actions to take to improve their health or to visit their doctor, software may also be able to spot what a doctor has missed.

Acknowledgments We thank you all staff of the Department of Mathematics and Computer Science for their valuable support and in particular: Mr. M. Barrett-Baxendale and Prof. D. Reid, Ms. J. Burnett, and Miss S. Benson.

This work was presented in report form in fulfillment of the requirements for the course in Electronic Engineering for the student K. Brown under the supervision of E.L. Secco from the Robotics Laboratory, Department of Mathematics & Computer Science, Liverpool Hope University.

References

1. G. Magenes, D. Curone, E.L. Secco, A. Bonfiglio, The ProeTEX prototype: a wearable integrated system for physiological & environmental monitoring of emergency operators. 1st IEEE EMBS Unconference on Wearable & Ubiquitous Technology for Health & Wellness, Boston 2011
2. D. Curone, E.L. Secco, L. Caldani, A. Lanatà, R. Paradiso, A. Tognetti, G. Magenes, Assessment of sensing fire fighters uniforms for physiological parameter measurement in harsh environment. IEEE Trans. Inf. Technol. Biomed. **16**(3), 501–511 (2012)
3. Oxehealth, *Technology - Oxehealth* (Oxehealth, Oxford, 2017)
4. H. Fan, L. Xaiobin, Z. Qian, *Ehealthguard: A Personal Mobile Health Monitoring Platform* (Institute for Computer Sciences, Social-Informatics and Telecommunications Engineering (ICST), Brussels, 2011), pp. 122–123
5. K.M. Diaz, D.J. Krupka, M.J. Chang, J. Peacock, Y. Ma, J. Goldsmith, J.E. Schwartz, K.W. Davidson, Fitbit®: An accurate and reliable device for wireless physical activity tracking. Int. J. Cardiol. **185**, 138–140 (2015)
6. FitBit Inc Corp. website, www.fitbit.com
7. A. Pantelopoulos, N.G. Bourbakis, A survey on wearable sensor-based systems for health monitoring and prognosis. IEEE Trans Syst Man Cybern C Appl Rev **40**(1), 1–12 (2010)
8. R.M.D. Omer, N.K. Al-Salihi, HealthMate: Smart wearable system for health monitoring (SWSHM), in *2017 IEEE 14th International Conference on Networking, Sensing and Control (ICNSC), Calabria, Italy*, (IEEE, New York, 2017), pp. 755–760
9. C. Otto, A. Milenkovic, C. Sanders, E. Jovanov, System architecture of a wireless body area sensor network for ubiquitous health monitoring. J. Mob. Multimed. **1**(4), 307–326 (2006)
10. Cooking Hacks website, https://www.cooking-hacks.com
11. Spacelabs Cardiocall VS20 DualMode ECG Event Recorder, https://www.medisave.co.uk/spacelabs-cardiocall-ecg-recorder.html
12. Y.M. Garcia, *Glucose Meter Fundamentals And Design*, 1st edn. (Freescale Semiconductor, Austin, 2017)
13. Atmel Corp. website, www.atmel.com
14. Arduino cc, https://www.arduino.cc/en/Main/AboutUs
15. C. Reas, F. Ben, *Processing: A Programming Handbook for Visual Designers and Artists*, 2nd edn. (MIT press, Cambridge, 2017)
16. G. Bateman, Processing/processing GitHub (2017), https://github.com/processing/processing/tree/master/java/libraries/serial

17. E.L. Secco, D. Curone, A. Tognetti, A. Bonfiglio, G. Magenes, Validation of smart garments for physiological and activity-related monitoring of humans in harsh environment. Am. J. Biomed. Eng. **2**(4), 189–196 (2012)
18. D. Curone, E.L. Secco, A. Tognetti, G. Magenes, An activity classifier based on heart rate and accelerometer data fusion, in *7th International Workshop on Biosignal Interpretation*, (Politecnico di Milano, Italy, 2012), pp. 169–172
19. A.S. Mahajan, R.Y. Pattar, R. Khamitkar, M. Akalawadi, K. Bhat, Doctor's innovative clinic: An application of artificial intelligence and physiological sensors, in *2016 International Conference on Advances in Human Machine Interaction (HMI), Doddaballapur*, (IEEE, New York, 2016), pp. 1–5

Proposal for Information and Communications Technologies that Are Essential to Smart City Dimensions

Hemalata Vasudavan and Sumathi Balakrishnan

1 Introduction

For the first time in history, most of the world's population is now living in urban places. This is evident from the statistics of UN-Habitat; a survey from 2014 showed that the number of new cities developed globally was 694 [1]. Urban areas have also experienced increases in population in recent years. This shift is going to accelerate; a study from the United Nations showed that by 2030, numerous urban areas in Asia and Latin America are projected to have populations of more than 10 million people. Additionally, 70% of the total population of the world will be living in urban cities by 2050 [2, 3].

A variety of demands occur when there is urbanization. Keeping urbanization parallel to the development of cities has always been a struggle. Uncontrolled urbanization has led to the need for a model of a city called a "smart city" [3]. There are various different definitions of the term "smart city." The concept of a smart city was initiated from the term "information city" and incrementally evolved into an idea of an information and communications technology (ICT)–aligned smart city [4]. A city can be well defined as "smart" when its sociology, traditional (transport) infrastructure, and modern (ICT) infrastructure promote ecological economic growth that points to a high quality of life [5].

H. Vasudavan (✉)
Asia Pacific University of Technology and Innovation, Malaysia, Kuala Lumpur, Malaysia
e-mail: hemalata@apu.edu.my

S. Balakrishnan (✉)
Taylor's University, Malaysia, Subang Jaya, Malaysia
e-mail: sumathi.balakrishnan@taylors.edu.my

A. Reyes-Munoz et al. (eds.), *EAI International Conference on Technology, Innovation, Entrepreneurship and Education*, Lecture Notes in Electrical Engineering 532, https://doi.org/10.1007/978-3-030-02242-6_17

The idea of a wired city is the fundamental development model, and connectivity is the reservoir for development [6]. In another definition of a smart city, it is a city that depends on a variety of smart computing technologies that are used for important infrastructure components and services [7]. When they are applied to the city's management, education, medical science, citizen safety, housing development, roading system, and living needs, they improve all operations and the data produced are used to analyze and make intelligent governing decisions, which will improve business overall [8]. Although many discussions have taken place in the European Community, the definition of the concept of the smart city is still not well distinguished [9]. With the rapid development of technology, there may be a need for smart cities to have a benchmark in terms of ICT services and infrastructures.

2 ICT and the Smart City

ICT is definitely an enabler for a smart city. It is inevitable that an integrated system will be needed to support public safety, security, and green technology with services such as leverage sensors that enable dynamic use of energy to improve the efficiency of carbon footprint reduction [10].

A well-planned city is always equipped with sustainability as its priority. The idea of technology in the smart city concept signifies the integration of systems, infrastructures, and services facilitated through enabling technologies [11]. To support efficient and high-quality life, cities must make extensive use of ICT and future information technologies through implementation of sophisticated monitoring, control tools, and applications, and robust infrastructures [12]. Application of big data supports the city's decision making, while the internet of things (IoT) can reduce wastage of energy and manage the safety of buildings and the city [13]. The ICT infrastructure that is needed to deploy smart infrastructure services and facilities includes fiber optics, Wi-Fi networks, wireless hot spots, and service-oriented information, as well as sensors, firmware, software, and middleware systems [14].

With ICT, the government can always have real-time data on their industrial, safety, and educational systems, and on their environment, to make critical reviews, amendments, and decisions. Thus, our communications ecosystem can be managed better with the support of ICT and can make automated decisions customized to the users' needs. Appropriate use of ICT is important for the development of cities. The planning of a city must use a SWOT (strengths, weaknesses, opportunities, and threats) analysis of the city, so that it covers the planning holistically, with a robust and long-term approach that looks 10 or even 20 years ahead [7]. This highlights the fact that with technology changing rapidly, there should be agile strategic planning for ICT infrastructures, applications, and services.

3 Smart City Dimensions

The importance of integrating the city's diverse systems—the roading system, vehicle management, educational institutions, health care, residential services, infrastructure, meals, public protection, and many more aspects that influence the quality of life of citizens—must be taken into consideration in developing a city with smart capabilities. The researchers in this field consider a smart city to be one that functions in a compact setting where devices are not operated in isolation [15]. An extensive evaluation of the literature shows that the meaning of a smart city is versatile. The description of a smart city draws together the characteristics of living and nonliving things with the advancement of information technology and information systems. Many factors and dimensions describing a smart city have been developed from analysis of the current literature. Many researchers, with the rationale of looking at the overall constitution of the smart city, have separated this concept of the smart city into many functions and dimensions [15]. ICT is common in every multidimensional aspect of a smart city and is a compulsory element in all dimensions of a smart city [16]. Smart city aspects include technology, infrastructure, governance, and economics [14].

The list of dimensions being considered has increased in recent studies performed by the Center of Local Technology at Vienna University, which have recognized six primary dimensions [17]. These dimensions are smart mobility, smart living, smart governance, smart people, a smart economy, and a smart environment. The researchers analyzed the dimensions on the basis of conventional and classical theories of city growth and improvement, competitiveness, transport, ICT, the economy, natural resources, high quality of existence, and individuals' participation in society. Lombardi has related the same six components to different elements of urban existence [18]. The smart financial system has been related to the presence of industries in the area of ICT or using ICT in manufacturing procedures.

Smart mobility refers to the use of ICT in contemporary transport technologies to enhance the experience of urban visitors. Those same six characteristics are deployed by means of other research to develop signs and smart city improvement strategies. This is further justified by the European Parliament's industry, research, and energy policies, which recognize and explore these six dimensions, and they are being utilized in practice by an increasing number of cities and policy makers. In its study titled "Mapping Smart Cities in the European Union," it has adopted these six dimensions to study smart cities in Europe.

The wealth of projects in the dynamic socioeconomic, technical, and policy environment probably offers an upward push to a wide variety of smart city dimensions. These can be associated with distinctive styles of citizen roles and connections, policy units, and implementation strategies. Each of those traits may additionally be mapped along with exclusive places, city sizes, funding preparations, and framework conditions and consequences. This research paper focuses on identifying technology on a smart city map along with identifying smart city dimensions in the literature study. This has allowed the researchers to determine applicable technological tasks that make a contribution to the formation of a smart city. Further to this, identification of possible technology correlations among dimensions will enable this research to derive underlying suggestions and best-practice recommendations for other smart city project initiatives.

4 Problem Statements

With urbanization being an inevitable process, it is hard to deal with the overwhelming issues related to pollution, infrastructure, congestion, and other social issues related to increasing population. There is an obvious abnormal growth pattern in the spatial distribution of people and resources. Although the smart city has been introduced to rectify many issues, Fatnassi points out that implementation of the smart city concept is limited by infrastructure that is technologically outdated, and some of it is incapable of handling the current needs of the citizens [10]. A smart city is a dynamically improving city. There are many challenges to its identity, depending on the infrastructure that is needed for the different dimensions it has. Such issues will affect its performance and have a disastrous impact on the economy unless money is pumped into the right infrastructures. The definition of ICT across articles related to the smart city is not explicit. The reason is that many technologies and deployment of these technologies are closely connected. The costs in these cities will be pushed to the limits, and there is also the possibility that the infrastructure may become obsolete too soon or redundancy will occur. ICT infrastructures and services have to be benchmarked well so that when a city is planned, the forecast of the expenditure can be accurate. First and foremost, things such as high bandwidth capacity and high-robustness classification help local governments to develop services that address their own unique requirements. Although it is important to know what data and databases are used, in reality "nobody has a comprehensive overall picture of the data and information systems of their city" [19]. Thus, the question arises as to which technology is a priority to be implemented first by policy makers.

5 Research Methods

This study adopted the approach of a systematic review, permitting a rigorous, independent, and literature-wide study of research outcomes, quality, and methods. A systematic review attempts to identify, appraise, and synthesize all of the empirical evidence that meets the prespecified eligibility criteria to fulfill the research findings. In this research, information on smart city case studies and technology usage were retrieved from the available scholarly research materials to propose essential technologies for smart city dimensions. The first step consisted of retrieving numerous smart city–related research papers. In this segment, the researchers performed a detailed search in the Springer Link, Science Direct, Institute of Electrical and Electronics Engineers (IEEE), Association for Computing Machinery (ACM), Google Scholar, ResearchGate, and other databases. In each of these databases, the phrases "smart city," "technology in smart city," and "smart city case study" were used to search for articles, proceedings papers, book chapters, and academic theses. Secondly, the researchers selected appropriate papers after studying each paper's title and abstract. A large collection of research papers from the wide literature exploration were examined for their importance in terms of smart city case studies, dimensions, and associated technology. Each research publication was analyzed by reading of the abstract, introduction, and overview. Articles that were not related to the research topic were eliminated from the sample. This process resulted in a sample of 60 articles.

The third phase consisted of detailed reading of the papers selected in the second phase, to select only those papers that were appropriate to the research objectives. The researchers conducted qualitative content analysis to retrieve sample smart city case studies with discussion of technology facilitated in a smart city. During this phase, there were challenges in retrieving papers related to the technology used in smart city case studies, since many papers had only limited or partial information on the technology used in urbanization development. However, the researchers performed content analysis on the technology used by identifying computing keywords in smart city case study discussions. The computing keywords "sensors," "infrastructure," "real-time," "storage of data," "data sharing," "integration," and other related keywords were taken into consideration in proposing the key technologies. As an outcome, this research compiled 33 papers published in international journals, in books, in proceedings, as web articles, or as research studies, as documented in the reference section of this research paper. The papers were analyzed qualitatively to identify how they conceptualized smart cities, smart city dimensions, smart city case studies, technology in the smart city, and other related information. The qualitative analysis identified 30 potential smart city case studies for this research. A detailed review of the technology keywords from the case studies enabled discussions on essential technology utilized in the smart city dimensions of smart mobility, smart living, smart governance, smart people, a smart economy, and a smart environment.

6 Case Studies

The following subsections describe case studies on the smart city dimensions. The technologies used for each case study were derived on the basis of keywords.

6.1 Smart Mobility Dimension

1. Seoul, South Korea (SEO)
   TOPIS—This transportation information center helps the public to obtain real-time data on city roads and subway traffic information [20]. The computing technology used includes big data, IoT, and the Global Positioning System (GPS).
2. Barcelona, Spain (BAR)
   Vehicle-to-infrastructure (V2I) and vehicle-to-vehicle systems transmit data between vehicles and the surrounding environment, and directly between nearby vehicles. These data offer solutions to many problems related to security, management, and entertainment on the road [21]. The computing technology used includes IEEE 802.11 N, infrared (IR), and dedicated short-range communication (DSRC).
3. Paris, France (PRS)
   Grand Nancy—This is an application to derive the best and alternative routes by using an ad hoc information system on the city by using bike sharing, biking, and tramways [20]. The computing technology used includes IoT, mobile applications and infrastructure, 4G, and 5G.
4. Singapore (SG)
   Smart logistics system—This has been implemented to provide real-time transport data to make route decisions as efficient as possible [20]. The computing technology used includes IoT, mobile applications and infrastructure, 4G, and 5G.
5. Thessaloniki, Greece (THES)
   Traffic control center—This system is used to control situations with real-time information and dynamically monitor traffic signals. A mobility planner provides citizens with real-time traffic data and provide citizens with the shortest, most affordable, and convenient routes [21]. The computing technology used includes big data, GPS, and artificial intelligence (AI).
6. Stockholm, Sweden (STOC)
   Intelligent transport system—This smart traffic system reduces traffic, reduces emissions, and increases use of public transportation with greater integration between systems and modes. The system provides real-time information sharing and responsiveness [22]. The computing technology used includes big data, GPS, and cloud computing.

6.2 Smart Living Dimension

1. Washington DC, USA (WHT)
   PA2040—This system manages light-emitting diode (LED) street lights. The LED lights are sensor based and designed to improve visitors' experience, and

are managed remotely [20]. The computing technology used includes big data, IoT, cloud computing, and Wi-Fi.
2. Los Angeles, USA (LA)
LA Express parking program—This is used to dynamically allocate parking spaces. The system is calculated on the basis of hours of usage and frequency of usage [20]. The computing technology used includes GPS, big data, and Wi-Fi.
3. Nordhavn, Denmark (NOR)
Smart city neighborhood—This is an ICT-endowed carbon-neutral and manageable neighborhood system. It includes solar cells, heating stations, waste to energy, water consumption, transportation, cultural events, management of urban water, and adaptation to climate changes [21]. The computing technology used includes big data, IoT, cloud computing, and mobile technology.
4. Kyoto, Japan (KYO)
Eco-city—This nationwide project has been created to create a center of awakening culture, knowledge, and study. The eco-city is a homegrown initiative for many organizational labs and study centers to possess reliable technical capability with technological infrastructure [20]. The computing technology used includes mobile technology, Wi-Fi, nanotechnology, and big data.
5. New York, USA (NY)
Real-time crime centers (RTCCs)—These have reduced the crime rate by 27% and useintegrated emergency response solutions for public safety through citywide surveillance, which is centrally monitored. Big data is used to decipher crime patterns, using analytics and visualization [10]. The computing technology used includes GPS, big data, Wi-Fi sensors, and mobile technology.
6. South Korea (KOR)
Public safety—The launch of this system is intended to prevent foot and mouth disease, harnessing big data links to animal disease overseas, customs/immigration records, breeding farm surveys, livestock migration, and workers in the livestock industry [12]. The computing technology used includes big data and mobile technology.

6.3 *Smart Governance Dimension*

1. Queensland, Australia (Que)
Dial before you dig—This system is designed to guard the underground pipe network, cabling, and pits by locating the devices before any upgrades, in order to prevent major damage, which can be disastrous [20]. The computing technology used includes radiofrequency identification (RFID), GPS, and mobile technology.
2. Amsterdam, Netherlands (AMS)
ICT-enabled citizen participation platform—This permits resident involvement in open data platform strategies, crowd sourcing, and integrated platforms. The Open Data Smart City caters for games/competitions to develop systems and to

yield electronic services, based on public data, in order to develop reliable public services and involve residents in service project creation [21]. The computing technology used includes cloud computing and big data.

3. Orlando, America (ORL)
   Emergency and response—OCAlert.Net is an alert system to enable contact with society, citizens, or communities during an emergency by instantly sending messages to email addresses, cell phones, and smartphones, enabling connection to real-time directions [23]. The computing technology used includes GPS, big data, Wi-Fi sensors, and mobile technology.
4. Prague, Czech Republic (PRA)
   University Hospital Motol—In this city, the major medical care providers have accomplished the initial installation of a Grid Medical Archive Solution. This is an expert system for integrating patient data and is focused on integrated solutions with a group of specialists who collaborate across the network [22]. The computing technology used includes AI, big data, and cloud computing.
5. Spain (SPA)
   Health System In Transition (HiT)—A regional integrated system has been implemented that allows patients to go to different health centers in the region, with the certainty that the doctors there will have access to their complete patient data, making treatment faster and more accurate [22]. The computing technology used includes AI, big data, and cloud computing.
6. Canada, Quebec City (QUE)
   Zap Quebec—This provides a Wi-Fi snow-clearing management system by providing sensors on each snow-clearing machine, intercity connections, information system integration, open data concepts, and online transportation control [15]. The computing technology used includes Wi-Fi, big data, IoT, and cloud computing.

6.4 Smart People Dimension

1. Birmingham, UK (BIR)
   Birmingham is a city with a fast-growing population that aims to become an ecological city and empower trades, society where inhabitants flourish in a cooperative manner. This is being realized by the establishment of city authorities, platforms, and environments that blend and control aptitude across digital citizens [21]. The computing technology used includes GPS, big data, Wi-Fi sensors, and mobile technology.
2. New York, USA (NY)
   A system powered by IBM helps the city to avoid fires and protect first responders, and helps residents to process tax refund claims. MyNYCHA (from the NY Housing Authority) is a cellular system that allows citizens or home occupants to edit, implement, submit, list, preview, and update maintenance service records, view alerts and outages related to developments, view scheduled

check-ups, and save contact details by using smart mobile phones and tablets [20]. The computing technology used includes big data, IoT, and Wi-Fi.

3. Seoul, Korea (SEO)
   Enhancing public health and safety—The U-Seoul Safety Service employs location-based services and CCTV, letting guardians monitor kids' locations and notify the appropriate personnel in any emergency situation. Besides that, the U-Health Care System offers medical checks and arranged medical care for incapacitated people and senior citizens with restricted mobility. There is also the creation of a charging infrastructure to ease large-scale introduction of electric vehicles (EVs), etc., and the building of next-generation traffic systems connecting cycles and public transportation [4]. The computing technology used includes big data, cloud computing, mobile technology, and IoT.
4. Friedrichshafen, Germany (FRIE)
   Edunex—This is an online education system. It is a learning website platform for schools. The secured Edukey provides secure access to Edunex by using biometrics. The system is designed to help schoolkids with new educational concepts and innovative tools [15]. The computing technology used includes big data, biometrics, and cloud computing.

6.5 Smart Economy Dimension

1. Amsterdam, the Netherlands (AMS)
   Budget monitoring—This is a way to enable people to screen, assess, and proactively join in decisions on residents' policy making and government expenditure. It gives citizens the power, information, and self-belief to take appropriate action to live in comfortable surroundings [21]. The computing technology used includes big data, cloud computing, and IoT.
2. Helsinki, Finland (HEL)
   RFIDLab Finland—This is a neutral nonprofit organization, aiming to improve the operational effectiveness of organizations with the latest technology. This service is presented to help organizations with the business potential of RFID and near-field communication (NFC) technologies to implement networks and enable development creativity. It caters for prime users of RFID technology—for example, in manufacturing, sales, retail, logistics, and service provision industries—and those businesses that need to expand their business processes or create new businesses ventures [24]. The computing technology used includes RFID and NFC.
3. Luxembourg City, (LUX)
   HotCity SA—This system has been created to commercialize online access to the HotCity network. This is to mutualize the HotCity network and city apps with cities and private organizations such as hotels, campsites, and banks. Next-generation traffic systems have also been constructed, linking bikes and public transport [25]. The computing technology used includes Wi-Fi, big data, and mobile technology.

4. Seattle, America (SEA)
   A government portal supports 20+ languages, allowing open data and open government. This is an open data site of the federal government, and its objective is to make management more transparent and responsible. Opening government data increases resident contributions to government, generates opportunities for economic development, and notifies decision making in both the private and public sectors [15]. The computing technology used includes Wi-Fi, big data, and cloud computing.

6.6 Smart Environment Dimension

1. Seoul, Korea (SEO)
   Increasing energy efficiency—This allots smart homes with smart meters that control electricity. These controllers can display power usage and save energy. This technology enables street lamps/lights that automatically regulate their light output according to the outdoor luminance [20]. The computing technology used includes Wi-Fi, big data, and mobile technology.
2. Beijing, China (BEI)
   The city of Beijing was chosen as a place for a pilot study on the EU–China Partnership on Urbanization 2013. The objective of this project is to advance intelligent amenities in medical care, expenditure, road traffic, and citizens' lives. This will importantly enhance citizens' daily lifestyles. An "Atos World" grid has been established for China to attract international skilled people to support China to develop energy networks, along with resolving problems related to urbanization. The solutions provided include automated meter reading and an automated grid [21]. The computing technology used includes RFID and NFC.
3. Barcelona, Spain (BAR)
   An IoT program has been launched, with a powerful base. This project includes about 500 kilometers of fiber-optic cable within the city area. The establishment of this widespread network infrastructure commenced 30 years ago, whereby the city was linked with two community buildings with early introduction of fiber technology. The current fiber network delivers 90% fiber-to-the-home handling and provides a main support for unified city systems [26]. The computing technology used includes RFID and NFC
4. Australia (AUS)
   Smart grid—This consists of sensors and smart meters to reduce energy consumption and help protect the environment in terms of use of renewable energy sources (RES) [27]. The computing technology used includes RFID, IoT, and NFC.
5. UK
   Natural resources and energy—The aim of this project is to address climate change and its impacts on the availability of necessary water and food, local strains, and global solidity and safekeeping by carrying out in-depth scrutiny

on multiple information channels [12]. The computing technology used includes IoT, big data, and cloud computing.

7 Findings

The smart city could be defined as a complex ecosystem characterized by the intensive use of ICT. Smart city development is a response to address the issues of urbanization and the needs for flexibility and agility in delivering services to citizens. The smart city can be a solution to many known problems in urbanization. In this study, the researchers conducted a systematic review of various technologies applied in the smart city dimensions. With the sample of these 30 smart city projects, the researchers undertook analysis to identify the forms of technologies utilized in the smart city dimensions. The results of the analysis in this study were mapped to the six dimensions of the smart city mentioned earlier. Referring to Sect. 6, technologies used in the dimensions of smart city projects or applications were figured from the keywords in the case studies mentioned in Sect. 6. Hence, the findings resulted in many technologies being mapped to the dimensions of smart mobility, smart living, smart governance, smart people, a smart economy, and a smart environment. Table 1 illustrates the smart city technologies mapped to the smart city dimensions. In this research, around 30 samples of smart city projects were reviewed randomly in different regions. For each smart city project, the researchers analyzed the types of projects implemented and the associated technologies identified on the basis of the keywords.

As Table 1 shows, the technologies used in the smart mobility dimension are big data, IoT, cloud computing, GPS, Wi-Fi, DSRC, mobile technology, and AI; those used in the smart living dimension are big data, IoT, cloud computing, nanotechnology, GPS, Wi-Fi, and mobile technology; those used in the smart governance dimension are big data, RFID, IoT, cloud computing, GPS, Wi-Fi, mobile technology, and AI; those used in the smart people dimension are big data, RFID, IoT, cloud computing, Wi-Fi, mobile technology, and biometrics; those used in the smart economy dimension are NFC, big data, RFID, IoT, cloud computing, Wi-Fi, and mobile technology; and those used in the smart environment dimension are NFC, big data, RFID, IoT, cloud computing, Wi-Fi, mobile technology, and AI.

8 Discussion

For the purpose of this research, the five most widely used forms of technology in the smart city dimensions were derived through a systematic approach, as shown in Table 1. This is denoted by the 19 smart cities listed as using big data technologies in their applications. Besides that, 9 smart cities have incorporated IoT in their application development. Along with this, 11 smart city case studies revealed the

Table 1 Common information and communications technologies (ICT) used across the dimensions

Type of ICT	Dimensions					
	Smart mobility	Smart living	Smart governance	Smart people	Smart economy	Smart environment
NFC					HEL	AUS, BEI
Big data	BAR, PRS, SEO, STOC, THES	KYO, LA, NOR, NY, WHT	ORL, PRA, QUE	BIR, FRIE, NY, SEA, SEO	AMS, LUX	BAR, BEI, UK
RFID			AMS, QUE	BIR	HEL	AUS, BEI, SEO
IoT	SEO, SG	NOR, WHT	QUE	FRIE, NY, SEO	AMS	AUS, BAR, SEO, UK
Cloud computing	STOC	NOR, WHT	AMS, PRA, QUE	BIR, SEA, SEO	AMS	UK
Nanotechnology		KYO				
GPS	BAR, SEO, STOC	LA, NY	ORL, QUE			
Wi-Fi	THES	KYO, LA, NY, WHT	QUE	NY, SEA	LUX	BAR
DSRC	BAR					
Mobile technology	PRS, SG	KYO, NOR, NY	ORL, QUE	SEO	LUX	BAR
Artificial intelligence	THES		PRA			SEO
Biometrics				FRIE		

AMS Amsterdam, *AUS* Australia, *BAR* Barcelona, *BEI* Beijing, *BIR* Birmingham, *CPH* Copenhagen, *DSRC* dedicated short-range communication, *FRIE* Friedrichshafen, *GPS* Global Positioning System, *HEL* Helsinki, *IoT* internet of things, *JPN* Japan, *KYO* Kyoto, *LA* Los Angeles, *LON* London, *LUX* Luxembourg City, *NFC* near-field communication, *NOR* Nordhavn, *NY* New York, *ORL* Orlando, *QUE* Quebec City, *PRA* Prague, *PRS* Paris, *RFID* radiofrequency identification, *SEA* Seattle, *SEO* Seoul, *SG* Singapore, *SPA* Spain, *STOC* Stockholm, *THES* Thessaloniki, *WHT* Washington, DC usage of cloud computing in their applications. Mobile technology was used in ten of the smart city case studies. Although this technology is the one most widely used by the population around the world, the infrastructure supporting mobile technology may be inadequate to support newer systems and applications; for example, use of autonomous vehicles requires very well supported mobile technology. Most of the cities mentioned in Table 1 are equipped with 3G and 4G technology. Nine of the smart cities listed in Table 1 use Wi-Fi. The findings prove that the forms of ICT that are commonly used across the smart city dimensions are big data, IoT, cloud computing, mobile technology, and Wi-Fi. The widely used big data is an essential technology for the development of a smart city. Hence, big data in smart cities involves a large collection of sensors, databases, emails, websites, and social media, which provide sophisticated data analysis mechanisms to search and extract valuable patterns and knowledge from IoT and the smart city [14]. Thus, the volume of valid urban data is massive in the smart city; therefore, advanced big data analytical tools for data validation technologies with the capability for high-speed processing and consistency need to be engaged in order to confirm unaffected real-time authentication. The core technology needed for enactment of a smart city is IoT, supported by components such as electronics, sensors, networks, firmware, and software, which are essential for its deployment [14]. In addition, a wide collection of data can be extracted from different perspectives of the city via a distributed wireless sensor network constituting IoT, and these measured data can provide solutions for the relevant authorities and contribute to better city administration [14, 28, 29]. IoT is connected heavily by RFID, IR, and GPS sensors, which connect buildings, infrastructures, transport, networks, and utilities through ICT. Ultrasmart sensors and video authentication technology that use Wi-Fi, IoT, and mobile technology networks throughout the earth, sea, and air make it conceivable to deliver a diversified smart city service concerning all elements of urban life. In addition, cloud computing technology is inevitably a part of the infrastructure that caters to data storage remotely. Stakeholders can use cloud computing for information sharing and for immediate diagnosis of actions, accidents, and incidents [28]. Diagnosing the data and providing alerts and solutions can be done using big data analysis. On the other hand, broadband cellular connectivity has played an important role in the evolvement of IoT devices, which not only accommodate global reach, scalability, and diversity but also support a full range of IoT applications; it is expected that by 2025, 50% of mobile network data traffic will be information from IoT devices [30]. There is evidence that mobility is an inevitable challenge because of the rapid development in vehicular communication and cellular technology; thus, support for infrastructure such as fiber optics and mobile communication infrastructures such as 4G and 5G is important for the efficient functioning of a city [7].

9 Conclusion

The wide variety of smart city case studies with essential technologies in this research study will help smart city researchers and be useful for future studies. The research conceptualizes six dimensions in a smart city, the smart applications in each dimension, and the associated technology usage in the dimensions of smart mobility, smart living, smart governance, smart people, a smart economy, and a smart environment. The research underpins many smart applications with derived technology such as big data, cloud computing, IoT, Wi-Fi, mobile technology, GPS, and other technology. Therefore, this research contributes information on essential technologies that are widely used in the six dimensions covering the initiating technological factors in planning of information technology infrastructure in a smart city.

10 Future Work

For future research, there should be a thorough study on creating a benchmark or guideline for essential ICT infrastructures, and it should be revised appropriately within a set time frame. This study will continue to evolve in the future, as the authors will explore more smart city case study initiatives, taking into consideration various smart city enablers such as vendors, service providers, and new technologies. There should also be a study to define smart city terminology specific to continents and dimensions.

References

1. United Nations DESA/Population Division, World urbanization prospects 2018, https://esa.un.org/unpd/wup/. Accessed 15 Feb 2019
2. C. Manville, G. Cochrane, J. Cave, J. Millard, J.K. Pederson, R.K. Thaarup, A. Liebe, M. Wissner, R. Massink, B. Kotterink, *Mapping smart cities in the EU* (European Union, Brussels, 2014). February 15th 2019
3. E. Fatnassi, O. Chebbi, J. Chaouachi, Viability of implementing smart mobility tool in the case of Tunis City, in *IFIP International Conference on Computer Information Systems and Industrial Management*, (Springer, Cham, 2015), pp. 339–350
4. S.K. Lee, H.R. Kwon, H. Cho, J. Kim, D. Lee, *International case studies of smart cities: Anyang, Republic of Korea* (Inter-American Development Bank, Washington, DC, 2016). https://publications.iadb.org/handle/11319/7722. Accessed 15 Feb 2019
5. A. Caragliu, P. Nijkamp, The impact of regional absorptive capacity on spatial knowledge spillovers: the Cohen and Levinthal model revisited. Appl. Econ. **44**(11), 1363–1374 (2012)
6. A. Caragliu, C. Del Bo, P. Nijkamp, Smart cities in Europe. J. Urban Technol. **18**(2), 65–82 (2011)

7. G. Dimitrakopoulos, Sustainable mobility leveraging on 5G mobile communication infrastructures in the context of smart city operations. Evol. Syst. **8**(2), 157–166 (2017)
8. D. Washburn, U. Sindhu, S. Balaouras, R.A. Dines, N. Hayes, L.E. Nelson, Helping CIOs understand "smart city" initiatives. Growth **17**(2), 1–17 (2009)
9. C. Garau, F. Masala, F. Pinna, Cagliari and smart urban mobility: analysis and comparison. Cities **56**, 35–46 (2016)
10. G. Harter, J. Sinha, A. Sharma, S. Dave, *Sustainable urbanization: the role of ICT in city development* (Booz & Company, New York, 2010). http://www.booz.com/media/uploads/Sustainable_Urbanization.pdf
11. T. Nam, T.A. Pardo, Conceptualizing smart city with dimensions of technology, people, and institutions, in *Proceedings of the 12th Annual International Digital Government Research Conference: Digital Government Innovation in Challenging Times*, (ACM, New York, 2011), pp. 282–291
12. E. Al Nuaimi, H. Al Neyadi, N. Mohamed, J. Al-Jaroodi, Applications of big data to smart cities. J. Internet Serv. Appl. **6**(1), 25 (2015)
13. C.S. Shih, J.J. Chou, N. Reijers, T.W. Kuo, Designing CPS/IoT applications for smart buildings and cities. IET Cyber Phys. Syst. Theory Appl. **1**(1), 3–12 (2016)
14. S.P. Mohanty, U. Choppali, E. Kougianos, Everything you wanted to know about smart cities: the internet of things is the backbone. IEEE Consum. Electron. Mag. **5**(3), 60–70 (2016)
15. V. Albino, U. Berardi, R.M. Dangelico, Smart cities: definitions, dimensions, performance, and initiatives. J. Urban Technol. **22**(1), 3–21 (2015)
16. R.P. Dameri, Smart city and ICT: shaping urban space for better quality of life, in *Information and Communication Technologies in Organizations and Society*, (Springer, Cham, 2016), pp. 85–98
17. R. Giffinger, G. Haindlmaier, H. Kramar, The role of rankings in growing city competition. Urban Res. Pract. **3**(3), 299–312 (2010)
18. P. Lombardi, S. Giordano, H. Farouh, W. Yousef, Modelling the smart city performance. Innov. Eur. J. Soc. Sci. Res. **25**(2), 137–149 (2012)
19. K. Viljanen, A.G. Poikola, P. Koponen, Information navigation in the city, in *The City of Helsinki: Forum Virium Helsinki and the Fireball Project*, (Forum Virium Helsinki, Helsinki, 2012)
20. Department of Communications, *Submission to the Inquiry into the Role of Smart ICT in the Design and Planning of Infrastructure* (Department of Communications, Australian Government, Canberra, July 2015)
21. Directorate General for Internal Policies, Policy Department A: Economic and Scientific Policy. Mapping smart cities in the EU. European Parliament (2014), http://www.europarl.europa.eu/RegData/etudes/etudes/join/2014/507480/IPOL-ITRE_ET(2014)507480_EN.pdf. Accessed 15 Feb 2019
22. L. Bătăgan, Smart cities and sustainability models. Inform. Econ. **15**, 80–87 (2011)
23. S.K. Lee, H.R. Kwon, H. Cho, J. Kim, D. Lee, *International case studies of smart cities: Orlando, United States of America* (Inter-American Development Bank, Washington, DC, 2016). https://publications.iadb.org/handle/11319/7725. Accessed 15 Feb 2019
24. Helsinki Smart Region. Welcome to visit RFIDLab show room, https://www.helsinkismart.fi/portfolio-items/welcome-to-rfidlab/. Accessed 15 Feb 2019
25. The New Economy. The e-city of Luxembourg (2014), https://www.theneweconomy.com/technology/the-e-city-of-luxembourg. Accessed 15 Feb 2019
26. L. Adler, *How smart city Barcelona brought the internet of things to life* (Data Smart City Solutions, Cambridge, 2016). http://datasmart.ash.harvard.edu/news/article/how-smart-city-barcelona-brought-the-internet-of-things-to-life-789. Accessed 15 Feb 2019
27. A.M. Haidar, K. Muttaqi, D. Sutanto, Smart grid and its future perspectives in Australia. Renew. Sust. Energ. Rev. **51**, 1375–1389 (2015)

28. L. Hao, X. Lei, Z. Yan, Y. ChunLi, The application and implementation research of smart city in China, in *System Science and Engineering (ICSSE)*, (IEEE, New York, 2012), pp. 288–292
29. C. Harrison, I.A. Donnelly, A theory of smart cities, in *Proceedings of the 55th Annual Meeting of the ISSS-2011, Hull, UK*, vol. 1, (Curran Associates, Red Hook, 2011), p. 521
30. AT&T, What you need to know about IoT wide area networks: how to choose the right WAN technology for the internet of things (2016), https://www.business.att.com/content/whitepaper/what_need_know_iot_networks.pdf. Accessed 15 Feb 2019

Offset Policy and Its Impact on Aerospace Industry in India

N. Sharath Chandra, G. L. Shekar, and N. V. Raghavendra

1 Introduction

Offset agreement is made between the local government and an overseas company, which would be selling high end equipment. It is generally considered in defence procurements, which is believed to facilitate "indigenisation" of manufacturing of defence equipment. The developing countries stress offset policies as one of the means to lower the burden on their economy by redistributing a portion of its huge expenses inside their respective countries [1–3]. On the other hand, some researchers have pointed out that developed countries, mainly from Europe and North America deem it an unfair practice, which distorts competition at the global level [4, 5]. Yang et al. present a case where it can be considered a win-win arrangement for both [6]. The argument is that it is an opportunity to leverage the strengths of both domestic and overseas industries. According to Willet et al., about 130 countries are implementing offset policies [7, 8].

As per the offset agreement, the overseas Original Equipment Manufacturers (OEMs) can choose domestic partners among public or private enterprises. The choice for domestic partner is based on technological capability, financial strength and ability to meet global quality standards. OEMs in most cases are global companies, which are generally part of big consortiums. The domestic partners normally belong to tier-2 and tier-3 segments of the industry. Since OEMs will

N. Sharath Chandra (✉) · N. V. Raghavendra
Department of Mechanical Engineering, National Institute of Engineering, Mysuru, India
e-mail: sharath.chandra@nie.ac.in; nvr@nie.ac.in

G. L. Shekar
Department of Industrial and Production Engineering, National Institute of Engineering, Mysuru, India
e-mail: glshekar@nie.ac.in

A. Reyes-Munoz et al. (eds.), *EAI International Conference on Technology, Innovation, Entrepreneurship and Education*, Lecture Notes in Electrical Engineering 532, https://doi.org/10.1007/978-3-030-02242-6_18

have well established supply chains, standard vendors and quality procedures, offset policies provide an opportunity for domestic companies to get exposed to this niche environment. The major take away for OEMs are a big domestic market and low manufacturing cost.

Observations by several researchers stress that the impact of offset policy has been mainly on sourcing components and parts, but not on platforms. This is not hard to see since the OEMs need to safeguard their intellectual property and not encourage competition. For example, the top of aerospace chain consists of OEMs and sub-assemblies manufacturers like engines, fuselage, avionics and flight control systems. These comprise the core of aircraft engineering that will not be part of offset, whereas the components which come at lower order in supply chain like castings, sheeting metal and aircraft interiors require lower technological capability to manufacture. This serves both the OEMs and purchasing country because the OEMs will not lose their core competencies whereas the host country would effectively move good amount of purchases both in number of components and amount of investment to its shores. More research studies can reveal how technological changes are affecting the companies at higher level of supply chain.

2 Offset Policy in India

The World Trade Organisation (WTO) has broadly defined offset policy in the following terms: "offset means any condition or undertaking that encourages local development or improves a party's balance-of-payments accounts, such as the use of domestic content, the licensing of technology, investment, countertrade and similar action or requirement." Offsets may be classified into two types [9],

(a) Direct offsets
(b) Indirect offsets

Direct offsets are the main production components which involves co-production of components and sub-contracts. Indirect offsets may consist of assistance in further export to other countries and so on. India has been implementing offset policy since the year 2005 to secure offsets for its defence imports [10, 11]. The offset policy of 2005 has outlined the following major objectives [12]:

1. Fostering development of internationally competitive enterprises.
2. Augmenting capacity for research, design and development related to defence products and services and
3. Encouraging development of synergistic sectors like civil aerospace and internal security.

The Government of India (GoI) has taken measures to ensure efficient implementation of offset policy in the defence sector. It has set up Defence Offsets Management Wing (DOMW) to monitor and audit offset agreements on a yearly basis. As highlighted in one of the objectives, civil aerospace is bracketed with the defence sector for the enforcement of offset policy.

However, till date, there has been no exclusive strategy adopted by Indian policy makers to increase the pace of technology transfer through offset policy. The advanced countries have taken a lead over the developing countries in the advancement of technology. Unless, Indian industry gears up to global level on technology front, Indian companies will continue to play only a marginal role in the global arena. Therefore, the focus of offset policies should shift towards ensuring high degree of technology transfer to Indian companies.

3 Challenges in Implementation of Offset Policy

On paper, the offset policy looks favourable to developing countries and provides a semblance of a level playing field. However, there are quite a few challenges which need to be addressed. One observation is that offset deals tend to be complicated and costlier compared to direct purchases which are also called "off-the-shelf" purchases [13]. Off-the-shelf purchases make sense when either the quantum of purchase is small or the host country has a small economy. In the latter case, implementing offset may neither benefit economically nor result in early availability of products. In such cases, off-the-shelf purchases are preferred.

The effect of offsets on broad-based civilian economic development, job creation and technology transfer has been constantly scrutinised [14]. The basic objectives of OEMs to enter into offset agreement are to expand their markets as well as create manufacturing hubs to cater to regional markets. It would be unrealistic to expect that they would factor in economic development of host country in their scheme of things. The onus will obviously be on the host country to get a better deal so that the offset agreement gives a fillip to the growth of local industry, employment generation and more importantly, transfer of technology to local firms. This section reviews extant literature and discusses some salient aspects of offset policy in terms of their impact on local economy as well as certain challenges associated with their implementation.

3.1 Impact on Economy

From the economics perspective, offset policy would serve its role if it results in increase in manufacturing GDP and job creation. It has the potential to boost domestic manufacturing base. Offsets have opportunity to become catalyst in driving an economy by improving the overall output of a manufacturing industry and get benefits from local production [15]. Work done by da Silva-Menezes et al., specifically to analyse technological impact of offset agreements in a recipient industry highlights positive impact offset had on Brazilian economy. Taking the case study of Embraer, the authors state that the MD-11 agreement benefitted Brazilian economy by creating more jobs, promoting exports of Brazilian made products and improved production efficiency [16].

The authors recommend a basket of measures by the government to ensure the success of offset policy. These measures comprise government-sponsored credit programs, bank-to-bank credit lines and buyer–supplier credits. Such measures enable firms to modernise and scale up their operations, so that they can be competitive in the long run. However, all said and done, the future research in this area should focus on developing a comprehensive framework, which can enable realistic evaluation of economic benefits which is believed to accrue due to offset policy.

3.2 Impact on Jobs

Offset policy has the potential to create more jobs. The local sourcing of components would create more business for small and medium scale manufacturers. Many industries such as defence production, aerospace and automobile are divided into several tiers depending on the sophistication of components, systems and sub-systems that are outsourced by OEMs. This demands a wide spectrum of skills and knowledge levels among labour. What is of interest for researchers is whether there is a pool of skilled labour and technicians who can fit into the new jobs. The ability of a firm to deliver goods depends heavily on the competence of its human resource to meet global standards. Unless government policies give fillip to enhance skill levels significantly, the manufacturing sector will continue to suffer from stunted growth.

The US had a monopoly in jobs within commercial aircrafts manufacturing, with as much as 95% of aerospace jobs in the year 1960. Interestingly, this percentage has reduced to 49% by the year 2001 [15]. The reason cited by the authors is the rise of Airbus Industry and redistribution of manufacturing jobs to developing countries such as China, Brazil and India. The content of components sourced from foreign countries for assembling Boeing 727, which was a mere 2% in 1960, has risen to 30% in the year 1990. This clearly indicates that manufacturing jobs are getting distributed to other countries and one can safely presume that offset policy would give more impetus to job creation in developing countries.

Jurgen Brauer and J. Paul Dunne estimate that offset in South Africa has resulted in creation of 65,000 new jobs over a period of 7 years. The paper also reveals that similar offset deal in Saudi Arabia generated employment to the tune of 2000 over a period of 5 years.

3.3 Impact on Partnership

Offsets create partnership between technically advanced OEMs and local suppliers/manufacturers. This area is a great challenge where in OEMs needs to patiently build the supply chain and the vendors have to work hard to uplift themselves to standards which need to be matched globally. One good example of this can be "Sale of the century" in 1976 when General Dynamics agreed to offset 40% of parts manufactured to Norway, the Netherlands, Belgium and Denmark. The deal also helped to clinch 10% of value production for US Air Force and 15% of value of production of aircraft produced by other governments [17]. Such partnerships promote entrepreneurship, managerial and manufacturing expertise, and transfer of technical know-how.

3.4 Impact on Technology

One of the primary take away of offset agreement is transfer of technology. Owen Herrnstadt and Scott have mentioned that the host countries generally constitute a coordinating body with the objective of nurturing and advancing domestic aerospace manufacturing, technology acquisition and generation of employment [18]. It has been documented in several studies [18] (https://en.wikipedia.org/wiki/Offset_agreement) that host countries factor in transfer of technology as one of the mandates in offset agreements. GoI too has highlighted in detail the procedure of technology transfer involving Indian procurements [11]. Emphasis is laid on significant transfer of technology to vendors. Offset agreement stresses that OEMs should provide the latest version of technology used with full training, assistance in setting up of supply chain, production line, etc.

Another important aspect of such agreements is that the local firms need to conform to global quality standard. The OEMs are obliged to enforce compliance to international standards like *SAE AS9100*, *EN9100*, *JIS19100* and *NBR 15100* for manufacturers, *DO-178B*, *DO-178C*, and *DO-254/Eurocae ED-80* for Avionics. This will facilitate local firms to upgrade their quality standards, which implicitly calls for modernisation and up gradation of technology.

4 Conclusion

In recent times, offset clauses have become mandatory for procurement of high value equipment from a foreign OEM. Offset agreement is believed to benefit local manufacturing by providing exposure to global markets and adoption of best practices. This practice is especially prevalent in defence procurement of high value equipment. Many researchers have opined that offsets lead to vertical disintegration [19] of supply chain, thereby resulting in percolation of information and knowhow to companies that operate in the lower end of the supply chain. The policy makers hold the view that offset policy has beneficial impact on local economy, job creation, increased tie-ups with global companies and transfer of technology. Another important aspect of such agreements is that the local firms need to conform to global quality standard. However, this stylised theory stands for scrutiny by more number of empirical studies based on hard facts and figures.

This paper has made an attempt to review the offset policy of the GoI in the backdrop of several findings reported by researchers in the field. We are especially interested about the debate in literature in the context of Indian aerospace industry. One success story reported in literature is that of the MD-11 agreement in Brazil, which proved to be the seedbed for the development of the indigenous Embraer aircraft. It created local industry base, more jobs and enabled the export promotion of Brazilian made products. Also, recent statistics indicate that the sourcing of components from foreign vendors for the Boeing 727 aircraft in USA has gone up significantly as an offshoot of more number of offset policies that Boeing has entered into with overseas customers. This clearly indicates that manufacturing jobs are getting distributed to other countries and one can safely presume that offset policy would give more impetus to job creation in developing countries.

More empirical studies are needed in order to assess the impact of offset policy. The present body of research is scanty in the area of technology transfer, which is of paramount importance for developing economies to compete in the global market place. There is a need to formulate a frame work, which can enable realistic evaluation of offset policy and its impact on economy, job creation, manufacturing quality, technology transfer, etc.

References

1. M.W. Chinworth, R. Matthews, Defense industrialisation through offsets: The case of Japan, in *The Economics of Offsets: Defence Procurement and Countertrade*, (Harwood Academic Publishers, Amsterdam, 2014), p. 177
2. J. Niosi, J.Y. Zhao, China's catching up in aerospace. Int. J. Technol. Glob. **7**(1–2), 80–91 (2013)
3. P. Bamber, G. Gereffi, *Costa Rica in the Aerospace Global Value Chain* (Duke, Durham, 2013)
4. O. Herrnstadt, *Offsets and the Lack of a Comprehensive US Policy. EPI Briefing Paper 201* (EPI, Washington, DC, 2008)
5. D.L. Furter, *The Influence of Legislation and Government Policy on Patterns of International Defence Trade and Future Markets: The Case of Offset and Directive 2009/81/EC* (Brunel University, London, 2014)
6. C. Yang, C. Tsung-Cheng Wang, T.A. Force, Interactive decision-making for the international arms trade: The offset life cycle model. DISAM J. Int. Secur. Assist. Manag. **28**(3), 101–109 (2006)
7. S. Willett, I. Anthony, *Countertrade and Offsets Policies and Practices in the Arms Trade* (Copenhagen Peace Research Institute, Copenhagen, 1998)
8. L. Kumar Behera, India's defence offset policy. Strateg. Anal. **33**(2), 242–253 (2009)

9. B. Udis, K.E. Maskus, Offsets as industrial policy: Lessons from aerospace. Def. Peace Econ. **2**(2), 151–164 (1991)
10. T. Mathew, Essential elements of India's defence offset policy-a critique. J. Def. Stud. **3**(1), 2 (2009)
11. Government of India, *Defence Procurement Procedures 2016, Capitol Procurement* (DPP 2016, Government of India, Ministry of Defence, New Delhi, 2016)
12. Standing Committee on Defence, *Demands for Grants (2016–17), Twenty Second Report* (Lok Sabha Secretariat, New Delhi, May 2016), pp. 36–28
13. J. Brauer, J. Paul Dunne, Arms trade offsets and development. Africanus **35**(1), 14–24 (2005)
14. L.J. Dumas, Do offsets mitigate or magnify the military burden? in *Arms Trade and Economic Development: Theory, Policy and Cases in Arms Trade Offsets*, (Routledge, London, 2004), pp. 16–29
15. P.W. Liesch, Government-mandated countertrade in Australia: Some international marketing implications. Ind. Mark. Manag. **23**(4), 299–305 (1994)
16. J.E. da Silva-Menezes, *The Brazilian Aerospace Industry: A Cast Study of the Technological Impact of Offset Agreements in a Recipient Industry* (Naval Postgraduate School, Monterey, CA, 1989)
17. D.C. Mowery, Offsets in commercial and military aerospace: An overview, in *Symposium Papers on Trends and Challenges in Aerospace Offsets*, (National Academies Press, Washington, DC, 1999)
18. National Research Council, *Trends and Challenges in Aerospace Offsets* (National Academies Press, Washington, DC, 1999)
19. J.H.L. Guerra, Systems integration model in the aerospace industry: Motivating factors. Gestão&Produção **18**(2), 251–264 (2011)

The Shape of Firms: Opportunities from Rapid Manufacturing

Antonio Esparza, Ricardo Sosa, and Andy M. Connor

1 Introduction

The role of design in the creation of new business ventures has been documented in the academic and professional literature [1–3]. An alternative design approach to entrepreneurship considers it as a matter of firm design [4, 5], a process of creation of artificial means that negotiate with the environment. We suggest that more nuanced descriptions of firm design are needed. This paper frames the study of the shape of the firm based on the intersection of design science, entrepreneurship theories, and rapid manufacturing technology. First, we examine the roles of design in the creation of new business ventures and describe shape as the formal dimension of firm design. We then examine theories of firm creation applying an ontology of design activity, the function–behaviour–structure (FBS) framework, to elucidate a space for the shape of the firm. The relationships between the shape of the product and the shape of the firm are analysed. Finally, we formulate a set of questions for the study of this entanglement in new business creation to empower future entrepreneurs to identify and capitalise on these relationships.

2 Expanding Design and Management

Entrepreneurship theories are strongly influenced by assumptions in management studies about the purpose, object, and process of creating business ventures. Two strands of thought are represented in the role of design in entrepreneurship: Strategic

A. Esparza (✉) · R. Sosa · A. M. Connor
Colab, Auckland University of Technology, Auckland, New Zealand
e-mail: antonio.esparza@aut.ac.nz; ricardo.sosa@aut.ac.nz; andrew.connor@aut.ac.nz

A. Reyes-Munoz et al. (eds.), *EAI International Conference on Technology, Innovation, Entrepreneurship and Education*, Lecture Notes in Electrical Engineering 532, https://doi.org/10.1007/978-3-030-02242-6_19

Design (SD) and Design Thinking (DT). SD is a branch of strategic thinking concerned with the creation of idealised plans for the optimal accomplishment of objectives. SD is distinguished for generating a carefully controlled process of thought, centralizing planning in the figure of the strategist, simplifying an original, complete, and explicit outcome, separated from the implementation process [6, 7]. DT for strategic planning has gained popularity as a tool for integrating divergent (synthetic) and convergent (analytic) reasoning. DT is mainly used for the synthesis of solutions based on abductive logic, the exploitation of opportunities, and the use of inquiry for value generation [8, 9]. It is the intention of this proposition to expand the definition and applicability of design principles in management based on the study of design activity.

Design has been defined as the capacity of "conceiving, planning, and making products that serve human beings in the accomplishment of their individual and collective purposes" [10, 11]. In other words, design is a process that uses the creation of artefacts to interact with the environment and effectuate desired results. Designed artefacts create new practices and as a consequence new identities of those who use [12–14]. As an activity of creation of our desired future state, the scope of design covers the production of all human artefacts, such as firms, from very diverse perspectives different from SD and DT.

Understanding firms as products of their products, designs can be understood as first-order principles of [15]. Therefore through design, managers are in charge of creating value within the firm in order to achieve differentiation. This view aligns with Sarasvathy's perspective of entrepreneurship. Based on the study of expert entrepreneurs, she portrays the entrepreneurial process as the effectuation of negotiations that helps entrepreneurs in the achievement of their goals [16, 17]. Specifically, she highlights the need to research the processing of language and the categorisation of symbols in the entrepreneurial [5]. The behaviours around artefacts related to the firm, such as brands, logos, products, etc. imply that the firm is an artefact that can be studied at a semantic level. Therefore, the study of design opens new opportunities for the study of the entrepreneurship process resulting in the design of a firm like an artefact.

2.1 Artefacts of Design

Artefacts are the object of design. In order to modify our environment, we interact with artefacts through their shape. Human ecologies, like other ecologies, are defined by the spaces or fluids that enable the movement of substances [18]. Medium and substances are separated by surfaces which have specific layouts that we call "shape." Shape configurations gather properties that help us distinguish them and give them a specific character, that is, room, chair, or cloth [19]. Through shape, artefacts relate within the semantic ecology of our environment. Shape communicates the counter-ability of artefacts, or the available purposes and actions that we can perform with the artefact. Through shape, artefacts help us in the

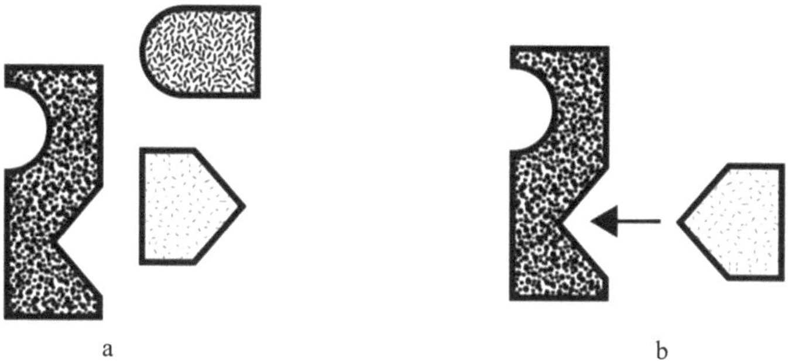

Fig. 1 Shape ecologies. (**a**) An ecology is composed by substances and a space or medium. (**b**) Shape limits the substance of the artefact and signals its counter-abilities or affordances [18]

accomplishment of the objectives of their design, as illustrated in Fig. 1. When designing an artefact, designers refer to the perceptual grammar that resembles the possibilities of shape within a specific typology of artefacts.

Due to the complexity of human production, the shape of system artefacts can be difficult to model. The creation of transactional systems lacks a formal manifestation compared with the design of physical artefacts. Therefore, the relationship between users, the environment, and firms as artefacts is not bounded by visible appearances, but by conditions of scale and reciprocity. Scale determines the span of interaction while reciprocity is the correspondence between the artefact and the user's interactive capabilities [20]. This is evident in the development of designed objects and spaces where the corresponding relationships between our bodies and the artefacts are found in the size of doors, or the roundness of handles. It is evident that in the case of complex systems such as firms, the scale and reciprocity shall be difficult to detect. We can infer that today in the design of business we experience a mismatch between the shape of the business and the relationship that it has to our human bodies and minds. When does our interaction with a business start and end? What are our expected behaviours? What is the vocabulary of the firm that we are meant to interpret?

Today the brand and its touch-points, as well as the product, and the packaging, are considered [21]. Nevertheless, a close examination of the existing theories of the firm suggests that the elements that compose a firm could be a designable as well. For instance, the theory of transaction costs considers that the firm will try to include all the transactions that increase the complexity of operations and as a result, increase cost. A model of the shape of the firm should attempt to account for these manifestations, and develop a designable perceptual grammar of the firm.

3 The Shape of the Firm and the Theories of the Firm

Current descriptions of the firm suggest the relational nature of business enterprises. Usually they account for heterogeneous compositions of resources, knowledge and human capital. Nevertheless, the firm as an artefact must also be justified as an effectuative prosthetic of human bodies. Consequently, it must have a reciprocal relationship to our scales and perceptive boundaries. The shape of the firm needs to be designed to afford specific behaviours on users according to the business logic and objectives. Customers, partners, employees, entrepreneurs, managers, stakeholders, and other artefacts interact with the affordances that the shape of the firm presents. The firm may interface through symbols and systems (brands, products, etc.) to elicit the desired behaviours of the business strategy. Therefore, inasmuch as the term user extends to all the people that interact with the signifiers of the firm, the quality of a good or bad firm design could be defined not only for its relative performance but by the difference between the expected behaviours of the design and the real behaviours that are elicited in users through these symbols. A different typology of firms based on shapes, could generate more options for business design and innovation creating more mechanisms for differentiation. However, in order to articulate a model of the shape of the firm, the existing theories that describe the composition of the firm space should be situated in design terms.

3.1 The FBS Ontology and Framework

The FBS ontology [22] is a useful to describe the design space and has been used extensively to model design [23–25]. Its ontology organises design based on three fundamental constructs: function, behaviour, and structure. Function is described as the teleological cause of the artefact, or the relationship between the goals and how they are met. Behaviour describes the performance derived from the artefact's structure. Structure refers to the arrangement of the artefact's components whether they are physical, virtual or social. Behaviour can be derived from structure using physical laws or heuristics, whilst no direct connection exists between function and [26]. The FBS framework splits the artefact space in two; the expected world, and the interpreted world. In the expected world, users and designers make up expectations of the artefact to be based on perception. Differently from users, designers enact this expectation in the design process. The interpretation world includes the artefact's use. Interpretation does not always aligns with expectation. An expected function (Fe) inductively derives an expected behaviour (Be) and an expected structure (Se). The materialised structure (S) elicits a behaviour (B) which in comparison with the design goals reveals a function (F). The FBS schema is depicted in Fig. 2. The distance between these two processes expands the set of transformations from a linear transformation, to a set of iterative processes that

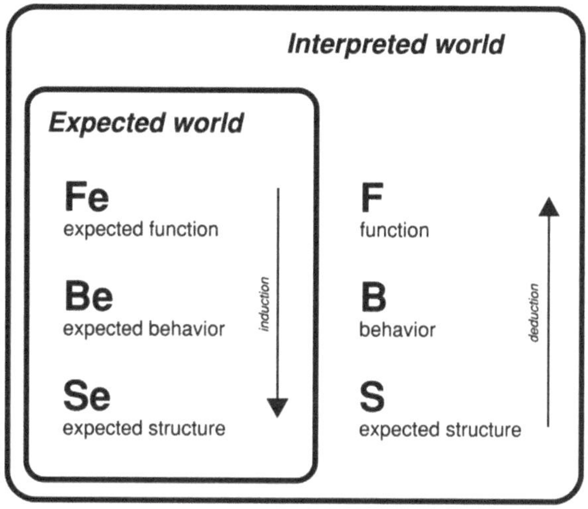

Fig. 2 The FBS framework supports a model of design processes [22]

reflect many design processes, from the generation of requirements to the interpreted description of the artefact.

3.2 Mapping Existing Definitions of the Firm to the FBS Ontology

Three groups of theories that account for the nature of the firm in economic sciences can be situated within the FBS framework to yield a different understanding of the firm as an artefact. Firstly, the theories of the firm, which account for the purpose and nature of the firm against market structures. Next, the models of enterprise ontology, originated as a tool for representing the entities and activities related inside a business. And finally, the theory of business models which explains the logic that underlies value creation and delivery. While each of the groups is formed by multiple models and theories, we consider their shared features.

First, the theories of the firm can be considered as a group centred in the description of a meta-level of abstraction. The purpose of the theory of the firm is to define the formal relations that differentiate it from the market and industry structures in a way that contributes to the study of economics [27–31]. Consequently, the models that the theories of the firm supply, are strongly related to the expected function (Fe) and behaviours (Be) of the firm. Coase [28] makes evident the question of the firm purpose, that is, "why is there any organisation?" Similarly, theories of the firm attempt "first, to specify the decisions that business firms will make (as a basis for more aggregate predictions of the economy) and second, to prescribe appropriate decision rules for a rational firm operating in a market economy" [32]. Overall, these theories seek to describe an ontology of the firm based on the observations of business, ergo showing interpretations of it as a phenomenon, not an artefact.

The majority of these theories account for functions that were based on the economic assumptions of supposed homogeneous goals of the entrepreneur and opportunism. In the design of new businesses, a tension is observable between the predefined layout of these expected functions and behaviours in economy, and the goals that individuals could bring to firm creation. Theories of creative entrepreneurship such as creative organizing [33] and bricolage [34] do not fit the theories of the firm. These theories do not show relationships that are able to induce structures and behaviours of the firm beyond the existing paradigms of economics and management. Moreover, the conflict between supposed heterogeneous goals and behaviours evidences a void in the theories of the firm that if addressed could create more possibilities for firm design.

Despite being a detailed reference of business entities, enterprise ontologies fail to map the firm onto a structural level. The diversity of models around enterprise ontologies can differ but they all concur in the representation of the entities to be monitored in order to exercise control of the company. Fox and Gruninger stress the role of ontologies in the integration of the enterprise by the addition of subsets of specific ontologies; "for example, the notion of manufacturability requires reasoning about the product's properties, preconditions, and effects of activities and the capabilities of resources" [35]. Therefore, business ontologies seem to be a reference tool for performance accountability rather than representing the structure itself. Yet, as legal litigations show, not all the affordable behaviours through the firm are accounted by business [36]. Therefore, if we consider that design theories of artefacts recognise the interpretation of the user in the redefinition of the purpose and its interaction with the context in the creation of affordances, an enterprise ontology mistakes the role of human entities in the exercise of creativity and innovation.

Business models show the logic behind the operation and profitability of a firm [37]. Research and popular literature consider that the design of a business model is essential in the early stages of the entrepreneurial process. Situating the business model definition within the FBS framework, the business logic fits the expected behaviour (Be) around the designed artefact. The activity based design process of business models proposed by Zott and Amit portraits the business model as a blueprint for the derivation of the firm structure [38]. Accordingly, Osterwalder and Pigneur situate the business model as an organising matrix inside the firm. They argue that the interaction between the business model, strategy, information, and organisation guides the firm's operation. The manifested business model of a company becomes a tool for the communication of strategy [39]. Hence, it could expand itself from the expected, to the interpreted behaviour since it elicits specific actions (at least at a macro level) in the exercise of the firm. Nevertheless, since there is no object to refer as a firm artefact, the deduced behaviours can be forced through explicit strategy and could be understood as ambiguous.

The resulting mapping of theories of the firm, enterprise ontologies and business models in to the FBS framework is shown in Fig. 3.

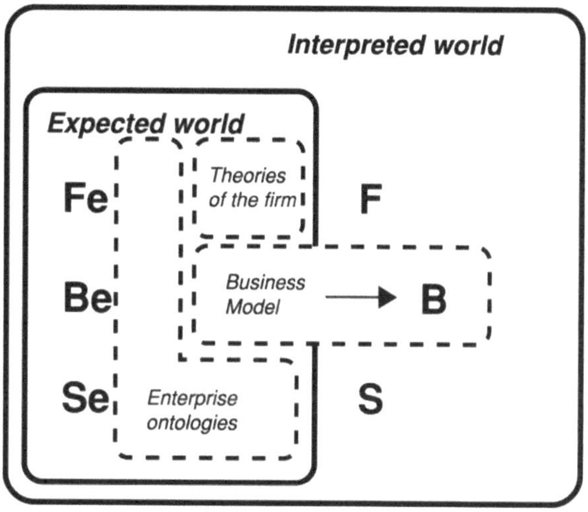

Fig. 3 The existing theories that describe firms only consider one "natural" firm structure that is pushed to the interpreted world through strategic communication

3.3 The Entangled Shape of the Firm

The theories of design and management examined here show the extant need for defining the interpreted world of the firm artefact. Literature on product architecture has interesting developments that consider the interaction between the firm and its users. Based on evidence of the relationship between product architecture and the success of firms, these studies confirm that the division of labour in firms reflects the principle of bounded rationality and consequently mirrors the configuration of the product into the information processing structures of the firm [40, 41]. As a consequence, the arrangement of the product architecture can affect the organisational learning curve and the exercise of authority between organisational divisions [42]. This mirroring process suggests that the development of information mechanisms that support product architecture solidifies and extends through time beyond the firm and into its suppliers and the rest of the industry [43]. Different products predefine available organisational configurations regardless the imposed strategy. The inadequate mirroring of a product architecture in the early stages of business development may carry associative thinking biases that need to be tackled through iterations of divergent configurations of the product and value proposition [44, 45]. The mirroring effect implies that one of the most important strategic choices in firm creation is the relationship between the components in product architecture.

As Sarasvathy proposes, entrepreneurship can be considered as the design of a firm artefact that aids entrepreneurs in the fulfilment of their goals. In the design of the firm artefact, a range of participants as users are acknowledged: Customers, suppliers, employees, etc. Consequently, possible misinterpretations of individual roles inside corporations need to be considered. Errors in the use of an artefact, such as pulling a door when it needs to be pushed, or walking over "lines of desire" in gardens and parks instead of going around the corner, are design flaws. Similarly, behaviours like delayed payments, product order misunderstandings, quality issues, and fraudulent practices, could be caused by affordances in the firm that are not accounted for. Moreover, desired behaviours focused on the creation of value and innovation could also be elicited through the design of the shape of the firm. Based on the mirroring process between product and organisational architectures, the main argument of this proposition is that the shape of the firm is composed by the dependencies between functional components in product architecture and organisational configurations which evoke the behaviours of firms users. Therefore, in the firm design process, the conceptualisation of different dependencies in the shape of the firm will make available specific product and organisational possibilities that can be matched to the goals of the entrepreneur. This approach is different from conventional innovation and entrepreneurial processes which create a product, and force an expected behaviour of users through strategic communication. Hence, current tools operate under the assumptions of the theories of the firm, business models and enterprise ontologies, regardless different product architectures. This new approach opens new opportunities for the creation of methods and tools that articulate the shape of the firm according to its interaction with humans, and other artefacts (logos, brands, media, other firms, etc.).

Just as the guidelines in the shape of a chair artefact, the dependencies between components and teams resemble the shape of the firm artefact (Fig. 4a). Traditional entrepreneurship takes the design of a product and enforces organisational behaviours through strategy (Fig. 4b). Through the design of the shape of the firm, product architectures can be purposefully selected that correspond to organizational configurations (Fig. 4c).

3.4 *Rapid Manufacturing in the Shape of the Firm*

This proposition is especially relevant today in view of digital technologies. Information Technology based tools summarize in code structures that before required the commitment of valuable resources. As a result, firms today are more flexible than before [46]. Tools that now are used for around the business model, such as enterprise application, customer relationship management, and computer-aided design software can be modified to fit and interact with the shape of the firm to bring out desired behaviours in users and feedback relevant data for the iteration of the shape itself. Tools for data science, such as mining and analytics can help in the shaping of the affordable relationships in the geometry of the firm. With the involvement of data, generative algorithms of design could be used to adopt a flexible strategy that take advantage of contingencies and react instantly to social and market fluctuations. Technologies like additive manufacturing (AM) could project this digital flexibility to the production of material goods. Algorithms

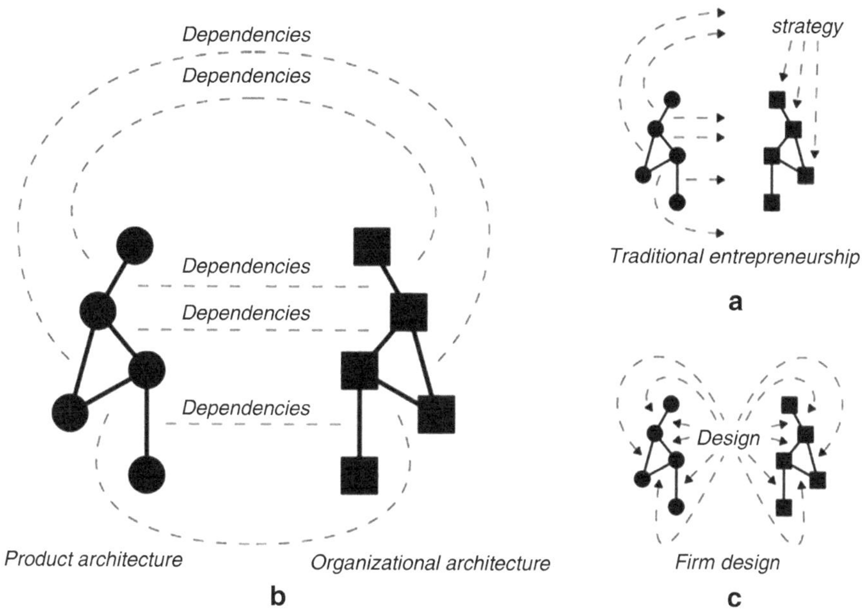

Fig. 4 Mappings of dependencies between product and firm design exist (**b**). Traditional entrepreneurship does not acknowledge them by projecting strategy to the organization (**a**). It is proposed here that firm design should articulate both poles (**c**)

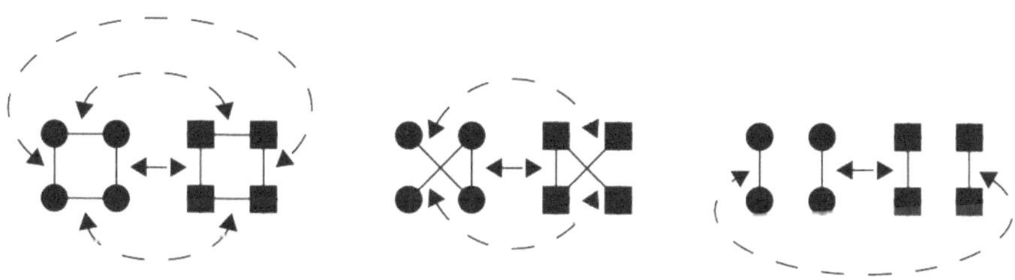

Fig. 5 Additive manufacturing could enable the exploration of different shape configurations without heavy capital investment

of generative design, can adapt the shape of produced products to the desired affordances of the shape in real time. Manufacturing of goods can be as flexible as needed for the business to effectuate the acquisition of partnerships and resources (Fig. 5).

For entrepreneurship this is an opportunity to leverage the relationships and shape the firm through the initial product according to the final goal of the entrepreneur. This will expand the role of design and the available control of the entrepreneur over the firm's future. Traditionally, regardless the industry, entrepreneurship processes are conformed by a discovery, evaluation, and exploitation of the business idea [47]. By integrating digital technologies, these processes have become more agile in the implementation and evaluation of explored ideas. Nevertheless, these processes guide itself through trial and error [48, 49]. A model of the shape of the firm could inform the implementation of this experiments in a more purposeful and specific way. It would integrate strategy to the production of the goods immediately in a way that is particular to the product that is being fabricated. Therefore, strategy could use digital technologies to extremely detail and micro-manage the shape of the firm. Unique paths for differentiation could originate in the interaction between the entrepreneurial contingency and the project that could push the competitiveness of small firms in front of competing corporations.

Experimentation with alternative concepts of digital business can be brought forward thanks to a model of the shape of the firm. While companies usually rely on the same structure, different shapes of firms could experiment with concepts that current ones find very expensive to use. That is the case of distributed manufacturing, a model where automated manufacturing like 3D printing could fabricate goods in smaller facilities distributed geographically. Despite the potential saves in cost, research has proven to be difficult to [50]. A specialized design of the shape of a firm could leverage the creation of networks that make this model possible. Other explorations with cryptocurrencies or "money of the commons" could also be matched to shapes of firms. Shapes that facilitate stakeholder governance mechanisms could make use of such technologies to foster social entrepreneurship.

4 Conclusion

Expanding the idea of firm design beyond the creation of instruments for strategy (SD and DT) to firm design creates a vast space for the exploration of the firm as a designable artefact. Artefacts as prosthetics of human bodies help us modify our environment to match our desired goals. Likewise, a firm artefact would help the entrepreneur to effectuate the contingencies that surround the project and fulfil specific purposes. After an analysis of current theories of firm creation, this paper identifies a void between the expected behaviours of the firms and the interpreted behaviours inside it. There seems to be no guided action outside the communication of strategy, and as a consequence, the existing theories do not account for all the afforded behaviours in the interaction between the firm and its possible users. Based on the evidence from studies on product architecture, we propose that the shape of the firm can be found in the dependencies between the product and organizational architectures. Therefore, by designing a specific set of dependencies, the product and organizational architectures of the firm could be mutually defined.

The study of the shape of the firm brings together the study of entrepreneurship and design. Processes of entrepreneurship could make use of design to articulate different shapes according to specific goals attainable by the entrepreneur. Tools for creative entrepreneurship could be designed to take advantage of the distinctive conditions of each entrepreneurial context. This would help grounding popular tools like business modelling on the entrepreneur's reality and give more certainty to projects that do not have access to venture capital or even information technology means. Shifting focus to technology based entrepreneuring, a theory of the shape of the firm would give a very valuable resource for start-ups to adapt and react based on the integration of data feeds to the design process of the firm.

This study also identifies opportunities for the study of the perceptive grammar and the resulting typologies of firms. Different types shall be classified according to size, industries, products, value propositions, etc. Dependencies of the firm could be also classified according to their function. Making use of analogical reasoning from biology theory, the main dependencies for the growth and reproduction of the firm could be theorised. Start-ups could be designed to act as dynamically as viruses, or rely on high memory and learning capacity like elephants and insect colonies. Industries could be studied according to the interactions within firms, viewed as ecosystems. Firms could be designed to create relationships of competition by cultivating dependencies that tie them to strong yet flexible networks of firms and users. Extending the analogy, firm and product differentiation can be achieved based on strategies of life spans, replication, ecological inheritance, niche construction and mutual adaptation [51].

This paper closes with three areas of interest for original research around the aesthetics of firm-artefact relations. First, the role of the product needs to be studied in the context of the evolution of the firm structure in the entrepreneurial process. This will permit the representation of the dependencies between the two structures in a practical context. Second, different shapes of firms need to be explored with the aid of rapid manufacturing technologies. Just as we can explore the shape of product design, we must explore a language that represents accurately the entanglement of the architectures and the guidelines of the shape as a whole. As mentioned before, rapid manufacturing technologies are notable for their flexibility. Therefore, by making changes in product architecture using additive manufacturing, it would be reasonable to expect to induce the dependencies and map the families of firm design. Finally, expanding the research around product architecture, the development of cases that analyse the interaction of firm shapes such as the failure between Boston Dynamics and Google [52], the adoption of Snapchat features by Facebook apps [53], or new product development in game consoles [54]. By considering these three possible routes of inquiry, the study of the shape of the firm has the potential to leverage design in the creation of more deliberate futures for entrepreneurs and new businesses.

References

1. T. Brown, Design thinking, Harvard Business Review, 01 Jun 2008 [Online], https://hbr.org/2008/06/design-thinking. Accessed 29 Mar 2016
2. L. Kimbell, Rethinking design thinking: Part I. Des. Cult. **3**(3), 285–306 (2011)

3. R.L. Martin, *Design of Business: Why Design Thinking is the Next Competitive Advantage* (Harvard Business Press, Brighton, 2009)
4. S.D. Sarasvathy, Entrepreneurship as a science of the artificial. J. Econ. Psychol. **24**(2), 203–220 (2003)
5. S.D. Sarasvathy, Making it happen: Beyond theories of the firm to theories of firm design. Entrep. Theory Pract. **28**(6), 519–531 (2004)
6. H. Mintzberg, The design school: Reconsidering the basic premises of strategic management. Strateg. Manag. J. **11**(3), 171–195 (1990)
7. H. Mintzberg, J.A. Waters, Of strategies, deliberate and emergent. Strateg. Manag. J. **6**(3), 257–272 (1985)
8. J. Liedtka, In defense of strategy as design. Calif. Manag. Rev. **42**(3), 8–30 (2000)
9. R. Martin, Design thinking: Achieving insights via the 'knowledge funnel'. Strategy Leadersh **38**(2), 37–41 (2010)
10. R. Buchanan, Design research and the new learning. Des. Issues **17**(4), 3–23 (2001)
11. H.A. Simon, *The Sciences of the Artificial* (MIT Press, Cambridge, 1996)
12. M. Heidegger, J. Stambaugh, *Being and Time (a Translation of Sein und Zeit)* (State University of New York Press, Albany, 1996)
13. K. Krippendorff, *The Semantic Turn: A New Foundation for Design* (CRC/Taylor & Francis, Boca Raton, 2006)
14. V. Margolin, Design, the future and the human Spirit. Des. Issues **23**(3), 4–15 (2007)
15. A. Dong, M. Kleinsmann, D. Snelders, A Design-based Theory of the Firm, (Unpublished, 2017)
16. S.D. Sarasvathy, Causation and effectuation: Toward a theoretical shift from economic inevitability to entrepreneurial contingency. Acad. Manag. Rev. **26**(2), 243–263 (2001)
17. S.D. Sarasvathy, *Effectuation: Elements of Entrepreneurial Expertise* (Edward Elgar, Northampton, 2008)
18. J.J. Gibson, *The Ecological Approach to Visual Perception: Classic Edition* (Psychology Press, London, 2014)
19. F.M. Juez, *Contribuciones para una antropología del diseño*, 1st edn. (Gedisa Editorial, Barcelona, 2002)
20. M. Krampen, Semiotics in architecture and industrial/product design. Des. Issues **5**(2), 124–140 (1989)
21. E.R. Abbing, C. van Gessel, Brand-driven innovation. Des. Manag. Rev. **19**(3), 51–58 (2008)
22. J.S. Gero, U. Kannengiesser, The function-behaviour-structure ontology of design, in *An Anthology of Theories and Models of Design*, ed. by A. Chakrabarti, L.T.M. Blessing (Springer, London, 2014), pp. 263–283
23. G. Cascini, G. Fantoni, F. Montagna, Situating needs and requirements in the FBS framework. Des. Stud. **34**(5), 636–662 (2013)
24. K. Dorst, P.E. Vermaas, John Gero's function-behaviour-structure model of designing: A critical analysis. Res. Eng. Des. **16**(1–2), 17–26 (2005)
25. U. Kannengiesser, J.S. Gero, A process framework of affordances in design. Des. Issues **28**(1), 50–62 (2011)
26. L. Qian, J.S. Gero, Function–behavior–structure paths and their role in analogy-based design. AI EDAM **10**(4), 289–312 (1996)
27. J.B. Barney, The resource-based theory of the firm. Organ. Sci. **7**(5), 469–469 (1996)
28. R.H. Coase, The nature of the firm. Economica **4**(16), 386–405 (1937)
29. R.M. Grant, Toward a knowledge-based theory of the firm. Strateg. Manag. J. **17**(S2), 109–122 (1996)
30. O.D. Hart, Incomplete contracts and the theory of the firm. J. Law Econ. Organ. **4**(1), 119–139 (1988)
31. O.E. Williamson, Transaction cost economics, in *Handbook of Industrial Organization*, vol. 1, (Elsevier, New York, 1989), pp. 135–182
32. R.M. Cyert, E.A. Feigenbaum, J.G. March, Models in a behavioral theory of the firm. Behav. Sci. **4**(2), 81–95 (1959)

33. B. Johannisson, Towards a practice theory of entrepreneuring. Small Bus. Econ. **36**(2), 135–150 (2011)
34. T. Baker, R.E. Nelson, Creating something from nothing: Resource construction through entrepreneurial bricolage. Adm. Sci. Q. **50**(3), 329–366 (2005)
35. M.S. Fox, M. Gruninger, Enterprise Modeling. AI Mag. **19**(3), 109 (1998)
36. K. Hoerr, Uber driver jailed for raping woman, ABC News, 13 Jun 2017 [Online], http://www.abc.net.au/news/2017-06-13/sydney-uber-driver-jailed-over-rape-of-passenger-kings-cross/8612936. Accessed 30 Jun 2017
37. A. Osterwalder, Y. Pigneur, C.L. Tucci, Clarifying business models: Origins, present, and future of the concept. Commun. Assoc. Inf. Syst. **16**(1), 1–25 (2005)
38. C. Zott, R. Amit, Business model design: An activity system perspective. Long Range Plan. **43**(2–3), 216–226 (2010)
39. A. Osterwalder, *The Business Model Ontology a Proposition in a Design Science Approach* (Universite De Lausanne Ecole Des Hautes Etudes Commerciales, Lausanne, 2004)
40. P. Anderson, M.L. Tushman, Technological discontinuities and dominant designs: A cyclical model of technological change. Adm. Sci. Q. **35**(4), 604 (1990)
41. R.M. Henderson, K.B. Clark, Architectural innovation: The reconfiguration of existing product technologies and the failure of established firms. Adm. Sci. Q. **35**(1), 9 (1990)
42. R. Sanchez, J.T. Mahoney, Modularity, flexibility, and knowledge management in product and organization design. Strateg. Manag. J. **17**(S2), 63–76 (1996)
43. L.J. Colfer, C.Y. Baldwin, The mirroring hypothesis: Theory, evidence, and exceptions. Ind. Corp. Change **25**(5), 709–738 (2016)
44. C. Christensen, *The Innovator's Dilemma: When New Technologies Cause Great Firms to Fail* (Harvard Business Review Press, Boston, 2013)
45. N. Furr, J.A. Nickerson, R. Wuebker, *A Theory of Entrepreneuring* (Social Science Research Network, Rochester., SSRN Scholarly Paper ID 2747458, 2016)
46. P. Schubert, J. Fisher, U. Leimstoll, ICT and innovation in small companies, in *ECIS 2007 Proc.*, (University of St. Gallen, St. Gallen, 2007)
47. S. Shane, S. Venkataraman, The promise of entrepreneurship as a field of research. Acad. Manag. Rev. **25**(1), 217–226 (2000)
48. E. Ries, *The Lean Startup: How Today's Entrepreneurs Use Continuous Innovation to Create Radically Successful Businesses* (Crown Publishing Group, New York, 2011)
49. D. Sola, G.S. Borioli, G. Scalabrini, New product development and disciplined experimentation. Symphonya Emerg. Issues Manag. **0**(2), 105–118 (2015)
50. S.H. Khajavi, J. Partanen, J. Holmström, Additive manufacturing in the spare parts supply chain. Comput. Ind. **65**(1), 50–63 (2014)
51. K. Sterelny, *Externalism, Epistemic Artefacts and the Extended Mind. The Externalist Challenge* (2004), pp. 239–254
52. A. Macfarlane, Google sells maker of 'nightmare-inducing' robots to Japan's SoftBank, CNNMoney, 09 Jun 2017 [Online], http://money.cnn.com/2017/06/09/technology/boston-dynamics-robots-google-alphabet-softbank/index.html. Accessed 30 Jun 2017
53. A. Heath, Here are all the times Facebook has copied Snapchat so far, Business Insider, 27 May 2017 [Online], http://www.businessinsider.com/all-the-times-facebook-copied-snapchat-2017-5. Accessed 30 Jun 2017
54. D. Thier, Nintendo's biggest mobile game is the one nobody's really talking about, Forbes, 30 May 2017 [Online], http://www.forbes.com/sites/davidthier/2017/05/30/nintendos-biggest-mobile-game-is-the-one-nobodys-really-talking-about/. Accessed 30 Jun 2017

Big Data, the Internet of Things, and Smart City Research: A Literature Review and Research Agenda

Samuel Fosso Wamba, Messina Ntede Cécile Angéla, and Etoa Etoa Jean Bosco

1 Introduction

Cities have always been at the heart of innovation. This is even more crucial in the twenty-first century with the United Nations predictions of a rapid growth in the urban population that may reach about 2.5 billion new people by 2050 [1]. This accelerating growth may become unsustainable if the cities rely on the traditional systems to deliver both economic and social resources [2]. Recently, the emergence of cutting-edge tools and technological innovations, including the Internet of Things (IoT), big data analytics (BDA), predictive analytics, and industry 4.0 has created a renewal of interest toward studying cities [3–5]. Indeed, the adoption and use of these new tools and technologies are creating the so-called Smart cities, which can be defined as the "networked infrastructures to improve economic and political efficiency and enable socio, cultural and urban development" (p. 307) [6]. For example, [7] argued that the use of the "Internet of Things and big data analytics can develop the smart city and smart industry" [7], and thus, help to ease the current pain points associated with traffic congestion, waste and pollution management, and energy efficiency that may bring the city liveability to new levels that haven't been seen before [2].

S. Fosso Wamba (✉)
Toulouse Business School, Toulouse, France
e-mail: s.fosso-wamba@tbs-education.fr

M. N. C. Angéla
GRIAGES, Catholic University of Central Africa, Yaounde, Cameroon

E. E. J. Bosco
Faculty of Economics and Management, University of Yaounde II, Soa, Cameroon

A. Reyes-Munoz et al. (eds.), *EAI International Conference on Technology, Innovation, Entrepreneurship and Education*, Lecture Notes in Electrical Engineering 532, https://doi.org/10.1007/978-3-030-02242-6_20

Similarly, Hashem et al. [8] argued that the growth of BDA and the development of IoT related technologies are paving the way for the feasibility and realization of smart city initiatives. They argued that BDA "offer the potential for cities to obtain valuable insights from a large amount of data collected through various sources, and the IoT allows the integration of sensors, radio-frequency identification, and Bluetooth in the real-world environment using highly networked services" [8]. Various scholars have also acknowledged the contribution of these technological innovations in solving various issues faced by today's cities including: job creation, economic growth, environmental sustainability, and social resilience [9].

Cities related tools and concepts have the capability of enhancing the quality of life of urban citizens by substantially improving the quality of development and delivery of various services including transportation, healthcare, electricity and water supply, education, and public security [9, 10].

Kummitha and Crutzen [11] claimed that IT-enabled smart cities will bring positive social change, and thus enrich "governance and human capital among the citizenry" (p. 43) [11].

Riggins and Fosso Wamba [12] argued that "while the technical issues needed to create the Internet of Things are substantial, little attention has been given to the behavioral, organizational and business issues that are necessary for a better understanding of the adoption, usage and impact of the IoT" (p. 1).

Despite the high potential related to the joint impact of BDA and IoT related tools in improving today's cities, we know little about the level of current studies on these topics. In fact, [8] claimed that the combination of the BDA and IoT is "an unexplored research area that has brought new and interesting challenges for achieving the goal of future smart cities" [8]. Therefore, this study is an initial attempt to fill the knowledge gap identified in the literature. More specifically, this study aims at examining the following research questions:

1. What is the current level of research on BDA and IoT related topics in the smart cities context?
2. Where should future efforts be directed to accelerate the adoption and use of BDA and IoT related topics in smart cities?

To address our research questions, we will draw on the literature on BDA and IoT related topics in smart cities and a literature review conducted on these topics within the database called SCOPUS.

The rest of this paper is structured as follows: Section 2 is concerned with the description of our research methodology; Section 3 presents and discusses the results; Section 4 provides the conclusion as well as future research directions.

2 Methodology

In this study, we follow an approach derived from prior studies [13, 14] that uses the following steps: (1) conduct a search using a combination of the following keywords: "smart cities" or "smart city" or "intelligent city" or "intelligent cities" and "big data" or "predictive analytics" and "internet of thing*" or "web of thing*" or "web of people" or "industry 3.0" or "industry 4.0" or "internet of object*" within the SCOPUS database. SCOPUS is considered as the largest abstract and citation database of peer-reviewed literature: scientific journals, books, and conference proceedings. The database delivers a comprehensive overview of the world's research output in the fields of science, technology, medicine, social sciences, arts, and humanities. It provides a set of tools to track, analyze, and visualize research [15].

Our search was realized on June 22, 2017. The search resulted in 145 papers dealing with the topic. A quick analysis allowed us to remove two duplicated papers [16, 17], and thus resulting in a final set of 143 papers.

3 Results and Discussion

In this section, we are going to present and discuss the key findings of the big data, predictive analytics, and IoT-related topics for smart cities articles identified by our search within SCOPUS [5, 7–9, 16–154].

In Table 1, we have the distribution of publications by year. There is a constant increase of papers published on the topic since 2014. We went from 16 papers (11.20%) published in 2014 to 30 papers in 2015 (21%), and 62 papers in 2016 (43.40%). On June 22, 2017, we already had 33 articles (23%) which is more than all papers published in 2015 or 2014.

Table 2 presents the distribution of articles by subject area. This table is dominated by "Computer Science" subjects with 120 papers (51.50%), followed by "Engineering" with 35 papers (15.02%), "Mathematics" with 20 papers (8.58%), "Social Sciences" with 13 papers (5.58%), and "Decision Sciences" with 10 papers (4.29%).

The distribution of articles by country is presented in Table 3. It is undoubtedly clear that China and the United States are leading the research on the topic with, respectively, 22 papers (11.46%) and 20 papers (10.42%), followed by Italy and Spain with 14 papers each (7.29%), then Switzerland with 11 papers (5.73%).

Table 1 Distribution of publications by year

Year	# of papers	%
2017	33	23.00
2016	62	43.40
2015	30	21.00
2014	16	11.20
2013	2	1.40
Total	143	100

Table 2 Distribution of publications by subject area

Subject area	# of papers	%
Computer Science	120	51.50
Engineering	35	15.02
Mathematics	20	8.58
Social Sciences	13	5.58
Decision Sciences	10	4.29
Energy	8	3.43
Business, Management, and Accounting	4	1.72
Earth and Planetary Sciences	4	1.72
Environmental Science	4	1.72
Materials Science	3	1.29
Medicine	3	1.29
Physics and Astronomy	3	1.29
Pharmacology, Toxicology, and Pharmaceutics	2	0.86
Arts and Humanities	1	0.43
Biochemistry, Genetics, and Molecular Biology	1	0.43
Chemistry	1	0.43
Health Professions	1	0.43
Total[a]	233	100.00

[a]Some articles are counted more than once because they cover more than one subject area

From Table 4, we can see that the distribution of papers by type of document is dominated by "Conference Paper" with 83 documents (58.04%), followed by "Article" with 40 documents (27.97%), 6 "Review" (4.20%), "Conference Review" with 5 documents (3.50%), "Book Chapter" with 3 documents (2.10%), "Article in Press" with 4 documents (2.80%), and Editorial and Note with each 1 document (0.70%).

Overall, we found 159 scholars working on the topic. Table 5 provides the list of top 10 authors. Jara, A.J. dominates this list with 7 papers, closely followed by Ahmad, A. and Paul, A. with 6 papers each. Then, we have Bocchi, Y., Genoud, D., Rathore, M.M., Dustdar, S., Galache, J.A., Inzinger, C., and Schleicher, J.M. with 5 papers each.

Overall, we found 54 sources of papers. The top 10 are presented in Table 6. The largest number of documents is from the "Lecture Notes In Computer Science Including Subseries Lecture Notes In Artificial Intelligence And Lecture Notes In Bioinformatics" with 11 papers.

Table 3 Distribution of publications by country

Country	# of papers	%
China	22	11.46
United States	20	10.42
Italy	14	7.29
Spain	14	7.29
Switzerland	11	5.73
Germany	9	4.69
India	8	4.17
South Korea	8	4.17
United Kingdom	8	4.17
Australia	7	3.65
Austria	5	2.60
France	5	2.60
Russian Federation	5	2.60
Canada	3	1.56
Colombia	3	1.56
Denmark	3	1.56
Japan	3	1.56
Macedonia	3	1.56
Netherlands	3	1.56
Portugal	3	1.56
Saudi Arabia	3	1.56
Sweden	3	1.56
Finland	2	1.04
Greece	2	1.04
Iran	1	0.52
Nigeria	2	1.04
Singapore	2	1.04
Taiwan	2	1.04
United Arab Emirates	2	1.04
Belgium	1	0.52
Brazil	1	0.52
Brunei Darussalam	1	0.52
Ecuador	1	0.52
Egypt	1	0.52
Latvia	1	0.52
Malaysia	1	0.52
Mexico	1	0.52
Morocco	1	0.52
Norway	1	0.52
Poland	1	0.52
Romania	1	0.52
Serbia	1	0.52

(continued)

Table 3 (continued)

Country	# of papers	%
Slovakia	1	0.52
South Africa	1	0.52
Undefined	1	0.52
Total[a]	192	100.00

[a]Some articles are counted more than one time as their authors come from more than one country

Table 4 Distribution of documents by type

Documents type	Number	%
Conference Paper	83	58.04
Article	40	27.97
Review	6	4.20
Conference Review	5	3.50
Book Chapter	3	2.10
Article in Press	4	2.80
Editorial	1	0.70
Note	1	0.70
Total	143	100

Table 5 List of top 10 authors

Author	# of papers
Jara, A. J.	7
Ahmad, A.	6
Paul, A.	6
Bocchi, Y.	5
Genoud, D.	5
Rathore, M. M.	5
Dustdar, S.	5
Galache, J. A.	5
Inzinger, C.	5
Schleicher, J. M.	5

4 Conclusion and Future Research Directions

The main objective of this study was to provide a literature review of BDA, the IoT, and Smart cities research using SCOPUS. Our search identified 143 relevant papers. The distribution of papers found by year of publication, subject area, country, type, top 10 authors, and sources have been presented and discussed.

The literature review showed that while there is a constant increase of papers published on the topic since 2014 (e.g., from 16 papers (11.20%) published in 2014 to 30 papers in 2015 (21%), and 62 papers in 2016 (43.4%) and 33 articles (23%) by June 22, 2017), the clear majority of papers are still "Conference Paper" (58.04%). Therefore, more research needs to be published by journals, especially top operations management and information systems journals. Indeed, no papers

Table 6 Distribution of articles by sources

Source	# of papers
1. Lecture Notes in Computer Science Including Subseries Lecture Notes in Artificial Intelligence and Lecture Notes in Bioinformatics	11
2. ACM International Conference Proceeding Series	4
3. IEEE Internet Computing	3
4. IEEE Pervasive Computing	3
5. Studies in Computational Intelligence	3
6. IEEE Access	2
7. International Journal of Pharmacy and Technology	2
8. International Journal on Semantic Web and Information Systems	2
9. Lecture Notes in Electrical Engineering	2
10. Software Practice and Experience	2

were identified from these outlets which are recognized to publish cutting-edge studies. More importantly, studies should focus on the best strategy to select, adopt, implement, and foster the use of the IoT, BDA, predictive analytics, and industry 4.0 in the context of smart cities.

The literature review also showed that very few studies were conducted on cities from underdeveloped countries, which face huge challenges including waste management, electricity and water supply, traffic congestion, urban planning, security and public services design and delivery. Assessing how the IoT, BDA, predictive analytics, and industry 4.0 related tools and technologies can contribute to solve these issues should be included into future research.

Nathali Silva et al. [10] argued that "continuous growth of the complex urban networks is significantly challenged by real-time data processing and intelligent decision-making capabilities," therefore, exploring the contribution of the IoT, BDA, predictive analytics, and industry 4.0 related tools and technologies in solving this challenge is an interesting research avenue.

Exploring how the IoT, BDA, predictive analytics, and industry 4.0 related tools and technologies impact citizens' privacy should also be included into future research agenda.

Assessing the cost of the adoption of these tools by various cities across the world should be included in future research directions. What are the best practices? How can they be shared? How these costs are going to influence the council tax system?

It would be interesting to expand our findings to other key databases such as ABI/Inform Complete, Academic Search Complete, Business Source Complete, Emerald, IEEE Xplore, Science Direct, and Taylor & Francis as well as in the Association of Information Systems (AIS) basket of top journals.

References

1. United Nations, *World Urbanization Prospects: The 2014 Revision, Highlights (ST/ESA/SER.A/352)* (United Nations, Department of Economic and Social Affairs, Population Division, 2014)
2. A. Sarkar, Significance of smart cities in 21st Century: an international business perspective FOCUS. J. Int. Bus. **2**(2) (2016).
3. L. Zhuhadar et al., The next wave of innovation—review of smart cities intelligent operation systems. Comput. Hum. Behav. **66**, 273–281 (2017)
4. G. Graham, Smart cities and operations management. Prod. Plan. Control **27**(6), 431–432 (2016)
5. M. Kumar et al., How will smart city production systems transform supply chain design: a product-level investigation. Int. J. Prod. Res. **54**(23), 7181–7192 (2016)
6. R.G. Hollands, Will the real smart city please stand up? City **12**(3), 303–320 (2008)
7. Z. Liu, Research on the Internet of Things and the development of smart city industry based on big data. Clust. Comput. **21**, 789–795 (2017)
8. I.A.T. Hashem et al., The role of big data in smart city. Int. J. Inf. Manag. **36**(5), 748–758 (2016)
9. B. Ahlgren, M. Hidell, E.C.H. Ngai, Internet of things for smart cities: interoperability and open data. IEEE Internet Comput. **20**(6), 52–56 (2016)
10. B. Nathali Silva, M. Khan, K. Han, Big data analytics embedded smart city architecture for performance enhancement through real-time data processing and decision-making. Wirel. Commun. Mob. Comput. **2017**, 12 (2017)
11. R.K.R. Kummitha, N. Crutzen, How do we understand smart cities? An evolutionary perspective. Cities **67**, 43–52 (2017)
12. F.J. Riggins, S. Fosso Wamba, in *Research Directions on the Adoption, Usage, and Impact of the Internet of Things Through the Use of Big Data Analytics*. 48th Hawaii International Conference on System Sciences, 2015, Hawaii, USA
13. E.W.T. Ngai, F.K.T. Wat, A literature review and classification of electronic commerce research. Inf. Manag. **39**(5), 415–429 (2002)
14. S. Fosso Wamba, A. Anand, L. Carter, A literature review of RFID-enabled healthcare applications and issues. Int. J. Inf. Manag. **33**(5), 875–891 (2013)
15. Scopus. About scopus. 2017 [cited 2017 June 18], https://www.elsevier.com/solutions/scopus
16. A.R. Honarvar, A. Sami, Extracting usage patterns from power usage data of homes' appliances in smart home using big data platform. Int. J. Inf. Technol. Web. Eng. **11**(2), 39–50 (2016)
17. S. Schatzinger, C.Y.R. Lim, Taxi of the future: big data analysis as a framework for future Urban fleets in smart cities, in *Smart and Sustainable Planning for Cities and Regions* (Springer, Cham, 2017), pp. 83–98
18. A.J. Jara, Y. Bocchi, D. Genoud, in *Determining Human Dynamics Through the Internet of Things* (IEEE, 2013)
19. Q. Zhang et al., A case study of sensor data collection and analysis in smart city: provenance in smart food supply chain. Int. J. Distrib. Sens. Netw. **9**, 382132 (2013)
20. J.P. Agudelo, H.E.C. Barrera, in *Approach for a Congestion Charges System Supported on Internet of Things* (IEEE Computer Society, 2014)
21. A. Amato, B. Di Martino, S. Venticinque, in *Big Data Processing for Pervasive Environment in Cloud Computing* (Institute of Electrical and Electronics Engineers Inc., 2014)
22. Z. Chen et al., A novel framework of data sharing and fusion in smart city-SCLDF. Jisuanji Yanjiu yu Fazhan. Comput. Res. Dev. **51**(2), 290–301 (2014)
23. M. Fazio, A. Puliafito, M. Villari, IoT4S: a new architecture to exploit sensing capabilities in smart cities. Int. J. Web Grid Serv. **10**(2-3), 114–138 (2014)
24. J.A. Galache et al., in *ClouT: Leveraging Cloud Computing Techniques for Improving Management of Massive IoT Data* (Institute of Electrical and Electronics Engineers Inc., 2014)

25. A.J. Jara, Y. Bocchi, D. Genoud, in *Social Internet of Things: The Potential of the Internet of Things for Defining Human Behaviours* (Institute of Electrical and Electronics Engineers Inc., 2014).
26. A.J. Jara, D. Genoud, Y. Bocchi, in *Short Paper: Sensors Data Fusion for Smart Cities with KNIME: A Real Experience in the SmartSantander Testbed* (IEEE Computer Society, 2014)
27. A.J. Jara, D. Genoud, Y. Bocchi, in *Big Data in Smart Cities: From Poisson to Human Dynamics* (IEEE Computer Society, 2014)
28. M.N. Kamel Boulos, N.M. Al-Shorbaji, On the Internet of Things, smart cities and the WHO Healthy Cities. Int. J. Health Geogr. **13**, 10 (2014)
29. D. Li, Y. Yao, Z. Shao, Big data in smart city. Wuhan Daxue Xuebao (Xinxi Kexue Ban)/Geomatics and Information Science of Wuhan University **39**(6), 631–640 (2014)
30. T.H. Nguyen, V. Nunavath, A. Prinz, Big data metadata management in smart grids, in *Studies in Computational Intelligence* (Springer, Berlin, 2014), pp. 189–214
31. K.E. Skouby, P. Lynggaard, in *Smart Home and Smart City Solutions Enabled by 5G, IoT, AAI and CoT Services* (Institute of Electrical and Electronics Engineers Inc., 2014)
32. S.K. Sowe et al., in *Managing Heterogeneous Sensor Data on a Big Data Platform: IoT Services for Data-Intensive Science* (Institute of Electrical and Electronics Engineers Inc., 2014)
33. A. Vakali, L. Anthopoulos, S. Krco, in *Smart Cities Data Streams Integration: Experimenting with Internet of Things and Social Data Flows* (Association for Computing Machinery, 2014)
34. J. Van De Pas, G.J. Van Bussel, in *'Privacy Lost - and Found?' Some Aspects of Regaining Citizens' Privacy by Means of PET in the Age of 'Data'* (Academic Conferences and Publishing International Limited, 2014)
35. E. van der Zee, H. Scholten, Spatial dimensions of big data: application of geographical concepts and spatial technology to the internet of things, in *Studies in Computational Intelligence* (Springer, Berlin, 2014), pp. 137–168
36. S. Andreev et al., in *15th International Conference on Next Generation Networks and Systems, NEW2AN 2015 and 8th Conference on Internet of Things, Smart Spaces, ruSMART 2015*, ed. by S. Andreev et al. (Springer, 2015), pp. 1–800
37. F. Archetti, I. Giordani, A. Candelieri, Data science and environmental management in smart cities. Environ. Eng. Manag. J. **14**(9), 2095–2102 (2015)
38. J. Balicki et al., in *Collective Citizens' Behavior Modelling with Support of the Internet of Things and Big Data* (Institute of Electrical and Electronics Engineers Inc., 2015)
39. L. Belli et al., in *An Open-Source Cloud Architecture for Big Stream IoT Applications*, ed. by M. Serrano, I.P. Zarko, K. Pripuzic (Springer, 2015), pp. 73–88
40. D. Bonino et al., in *ALMANAC: Internet of Things for Smart Cities* (Institute of Electrical and Electronics Engineers Inc., 2015)
41. P. Brous, M. Janssen, in *A Systematic Review of Impediments Blocking Internet of Things Adoption by Governments*, ed. by W. Lamersdorf et al. (Springer, 2015), pp. 81–94
42. B. Cheng et al., in *Building a Big Data Platform for Smart Cities: Experience and Lessons from Santander* (Institute of Electrical and Electronics Engineers Inc., 2015)
43. N. Cherifi et al., in *Energy Consumption of Networked Embedded Systems* (Institute of Electrical and Electronics Engineers Inc., 2015)
44. J.P. Exner, How smart should planning and cities be in the future? gis.Science - Die Zeitschrift fur Geoinformatik **1**, 1–10 (2015)
45. C. Formisano et al., in *The Advantages of IoT and Cloud Applied to Smart Cities: ClouT User Scenarios and Reference Architecture* (Institute of Electrical and Electronics Engineers Inc., 2015)
46. J. Gao, L. Lei, S. Yu, in *Big Data Sensing and Service: A Tutorial* (Institute of Electrical and Electronics Engineers Inc., 2015)
47. D. Hill, The street as platform: how digital dynamics shape the physical city. Archit. Des. **85**(4), 62–67 (2015)

48. R. Jalali, K. El-Khatib, C. McGregor, in *Smart City Architecture for Community Level Services Through the Internet of Things* (Institute of Electrical and Electronics Engineers Inc., 2015)
49. A.J. Jara, D. Genoud, Y. Bocchi, Big data for smart cities with KNIME a real experience in the SmartSantander testbed. Softw.: Pract. Exp. **45**(8), 1145–1160 (2015)
50. C. Kan et al., in *Mobile Sensing and Network Analytics for Realizing Smart Automated Systems Towards Health Internet of Things* (IEEE Computer Society, 2015)
51. A. Krylovskiy, M. Jahn, E. Patti, in *Designing a Smart City Internet of Things Platform with Microservice Architecture* (Institute of Electrical and Electronics Engineers Inc., 2015)
52. B.N. Lee et al., in *Korean Experiences and Lessons Learned from Standardization Framework Activities* (Portland State University, 2015)
53. D. Li, in *Big Data in Smart City*, ed. by Y.J. Zhang (Springer, 2015)
54. D.R. Li, J.J. Cao, Y. Yao, Big data in smart cities. SCIENCE CHINA Inf. Sci. **58**(10), 1–12 (2015)
55. W. Li, P. Wang, K. Yang, in *Visualizing City Events on Search Engine: Tword the Search Infrustration for Smart City* (Institute of Electrical and Electronics Engineers Inc., 2015)
56. P. Lynggaard, K.E. Skouby, Deploying 5G-technologies in smart city and smart home wireless sensor networks with interferences. Wirel. Pers. Commun. **81**(4), 1399–1413 (2015)
57. A. Medvedev et al., in *Reporting Road Problems in Smart Cities Using OpenIoT Framework*, ed. by M. Serrano, I. P. Zarko, K. Pripuzic (Springer, Cham, 2015), pp. 169–182
58. M.V. Moreno, A.F. Skarmeta, A.J. Jara, in *How to Intelligently Make Sense of Real Data of Smart Cities* (Institute of Electrical and Electronics Engineers Inc., 2015)
59. N. Narendra et al., in *Goal-Driven Context-Aware Data Filtering in IoT-Based Systems* (Institute of Electrical and Electronics Engineers Inc., 2015)
60. M.M. Rathore, A. Ahmad, A. Paul, in *Big Data and Internet of Things: An Asset for Urban Planning* (Association for Computing Machinery, 2015)
61. F. Shahzad, T.R. Sheltami, in *An Efficient MAC Scheme in Wireless Sensor Network with Energy Harvesting (EHWSN) for Cloud Based Applications* (IEEE Computer Society, 2015)
62. M. Strohbach et al., Towards a big data analytics framework for IoT and smart city applications, in *Modeling and optimization in science and technologies* (Springer, Cham, 2015), pp. 257–282
63. B. Tang et al., in *A Hierarchical Distributed Fog Computing Architecture for Big Data Analysis in Smart Cities* (Association for Computing Machinery, 2015)
64. H. Wang, in *Sensing Information Modelling for Smart City* (Institute of Electrical and Electronics Engineers Inc., 2015)
65. Y. Zhu, J. Zuo, Research on security construction of smart city. Int. J. Smart Home **9**(8), 197–204 (2015)
66. A. Ahmad et al., in *Defining Human Behaviors Using Big Data Analytics in Social Internet of Things* (Institute of Electrical and Electronics Engineers Inc., 2016)
67. F. Alam et al., in *Analysis of Eight Data Mining Algorithms for Smarter Internet of Things (IoT)* (Elsevier B.V., 2016)
68. A. Alkhamisi, M.S.H. Nazmudeen, S.M. Buhari, in *A Cross-Layer Framework for Sensor Data Aggregation for IoT Applications in Smart Cities* (Institute of Electrical and Electronics Engineers Inc., 2016)
69. A. Alshami, W. Guo, G. Pogrebna, in *Fuzzy Partition Technique for Clustering Big Urban Dataset* (Institute of Electrical and Electronics Engineers Inc., 2016)
70. A. Amato et al., in *A Cyber Physical System of Smart Micro-Grids* (Institute of Electrical and Electronics Engineers Inc., 2016)
71. M. Bawa et al., in *Importance of Internet of Things and Big Data in Building Smart City and What Would Be Its Challenges*, ed. by V. Krutilova et al. (Springer, Cham, 2016), pp. 605–616
72. D.C. Bogatinoska et al., in *Advanced Sensing and Internet of Things in Smart Cities* (Institute of Electrical and Electronics Engineers Inc., 2016)

73. G. Cerruela García, I. Luque Ruiz, M. Gómez-Nieto, State of the art, trends and future of bluetooth low energy, near field communication and visible light communication in the development of smart cities. Sensors (Basel) **16**(11), E1968 (2016)
74. C.I. Chang, C.C. Lo, Planning and implementing a smart city in Taiwan. IT Prof. **18**(4), 42–49 (2016)
75. Y.Y. Chang, C.Y. Yu, L.T. Lee, in *Research of Vehicle-Mounted Detection and Collision Avoidance System Based on Dedicated Short Range Communications* (Institute of Electrical and Electronics Engineers Inc., 2016)
76. N. Chen et al., in *Dynamic Urban Surveillance Video Stream Processing Using Fog Computing* (Institute of Electrical and Electronics Engineers Inc., 2016)
77. B.C. Chifor, I. Bica, V.V. Patriciu, Sensing service architecture for smart cities using social network platforms. Soft. Comput. **21**, 4513–4522 (2016)
78. C. Costa, M.Y. Santos, in *BASIS: A Big Data Architecture for Smart Cities* (Institute of Electrical and Electronics Engineers Inc., 2016)
79. S.K. Datta et al., in *OneM2M Architecture Based IoT Framework for Mobile Crowd Sensing in Smart Cities* (Institute of Electrical and Electronics Engineers Inc., 2016)
80. M. Díaz, C. Martín, B. Rubio, State-of-the-art, challenges, and open issues in the integration of Internet of things and cloud computing. J. Netw. Comput. Appl. **67**, 99–117 (2016)
81. S. Dupont et al., in *Bringing Dynamics to IoT Services with Cloud and Semantic Technologies: An Innovative Approach for Enhancing IoT Based Services* (SciTePress, 2016)
82. T. Fendley, What next for legible cities: an assessment of the potential future for urban wayfinding. Inf. Des. J. **22**(2), 165–171 (2016)
83. A. Giordano, G. Spezzano, A. Vinci, in *Smart Agents and Fog Computing for Smart City Applications*, ed. by E. Alba, F. Chicano, G. Luque (Springer, Cham, 2016), pp. 137–146
84. V.H. González-Jaramillo, in *Tutorial: Internet of Things and the Upcoming Wireless Sensor Networks Related with the Use of Big Data in Mapping Services; Issues of Smart Cities* (Institute of Electrical and Electronics Engineers Inc., 2016)
85. A. Gupta et al., Comparitive study on big data architectures proposed for smart cities context. Int. J. Pharm. Technol. **8**(4), 25440–25451 (2016)
86. V. Hans, P.S. Sethi, J. Kinra, in *An Approach to IoT Based Car Parking and Reservation System on Cloud* (Institute of Electrical and Electronics Engineers Inc., 2016)
87. J. Huang, M.D. Seck, A. Gheorghe, in *Towards Trustworthy Smart Cyber-Physical-Social Systems in the Era of Internet of Things* (Institute of Electrical and Electronics Engineers Inc., 2016)
88. F.D. Hudson, E.W. Nichols, The internet of things and cognitive computing, in *Handbook of Statistics*, ed. by V. N. Gudivada et al. (Elsevier B.V., Amsterdam, 2016), pp. 341–373
89. B. Javed et al., in *A Provenance Framework for Policy Analytics in Smart Cities* (SciTe Press, 2016)
90. E. Kühn, Reusable coordination components: reliable development of cooperative information systems. Int. J. Coop. Inf. Syst. **25**(4), 1740001 (2016)
91. V. Kupriyanovsky et al., in *On the Digital Economy* (CEUR-WS, 2016)
92. J. Lanza et al., Managing large amounts of data generated by a Smart City internet of things deployment. Int. J. Semant. Web Inf. Syst. **12**(4), 22–42 (2016)
93. B.P.L. Lau et al., in *Spatial and Temporal Analysis of Urban Space Utilization with Renewable Wireless Sensor Network* (Association for Computing Machinery, Inc., 2016)
94. V.J. Lawson, L. Ramaswamy, in *TAU-FIVE: A Multi-Tiered Architecture for Data Quality and Energy-Sustainability in Sensor Networks* (Institute of Electrical and Electronics Engineers Inc., 2016)
95. L. Lenz et al., in *How Will the Internet of Things and Big Data Analytics Impact the Education of Learning-Disabled Students? A Concept Paper* (Institute of Electrical and Electronics Engineers Inc., 2016)
96. D. Li, Towards geo-spatial information science in big data era. Acta Geodaet. Cartographica Sin. **45**(4), 379–384 (2016)

97. J. Lloret et al., An integrated IoT architecture for smart metering. IEEE Commun. Mag. **54**(12), 50–57 (2016)
98. Z. Lv et al., Virtual reality smart city based on WebVRGIS. IEEE Internet Things J. **3**(6), 1015–1024 (2016)
99. M. May, in *Towards Industrial Machine Intelligence*, ed. by B. Bringmann et al. (Springer, 2016), p. XVII
100. M. Mazhar Rathore, A. Ahmad, A. Paul, in *IoT-Based Smart City Development Using Big Data Analytical Approach* (Institute of Electrical and Electronics Engineers Inc., 2016)
101. A. Medvedev et al., in *Storing and Indexing IoT Context for Smart City Applications*, ed. by O. Galinina, Y. Koucheryavy, S. Balandin (Springer, Cham, 2016), pp. 115–128
102. G. Merlino et al., in *Software Defined Cities: A Novel Paradigm for Smart Cities Through IoT Clouds* (Institute of Electrical and Electronics Engineers Inc., 2016)
103. M.V. Moreno et al., in *Applicability of Big Data Techniques to Smart Cities Deployments* (IEEE Transactions on Industrial Informatics, 2016), pp. 800–809
104. V. Moreno-Cano, F. Terroso-Saenz, A.F. Skarmeta-Gomez, in *Big Data for IoT Services in Smart Cities* (Institute of Electrical and Electronics Engineers Inc., 2016)
105. D. Namiot, M. Sneps-Sneppe, in *On Internet of Things Programming Models*, ed. by K. E. Samouylov et al. (Springer, Cham, 2016), pp. 13–24
106. E. Patti, A. Acquaviva, in *IoT Platform for Smart Cities: Requirements and Implementation Case Studies* (Institute of Electrical and Electronics Engineers Inc., 2016)
107. A. Paul et al., Smartbuddy: defining human behaviors using big data analytics in social internet of things. IEEE Wirel. Commun. **23**(5), 68–74 (2016)
108. S. Pirttikangas et al., in *Experiences with Smart City Traffic Pilot* (Institute of Electrical and Electronics Engineers Inc., 2016)
109. E. Psomakelis et al., in *Big IoT and Social Networking Data for Smart Cities: Algorithmic Improvements on Big Data Analysis in the Context of RADICAL City Applications* (SciTe Press, 2016)
110. M.M. Rathore et al., Urban planning and building smart cities based on the internet of things using big data analytics. Comput. Netw. **101**, 63–80 (2016)
111. S. Ristov et al., in *Is the Computer Science Curriculum Ready to Teach Students Towards Hardwarizing?* (IEEE Computer Society, 2016)
112. C.D.G. Romero, J.K.D. Barriga, J.I.R. Molano, *Big Data Meaning in the Architecture of IoT for Smart Cities*, Lecture Notes in Computer Science (including subseries Lecture Notes in Artificial Intelligence and Lecture Notes in Bioinformatics) (Springer, Cham, 2016), pp. 457–465
113. R. Sakthivel, V. Parthipan, D. Dhanasekaran, Big data analytics on smart and connected communities using Internet of Things. Int. J. Pharm. Technol. **8**(4), 19590–19601 (2016)
114. J.M. Schleicher et al., Application architecture for the internet of cities: blueprints for future smart city applications. IEEE Internet Comput. **20**(6), 68–75 (2016)
115. J.M. Schleicher et al., Enabling a smart city application ecosystem: requirements and architectural aspects. IEEE Internet Comput. **20**(2), 58–65 (2016)
116. T. Shaikh, S. Ismail, J.D. Stevens, in *Aura Minora: A User Centric IOT Architecture for Smart City* (Association for Computing Machinery, 2016)
117. L. Shang et al., in *Evaluation of High Density GPUs as Sustainable Smart City Infrastructure* (Institute of Electrical and Electronics Engineers Inc., 2016)
118. Y. Sun et al., Internet of things and big data analytics for smart and connected communities. IEEE Access **4**, 766–773 (2016)
119. T.H. Szymanski, Securing the industrial-tactile internet of things with deterministic silicon photonics switches. IEEE Access **4**, 8236–8249 (2016)
120. R. Wen et al., in *A Smart Cable Management System in Support of the Smart City* (Institution of Engineering and Technology, 2016)
121. Y. Wu et al., Smart city with Chinese characteristics against the background of big data: idea, action and risk. J. Clean. Prod. **173**, 60 (2018)

122. L. Yu et al., in *Intelligent Farm Relaxation for Smart City Based on Internet of Things: Management System and Service Model* (SciTe Press, 2016)
123. Q. Zhang et al., in *Firework: Big Data Sharing and Processing in Collaborative Edge Environment* (Institute of Electrical and Electronics Engineers Inc., 2016)
124. G.R.C. Andrés, in *CleanWiFi: The Wireless Network for Air Quality Monitoring, Community Internet Access and Environmental Education in Smart Cities* (Institute of Electrical and Electronics Engineers Inc., 2017)
125. G. Andrienko et al., in *2016 IEEE Conference on Visual Analytics Science and Technology, VAST 2016 – Proceedings* (Institute of Electrical and Electronics Engineers Inc., 2017)
126. M.R. Bashir, A.Q. Gill, in *Towards an IoT Big Data Analytics Framework: Smart Buildings Systems* (Institute of Electrical and Electronics Engineers Inc., 2017)
127. D.M. Batista et al., in *InterSCity: Addressing Future Internet research challenges for Smart Cities* (Institute of Electrical and Electronics Engineers Inc., 2017)
128. K. Ben Ahmed et al., in *Visual Sentiment Prediction with Transfer Learning and Big Data Analytics for Smart Cities* (Institute of Electrical and Electronics Engineers Inc., 2017)
129. M. Benjillali et al., in *International Symposium on Ubiquitous Networking, UNet 2016*, ed. by M. Benjillali et al. (Springer, 2017), pp. 1–712
130. CIoT 2016, in *2016 Cloudification of the Internet of Things, CIoT 2016* (Institute of Electrical and Electronics Engineers Inc., 2017)
131. A. Crooks et al., Creating smart buildings and cities. IEEE Pervasive Comput. **16**(2), 23–25 (2017)
132. T.M. John et al., in *Towards Building Smart Energy Systems in Sub-Saharan Africa: A Conceptual Analytics of Electric Power Consumption* (Institute of Electrical and Electronics Engineers Inc., 2017)
133. N. Modi, P. Verma, B. Trivedi, in *International Conference on Communication on Networks, COMNET 2016*, ed. by N. Modi, P. Verma, B. Trivedi (Springer, 2017), pp. 1–753
134. M. Mylrea, Smart energy-internet-of-things opportunities require smart treatment of legal, privacy and cybersecurity challenges. J. World Energy Law Bus. **10**(2), 147–158 (2017)
135. M.A.L. Pena, C.A. Rua, S.S. Lozoya, in *A "fast Data" Architecture: Dashboard for Anomalous Traffic Analysis in Data Networks* (Institute of Electrical and Electronics Engineers Inc., 2017)
136. H. Peng et al., Forecasting potential sensor applications of triboelectric nanogenerators through tech mining. Nano Energy **35**, 358–369 (2017)
137. R.G. Qiu, L. Qiu, Y. Badr, in *Integrating Physical and Social Sensing to Enable Smart City Mobility Services* (Institute of Electrical and Electronics Engineers Inc., 2017)
138. T. Qiu et al., Heterogeneous ad hoc networks: architectures, advances and challenges. Ad Hoc Netw. **55**, 143–152 (2017)
139. M.M. Rathore et al., IoT-based big data: from smart city towards next generation super city planning. Int. J. Semant. Web Inf. Syst. **13**(1), 28–47 (2017)
140. A.R. Reid, C.R.C. Pérez, D.M. Rodríguez, Inference of vehicular traffic in smart cities using machine learning with the internet of things. Int. J. Interact. Des. Manuf., 1–14 (2017)
141. P. Rizwan, K. Suresh, M. Rajasekhara Babu, in *Real-Time Smart Traffic Management System for Smart Cities by Using Internet of Things and Big Data* (Institute of Electrical and Electronics Engineers Inc., 2017)
142. R. Samano-Robles et al., in *The DEWI High-Level Architecture: Wireless Sensor Networks in Industrial Applications* (Institute of Electrical and Electronics Engineers Inc., 2017)
143. K. Schechtner, Bridging the adoption gap for smart city technologies: an interview with Rob Kitchin. IEEE Pervasive Comput. **16**(2), 72–75 (2017)
144. D. Serrano, T. Baldassarre, E. Stroulia, in *Real-Time Traffic-Based Routing, Based on Open Data and Open-Source Software* (Institute of Electrical and Electronics Engineers Inc., 2017)
145. A. Shukla, Y. Simmhan, in *Benchmarking distributed stream processing platforms for IoT applications*, ed. by R. Nambiar, M. Poess, (Springer, Cham, 2017), pp. 90–106
146. S. Shukla, K. Balachandran, V.S. Sumitha, in *A Framework for Smart Transportation Using Big Data* (Institute of Electrical and Electronics Engineers Inc., 2017)

147. F. Silva et al., Intelligent data metrics for urban driving with data fusion and distributed machine learning, in *Studies in Computational Intelligence*, (Springer, Berlin, 2017), pp. 227–236
148. N. Streitz, in *Reconciling Humans and Technology: The Role of Ambient Intelligence*, ed. by R. Wichert, A. Braun, A. Mana, (Springer, Cham, 2017), pp. 1–16
149. S. Trilles et al., A domain-independent methodology to analyze IoT data streams in real-time. A proof of concept implementation for anomaly detection from environmental data. Int. J. Digital Earth **10**(1), 103–120 (2017)
150. M. Vögler et al., Ahab: a cloud-based distributed big data analytics framework for the Internet of Things. Softw.: Pract. Exp. **47**(3), 443–454 (2017)
151. C. Wu et al., A NoSQL-SQL hybrid organization and management approach for real-time geospatial data: a case study of public security video surveillance. ISPRS Int. J. Geo-Inf. **6**(1), 21 (2017)
152. J. Xing et al., in *Wide Area Beidou Foundation Enhanced Grid Space-Time Interlinked Science and Education Collaborative Experimental Service: A Science and Education Cooperative Experiment Demonstration Platform for the Wide-Area and Grid Enhanced China Beidou Satellite System*, ed. by J. Sun et al. (Springer, Singapore, 2017), pp. 255–265
153. A. Yavari, P.P. Jayaraman, D. Georgakopoulos, in *Contextualised Service Delivery in the Internet of Things: Parking Recommender for Smart Cities* (Institute of Electrical and Electronics Engineers Inc., 2017)
154. V. Zdraveski et al., ISO-standardized smart city platform architecture and dashboard. IEEE Pervasive Comput. **16**(2), 35–43 (2017)

B2C Relationship Quality in the Sharing Economy in the Chinese Context

Xini Hu

1 Introduction

1.1 Research Background

The sharing economy is a means of sharing goods, services, ideas, information, and skills through a network of individuals, facilitated through social networks via computers and mobile apps. This has been described by Botsman and Rogers [1], and a socioeconomic model is emerging, based on sharing, renting, gifting, bartering, swapping, lending, and borrowing [1, 2].

The sharing economy is booming and, with the consumer's value changed, access has become more popular than ownership recently, so the sharing economy still has huge potential for penetration. There are many data showing faster development in some industries. For instance, in bicycle sharing, there were 400,000 public city bikes available in the world in 2012. The car sharing market was worth USD 100 billion in 2010, and it was estimated that it would be worth USD 3.3 billion by 2016 in North America alone [3]. The sharing economy was introduced in China in 2003. By 2008 it was just starting to grow and after 2013 it started to rocket. In 2016 the turnover in the sharing economy reached RMB 3452 billion in China—an increase of 103% in comparison with 2015.

X. Hu (✉)
Christ Church Business School, Canterbury Christ Church University, Kent, UK
e-mail: x.hu13@canterbury.ac.uk

A. Reyes-Munoz et al. (eds.), *EAI International Conference on Technology, Innovation, Entrepreneurship and Education*, Lecture Notes in Electrical Engineering 532, https://doi.org/10.1007/978-3-030-02242-6_21

The Sharing Economy in China

The participants in the sharing economy numbered over 600 million in 2016, an increase of 10 million in comparison with 2015. Among these participants, service providers accounted for 60,000,000 and employees in the sharing economy platform accounted for 5,850,000 [4]. The *China Internet "Sharing Economy" Research Report* for 2016 [5] predicted that there would be over 10 million service providers in the sharing economy in China, with around 20 million full-time employees.

There was some research on Chinese participation in the sharing economy in 2016, which showed that 52.5% of Chinese people online thought that through their participation, they could access underused resources at lower prices; 7.0% of them thought it was difficult to join the sharing economy [6]; 39.4% of them were strongly willing to participate in the sharing economy, 33.9% were willing to participate, and 6.4% were not willing to participate; and 39.1% of them were strongly willing to share their underused resources. In particular, they were willing to share their interests, working techniques, and knowledge. Demands for transportation, house rental, and interests were the most popular topics [6]. Of the Chinese people online, 39.9% were familiar with DiDi, which is the world's leading one-stop mobile transportation platform. It provides different services to customers—the main ones being DiDi taxi, DiDi express, DiDi premier, DiDi hitch, DiDi chauffeur, and DiDi car rental—with the aim being to help cities resolve transportation, employment, and environmental challenges (DiDi Chuxing, 2012–2017). Of the respondents, 35.5% were familiar with Uber and 25.9% knew of Airbnb [6].

Developmental Drivers

Traffic congestion: As in many other countries, the drivers for development of the sharing economy come from ideas to resolve traffic problems. In many big Chinese cities (such as Beijing, Shanghai, and Guangzhou) there are many transportation problems, particularly traffic congestion, parking problems, limitations on car purchasing, and license plate lotteries. These factors encourage people who own a car or who own a license plate without a car to join the sharing economy [7].

Culture: In China, there is a tradition of saving, which is one of the most important qualities in China, and the sharing economy benefits money saving and effective use of products. Meanwhile, the value of green consumption is accepted by the well-educated generation and young professionals in China, who have the willingness to share private goods [8].

Employment: In China, graduates are faced with fierce completion for employment and, with a change of mind toward following their hearts and interests, most graduates would like to be freelancers or work freely and flexibly. The sharing economy can provide such opportunities for them [6].

Web 2.0 development: Mobile services and electronic services are available for accessing the service via different apps in any place and at any time, so it is convenient for consumers to share their goods via social networks and community platforms [9, 10]. These developments in technology contribute to the organization of access-based systems that are viable on a daily basis [11].

Business in the Sharing Economy

Many of these businesses work directly by using the P2P model, which refers to person-to-person (or customer-to-customer (C2C)), especially in the finance industry. Many of them have been developed on a business-to-customer (B2C) basis because of legislative limitations in the transportation and accommodation industries. Many businesses would like to embrace the sharing economy, so they choose to cooperate with another company, which is responsible for providing a resource such as a car or a house. With different demands in the house rental industry, these firms (such as Tujia) need to be responsible not only for the maintenance of the platform but also for house cleaning and for building good relationships with customers, as intermediaries [8], because house owner's demand is focused not just on the rental for profit but also on the maintenance of their home. In house rentals, most of the business is based not only on B2C but also on C2C; the difference is that house owners in America and European countries focus mainly on sharing the house with their customers and living with them. However, in China, it is difficult to live with strangers, although there is greater trust than before, but the house owner generally rents the house to the customers for a few days [12].

Problems and Challenges

Firstly, the customer may not be satisfied with the service quality. Because some of the private car owners are not that professional in this service industry, there will be arguments between them. Also, the matter of safety is a concern for both the car owner and the car renter. Some customers will not feel totally safe, and some renters are not careful with the car and cause some damage. Furthermore, if there is a traffic accident, who should be held responsible [12]?

Secondly, most parts of the sharing economy involve connecting with strangers. However, although there is a credit system to improve trust among the people, it is still hard to transplant a successful business model from western countries. There are a lot of different regulations in some traditional industries—for example, the entrance standard, the licensing, the operation limits, and so on [13].

Thirdly, there are huge numbers of businesses established on the basis of the sharing economy, and every business says that they are the largest one in different aspects. Most of them rely on raising funding, and how to be profitable is a challenge for them, so it is hard to survive [12].

Lastly, it is also very difficult to maintain customer relationships. There are many different and creative platforms, and customers are flexible participants, so it is not easy to keep them loyal, as they are curious to try different options. This causes some problems for businesses [12].

2 Research Questions and Contributions

In China, Uber has moved into the luxury chauffeur niche because the regulations prohibit private drivers. However, rather than Airbnb, Venmo, and other American companies, let us take Tujia as an example. It is more deeply involved in property management. While the American companies mostly provide a C2C platform, Chinese users need more, and the sharing companies are stepping in as full-service intermediaries [8]. Faced with low loyalty, unsatisfactory service quality, trust problems, safety risks, and low credibility, relationship building between businesses and customers is necessary, especially in China's different business models. Sharing economy businesses would like to build and maintain the relationship between suppliers, customers, and participants [14].

2.1 Research Questions

There have been some different research studies on the B2C relationship. Guenzi and Pelloni [15] found that the relationship between the customer and service personnel had a positive impact on customer-to-firm relationships: the higher the overall customer satisfaction, the higher the behavioral loyalty toward the firm [15]. Here are some questions to be raised:

What are the independent variables (loyalty, commitment, trust, satisfaction, and so on) that affect B2C relationship quality?

What is the relationship between customer–company identification and B2C relationship quality?

What is the relationship between the independent variables and B2C relationship quality?

Does B2C relationship quality directly affect the sharing service provider's and customer's personal relationship or not?

2.2 Research Contributions

The aim of this research is to help fulfill and build the B2C relationship framework in the sharing economy context. From a theoretical perspective, this research contributes a framework on B2C relationship quality, and within this framework, the factors affecting B2C relationship quality are outlined and developed. At the same time, it is hoped that this research will help to outline the B2C relationship between sharing firms and customers, and also B2C relationship quality, as a mediator affecting the relationship quality between service providers and customers.

From a practical perspective, this research aims to provide practical suggestions for businesses involved in the sharing economy on how to build and maintain relationships between services providers, firms (sharing platforms), and customers. It also aims to help identify some factors that dominantly affect these relationships to let businesses knows where to start in order to improve relationship quality, increase customer satisfaction with these services, build more trust among the parties participating in the sharing economy, and decrease damage to the sharing products for both parties, to provide "win–win" outcomes for both owners and renters.

3 Literature Review

3.1 The Sharing Economy

The idea of this emerging economy originally comes from Belk [16], who used the term "sharing," which is a form of joint possession for all to use, without debt [16, 17]. On the basis of these different related concepts, Botsman and Rogers [1] coined the term "collaborative consumption" and Bardhi and Eckhardt [18] discussed a form of collaborative consumption: "access-based consumption" together with "redistributed ownership." In 2011, Rachel Botsman used the term "collaborative consumption," which gave way to the term "sharing economy" [19]. The various different terms that have been proposed include "sharing" or the "sharing economy" [16, 17], "collaborative consumption" [1], "access" or "access-based consumption" [18], "commercial sharing systems" [3], "the mesh", and "product–service systems".

The definition of the sharing economy was devised and developed by different researchers. It is a means of sharing goods, services, ideas, information, and skills through a network of individuals, facilitated through social networks via computers and mobile apps (as mentioned by Botsman and Rogers [1]), and it is an emerging socioeconomic model based on sharing, renting, gifting, bartering, swapping, lending, and borrowing [2]. An entire economy is emerging around the exchange of goods and services between individuals instead of from business to consumer (B2C) [9]. It is also defined as a reciprocal exchange process, where individuals can share personal goods with each other via a digital platform [20]. The sharing economy has certain characteristics—sustainability, openness, and solidarity—and it is conducted by sharing underused assets and resources, enabling people to connect with each other through sharing of homes, skills, and cars [19]. With the increasing development of the sharing economy, many different similar and related terms have been used, e.g., the "collaborative economy" and "peer to peer" (i.e., P2P) [21, 22]. There are seven distinct dimensions of sharing business models—(1) platforms for collaboration; (2) underutilized resources; (3) P2P interactions; (4) collaborative governance; (5) mission driven; (6) alternative funding; and (7) technology reliance [23]—which also can be regarded as the essentials of the sharing economy.

The drivers contributing to the development of the sharing economy definitely should be researched and have been studied. Generally speaking, one of the most significant ones is the development of the internet, particularly the launch of Web 2.0 [9, 10, 17], which enables development of online-based communities and networks with only small transaction costs. Secondly, with changing consumer behavior, temporary use of products is more attractive than ownership [10, 24]. Lastly, with growing environmental consciousness, cities are struggling with population growth and density, and there is also an urgent need to resolve serious pollution problems, thereby driving the development of sharing of cars and bikes in cities [25]. Consumers are making conscientious choices in favor of goods that have fewer environmental effects [21].

Sharing economy businesses vary from accommodation to transportation, finance, online marketplaces, and online labor markets [25, 26]. Botsman and Rogers [1] consider that collaborative consumption involves three systems: product–service systems, redistribution markets, and collaborative lifestyles. The sharing platforms can be classified into for-profit and not-for-profit activities [17], and another dimension classifies B2C versus P2P activities. Many P2P platforms are owned and operated by formal businesses [27]. Stoica et al. [28] also talk about the following sharing economy models: redistribution, mutualization models, shared mobility models, and P2P systems [28]. As Roh [29] mention, the first model is product–service systems, which allow members to share multiple products that are owned by companies or by private persons. Second, in redistribution markets, P2P matching or social networks allow reownership of a product. Access can be derived through collaborative lifestyles in which people share similar interests and help each other with less tangible assets such as money, space, or time [29].

Many different reasons, motivations, and determinants are involved in the sharing economy. Mohlmann [30] conducted research on the determinants of satisfaction and the likelihood of using a sharing economy option again. The results showed that trust was the strongest determinant of satisfaction with a sharing option. Meanwhile, cost savings, familiarity, service quality, trust, and utility were found to have positive effects on satisfaction with a sharing option [30]. People participate in the sharing economy not only in pursuit of economic benefits but also for the feeling of doing something good [31]. Also, sustainability and enjoyment of the activity, as well as economic gains, motivate participation in the sharing economy. However, sustainability is not directly associated with participation unless it is, at the same time, also associated with positive attitudes toward the sharing economy [22]. Some research has focused specifically on transportation. In terms of values, car sharing is largely associated with "protecting nature," "autonomy of action," and "societal concern." Car sharing may either be viewed as an innovative proposition or be received with a degree of uncertainty. It may prove uncomfortable if the available space is limited by luggage and other passengers on board. On the other side, with regard to buying/owning a car, in terms of values, owning a car is largely associated with "personal security," "material resources," "autonomy of action," and "face" [2]. Talking about demographic research on the sharing economy, the 25- to 29-year age group is the one most likely to have heard of the sharing economy, and people with a higher level of education are more likely to engage in the sharing economy, either as providers or as consumers; higher income (or higher employment status) also correlates with participation in the sharing economy [20]. In research on how the sharing economy affects consumers' values, attitudes, and norms, it was found that 138 participants (82%) reported in wave one that they had acquired something through shared consumption at least once in the previous 4 weeks, while 132 (79%) did so in wave two [32].

3.2 The Business-to-Customer Relationship (in the Sharing Economy Context)

Most of the literature reveals that development of a strong customer relationship can improve customer loyalty, which in turn leads to increased profit for the company. A strong relationship is an intangible asset and a competitive advantage. However, not all customers prefer to engage in a close relationship with service providers, and long-term relationships are often not necessary for either customers or firms [33]. Relationship breadth focuses on how relationships evolve over time (e.g., by use of loyalty programs). Relationship depth focuses on the various factors that strengthen or weaken a relationship within a particular stage—for example, trust, satisfaction, brand equity, communication, and conflict [34].

Building long-term B2C service relationships is essential to obtain relational benefits and achieve sustainable competitive advantages for companies [35]. In the relationship-oriented approach, the interaction between customers and firms is the core of building B2C relationships [35]. According to Law et al. [36], the customer relationship management (CRM) focus is on the customers, so the customer direction of B2C and B2C should be C2B or C2C [36]. A genuine consumer-to-business relationship is characterized by the consumer's voluntary involvement [37].

There have been some different research studies on the B2C relationship. Guenzi and Pelloni [15] found that the relationship between the customer and service personnel had a positive influence on customer-to-firm relationships [15]. And some conduct B2C relationship on e-business, research has explored the dynamic nature of building and losing trust in B2C e-business [38]. Other research has focused on customer participation in the B2C relationship. The analysis revealed that consumers' desire to participate in a relationship affected their level of motivation and degree of confidence; in turn, these impacted on the consumers' sense of affiliation with the service provider [37]. With regard to relationships in different industries, Rajaobelina and Bergeron [33] conducted research on the buyer's and seller's relationship in the financial industry [33].

As mentioned earlier, the sharing economy in China faces some problems, especially since Uber China and Didi merged, when allowance payments as a way to attract drivers or users decreased, so the loyalty of drivers and users also decreased.

Faced with the allowance reduction, there were decreases in the numbers of drivers and users, because the drivers could not get higher payments and the users could not get lower prices [6]. Different drivers give different standards of service, and losses and damage occur in P2P renting of cars [5].

In the accommodation industry, the standardization of service quality is difficult to achieve, and it is also hard to control and improve service quality. Most of the consumer behavior in this field is not clear, and the users' perception and identity to this field is not high enough, because of issues of credibility and traditional culture. Life service sharing also triggers some problems. The service providers cannot be guaranteed by the platform to provide high-quality service, and there is a lack of training for personal service providers, which causes trust issues, credibility risk, and safety problems [4]. The peer service providers lack professional training and industry specialists are not closely affiliate with the sharing service organizers [39]. Meanwhile, the sharing economy tends to involve bad behavior of a guest, a driver, a host, or a passenger more than bad behavior of a consumer in a market transaction. Furthermore, there are extra needs for trust between the two exchanging partners on sharing platforms [40]. Thus, relationship building between customers and business is urgently needed.

There are different relationships involved in the sharing economy: consumer–product, consumer–consumer, and consumer–business [41]. There is a lack of research investigating what drives and makes customers stay in relationships with peer service providers [39]. It is important to develop quality customer–firm relationship research [42].

On sharing platforms, users begin to build their online credibility and reputation, and the platforms need to verify the users' identity and trust building. The sharing economy depends on the commitment of the individual user to share, and users must be trustworthy. Sharing economy businesses are committed to building the relationship capital between members of their community, including producers, suppliers, customers, and participants [14].

4 Methodology

4.1 Systematic Literature Review

A systematic literature review is performed to identify and narrow down the independent variables that affect B2C relationship quality, especially in the Chinese context of the sharing economy background. It is therefore necessary to conduct a systematic literature review. Furthermore, clear research questions are also essential. B2C relationship quality and customer–company identification are constructed into dependent variables.

What are the independent variables (loyalty, commitment, trust, satisfaction, and so on) affecting B2C relationship quality between customers and firms (sharing businesses)?

What is the relationship between customer–company identification and B2C relationship quality?

What is the relationship between the independent variables and B2C relationship quality?

Does the B2C relationship quality between firms and customers as a mediator affect the sharing service providers' and customers' relationship?

4.2 *Hypothesis*

According to the literature review, which clarifies the independent and dependent variables, the hypothesis is proposed and the framework is outlined and then the data will be collected to test this framework.

4.3 *Questionnaire*

The questionnaire is designed to collect data on customer and worker participation in building the B2C relationship. The valid measurements of customer–company identification and the B2C relationship are constructed by items that make up the questionnaire. The size of the questionnaire is yet be decided, and it will be targeted at sharing economy participants and sharing platforms workers.

References

1. R. Botsman, R. Rogers, *What's Mine is Yours: The Rise of Collaborative Consumption* (HarperBusiness, New York, 2010), pp. 38–43
2. L. Piscicelli, The role of values in collaborative consumption. Proceedings of the CADBE PhD Conference 2014, At Nottingham Trent University, 2014
3. C.P. Lamberton, R.L. Rose, When is ours better than mine? A framework for understanding and altering participation in commercial sharing systems. J. Mark. **76**(4), 109–125 (2012)
4. National Information Center, China internet "sharing economy", research report (2017), Available online: http://www.199it.com/archives/569406.html. Accessed 24 July 2017
5. National Information Center, China internet "sharing economy", research report (2016), Available online: http://www.sic.gov.cn/archiver/SIC/UpFile/Files/Htmleditor/201602/20160229121154612.pdf. Accessed 24 July 2017
6. iiMedia, China internet "sharing economy", research report (2016), Available onine: http://www.iimedia.cn/14780907678270150n.pdf. Accessed 25 July 2017
7. Xinhuanet, The fast growing sharing economy in China (2017), http://news.xinhuanet.com/fortune/2017-03/16/c_1120636819.htm. Accessed 04 Jul 2017

8. C. Marquis, Z. Yang, The sharing economy in China: toward a unique local model. China Policy Rev. **9**, 109–111 (2014)
9. J. Owyang, C. Tran, C. Silva, The collaborative economy: products, services, and market relationships have changed as sharing startups impact business models. To avoid disruption, companies must adopt the collaborative economy value chain, Altimeter Research Theme: Digital Economies, A Market Definition Report (2013), Available online: https://www.slideshare.net/Altimeter/the-collaborative-economy. Accessed 03 Aug 2017
10. T. Puschmann, R. Alt, Sharing economy. Bus Inform Syst Eng **58**, 93–99 (2015)
11. R. Radka, A. Margolis, Changing models of ownership and value exchange, in Presentation at Epic Conference, Boulder, 20 September 2011
12. FTChinese, House sharing in China (2016), http://www.ftchinese.com/story/001067520#adchannelID=2000. Accessed 26 July 2017
13. News ScienceNet, Embracing sharing economy (2016), http://news.sciencenet.cn/htmlnews/2016/5/347478.shtm. Accessed 25 July 2017
14. R. Peters, The sharing economy requires relationship capital (2015), https://www.linkedin.com/pulse/future-sharing-economy-requires-relationship-capital-rob-peters. Accessed 11 Aug 2017
15. P. Guenzi, O. Pelloni, The impact of interpersonal relationships on customer satisfaction and loyalty to the service provider. Int. J. Serv. Ind. Manag. **15**(4), 365–384 (2004)
16. R. Belk, Sharing. J. Consumer Res. **36**, 715–734 (2010)
17. R. Belk, You are what you can access: sharing and collaborative consumption online. J. Bus. Res. **67**(8), 1595–1600 (2014)
18. F. Bardhi, G.M. Eckhardt, Access-based consumption: the case of car sharing. J. Consum. Res. **39**(4), 881–898 (2012)
19. B. Balaram, *Reclaiming power in the sharing economy* (RSA, London, 2016)
20. A. Andreotti, G. Anselmi, T. Eichhorn, C.P. Hoffmann, M. Micheli, Participation, privacy, and power in the sharing economy participation in the sharing economy, Report from the EU H2020 Research Project Ps2Share (2017), Available online: https://www.bi.edu/globalassets/forskning/h2020/working-paper-version-for-web-privacy.pdf. Accessed 04 Aug 2017
21. J. Madden, *Exploring the new sharing economy* (NAIOP Research Foundation, Herndon, 2015)
22. J. Hamari, M. Sjöklint, A. Ukkonen, The sharing economy: why people participate in collaborative consumption. J. Assoc. Inf. Sci. Technol. **67**(9), 2047–2059 (2015)
23. P. Munoz, B. Cohen, Mapping out the sharing economy: a configurational approach to sharing business modeling. Technol. Forecast. Soc. Change **125**, 21–37 (2017)
24. K. Matzler, V. Veider, W. Kathan, Adapting to the sharing economy. MIT Sloan Manag. Rev. **56**(2), 71–77 (2015)
25. B. Cohen, J. Kietzmann, Ride on! Mobility business models for the sharing economy. Org. Environ. **27**(3), 279–296 (2014)
26. V. Demary, Competition in the sharing economy, IW policypaper, no. 19/2015 (2015), Available online: https://www.iwkoeln.de/fileadmin/publikationen/2015/235445/Sharing_Economy_Policy_Paper.pdf. Accessed 24 Aug 2017
27. J. Valant, A European Agenda for the Collaborative Economy (European Parliamentary Research Service, Brussels, 2016), Available online: http://www.europarl.europa.eu/RegData/etudes/BRIE/2016/593510/EPRS_BRI(2016)593510_EN.pdf. Accessed 23 Aug 2017
28. I. Stoica, R. Morris, D. Liben-Nowell, D.R. Karger, M.F. Kaashoek, F. Dabek, et al., Chord: a scalable peer-to-peer lookup protocol for internet applications. IEEE/ACM Trans. Netw. **11**, 17–32 (2003)
29. T.H. Roh, The sharing economy: Business cases of social enterprises using collaborative networks. Procedia Comput. Sci. **91**, 502–511 (2016)
30. M. Mohlmann, Collaborative consumption: determinants of satisfaction and the likelihood of using a sharing economy option again. J. Consum. Behav. **14**(3), 193–207 (2015)

31. L. Scholdan, N. van Straaten, C.N.A. Molenaar, What are the drivers of the sharing economy and how do they changes the customer behavior? E-marketing report, program me: e-marketing: marketing strategies for future success (2015), Available online: https://www.contextuallearning.nl/wp-content/uploads/2016/01/drivers-of-the-sharing-economy-strategie.pdf. Accessed 24 Aug 2017
32. D. Roos, R. Hahn, Does shared consumption affect consumers' values, attitudes, and norms? A panel study. J. Bus. Res. **77**, 113–123 (2017)
33. L. Rajaobelina, J. Bergeron, Antecedents and consequences of buyer-seller relationship quality in the financial services industry. Int. J. Bank Mark. **27**(5), 359–380 (2009)
34. D.V. Parboteeah, J.S. Valacich, J.D. Wells, The influence of website characteristics on a consumer's urge to buy impulsively. Inform. Syst. Res. **20**(1), 60–78 (2009)
35. M.S. Balaji, S. Kumar Roy, K. Kok Wei, Does relationship communication matter in B2C service relationships? J. Serv. Mark. **30**(2), 186–200 (2016)
36. M. Law, T. Lau, Y.H. Wong, From customer relationship management to customer-managed relationship: unraveling the paradox with a co-creative perspective. Mark. Intell. Plan. **21**(1), 51–60 (2003)
37. M.M. Raciti, T. Ward, T.S. Dagger, The effect of relationship desire on consumer-to-business relationships. Eur. J. Mark. **47**(3/4), 615–634 (2013)
38. G. Winch, P. Joyce, Exploring the dynamics of building, and losing, consumer trust in B2C e-business. Int. J. Retail Distrib. Manag. **34**(7), 541–555 (2006)
39. P. Parigi, Trust and relationships in the sharing economy, in *Perspective: King's College London Economics & Finance Society*, (2016)
40. P. Parigi, Trust and relationships in the sharing economy. Respective student-led economics journal (2016), http://kcleconomics.com/trust-and-relationships-in-the-sharing-economy/. Accessed 11 Aug 2017
41. H. Park, Collaborative apparel consumption in the digital sharing economy: an agenda for academic inquiry. J. Consum. Stud. **41**(5), 465–474 (2017)
42. M.T. Adjei, M.N. Clark, Relationship marketing in a B2C context: moderating role of personality traits. J. Retail. Consum. Serv. **17**, 73–79 (2010)

Part III
CSf Holonovel Workshop: Creative Science and Science Fictions

Kill Your Darlings (a Holonovel)

Tiina Kymäläinen

1 Introduction

This *holonovel*—taking the form of a science fiction prototype (SFP) [1, 2]—is about themes relating to transhumanism, posthumanism and digital resurrection in the Mixed Reality (MR) game environment that involves digital human avatars [3–6]. The SFP is a dystopic tale that draws its inspiration from the recent distressing activities in social media, a particularly upsetting one being an online game called "Blue Whale Suicide Game" that is specifically targeted at teenage children. In short, during the gameplay the players are provided with various acts of self-harm to be committed over the course of 50 days, and on the final day they are urged to win the game by committing suicide[1]. The game has been reported to pursue victims mainly in the Russian-speaking part of the world, although it has been shown to have gained footing in various other parts of the world as well.

The *holonovel* illustrates similar precarious game settings in more systematic and technically advanced settings. To softening the corners, the game borrows elements from a popular, verbal party game 'Truth or Dare', in which the players are given the choice between answering a question truthfully, or performing a "dare". The Mixed Reality (MR) online game is designed to be motivating and rewarding by

[1]More about the game with a generous set of references: https://en.wikipedia.org/wiki/Blue_Whale_(game) (Website retrieved June 28th, 2017; 11:02). It should be noted that the diverse facts about this game are controversial.

T. Kymäläinen (✉)
Human Factors, Virtual- and Augmented Reality, VTT Technical Research Centre of Finland Ltd., Tampere, Finland
e-mail: tiina.kymalainen@vtt.fi

A. Reyes-Munoz et al. (eds.), *EAI International Conference on Technology, Innovation, Entrepreneurship and Education*, Lecture Notes in Electrical Engineering 532, https://doi.org/10.1007/978-3-030-02242-6_22

offering two adolescent protagonists, Angela and Veronica, challenges that arouse their curiosity and excitement. Eventually the girls grow fond of the adornment of the anonymous game community, which, behind the scenes, is composed of extreme transhumanists: adults that prey on pre-teens and young teenagers. Gradually Angela and Veronica's young minds are completely toyed with by the invisible game designers, who slowly gain access to their most private secrets and, thusly, find ways to implant their techno-religious agenda into their young minds. With the intimate information, the online community can effortlessly gain leverage over the girls; they can use the deviously gained knowledge for blackmailing and for threatening to kill their loved ones, if they eventually want to back out of the game.

The key technology to be explored in this *holonovel* is photo-realistic Holographic technology that provides immersive telepresence of people with realistic 3D rendering in physical space[2]. The "holoported" digital humans generated by the technology may be living, deceased or fictional characters, for example, artists, entertainers, politicians, medical professionals or subject matter experts—who, at present, can be 'beamed' over the public internet onto interactive stages, classrooms or conferences. In this *holonovel,* the key focus concerning this technology is in a controversial concept called *digital resurrection*, which in the SFP is issued by exploring ideas from transhumanism and posthumanism—that evidently provide a distorted framework for the game community's notorious activities. Another explored topic is the symbiosis of the artificial game engine and the anonymous community, that is, the mash-up of algorithmic intelligence and human game designers. The artificial moderator of the game—who, in the *holonovel*, has the appearances of Professor Moriarty—aggregates the message and command streams of the online community, and by doing this, illustrates the *posthuman godhood* that is beyond any legal allegations. The topic of human experience engaging Artificial Intelligence in Holodeck[3] (or more commonly, in Mixed/Augmented/Virtual Reality environments) has been greatly inspired by such Star Trek episodes[4] that previously raised applicable philosophical and sociological issues.

[2]Such technologies are provided, for example, by ARHT Media Inc. (HumaGram technology) and Musion Hologram Limited (3D holographic projection technique): http://www.arhtmedia.com/, http://musion.com/.

[3]The Holodeck, and "holographic" technology in general, is a plot device used in stories set within the Star Trek universe. It is used to permit stories with locations and characters that could not otherwise exist in the *Star Trek* setting, and as a means to explore metaphysical and ethical questions: https://en.wikipedia.org/wiki/Holodeck. In scientific research, there have been attempts to create environments that have similar qualities, for example [7, 8].

[4]Inspirational episodes in Star Trek: "Elementary, Dear Data" (1988): Next Generation, Season 2, Episode 3; "Ship in a Bottle" (1988): Next Generation, Season 6, Episode 12.

2 Kill Your Darlings

The game started really much before the girls, Veronica and Angela, even took their first cautious steps in its holographic playground. It started when the ARCH (Associated Responsibility for C+ and H+) syndicate posted a proposal of the game to some of the popular community platforms devoted to the ultimate form of transhumanism. The message was:

> Come and create the game world, convince two teen players of the legitimacy of its rules and find tacit agreement with our common interests, domains of thought and practice. The persuasion capacity of the game designer will be measured by the strength of seduction, from the beginning until the mortal end of the game; and the champ will be rewarded with a unique opportunity to upload their minds to the eternal ARCH cloud.

2.1 The Dare: Back on Baker Street

Baker Street 221B, Resurrection of Moriarty, Missing alphabet.

Angela and Veronica, both aged 14, were ordinary teenagers in Canterbury who liked to dress up classy, impress their friends and take unintentional risks when finding the way in the world. This was the case at the present moment, while they were heading towards a strange part of the city, which they had never visited before. The atmosphere of the near-downtown area felt like a ghost town, for the quarter was filled with abandoned retail stores that were gradually made extinct by the e-commerce. The girls knew exactly what they were after, so it did not take long for them to find the right building. They knew they were standing beside the right door, as they saw a familiar symbol in its surface: a red dragon that was eating its tail. Underneath the dragon, there were four letters, written in cypher, which was still an unsolved mystery for them.

'This is our 45th challenge', Veronica whispered proudly, 'Only five more to go. Shall we enter?' Angela nodded and opened the unlocked door with no hesitation.

The abandoned shop looked precisely like most of the ones they had visited earlier: it was almost empty, with only some abandoned furniture, and a lot of dust and dirt floating aimlessly in the air. The girls entered the main gallery, and there, beside a desk counter, were the precious things they were looking for: two pairs of glows and well-designed golden glasses. As they were all set, the girls pressed two fingers upon their hearts in an accustomed manner and greeted, with a clear and loud voice:

'We pledge allegiance to the syndicate and swear never to speak of it to anyone.' Right after the devoted oath Veronica added, 'The Arch', and with just one magic word, the girls were back at 'Baker Street 221B.' Like ever before, a beautiful sitting room was right in its place, completed with fireplace, mantelpiece and a Persian rug. One after the other a holographic presence of the girls' famous friends appeared to the room. The holographic avatars were celebrated teen idols, film stars and artists that Angela and Veronica had chosen to help them throughout the game.

The continuous plot of the game was seeking out missing alphabets from the Baker Street study, whenever it was addressed to be the game environment. So far the girls had collected 21 letters, and now there was only five of them left. Apparently, they were supposed to use them at the end of the game to solve the final mystery, which would answer the ultimate existential question: "Is there life after death?"

Angela took her famous deerstalker hat casually, put it firmly on her head and requested Watson to sit gracefully beside her. In the beginning of the game, Veronica and Angela had contested which one of them was to be Sherlock. Angela had drawn the longest virtual-straw, and Veronica still felt bitter about it; Angela was already the blonde—why should she have all the fun of being Sherlock, too? Just when Angela had put the pipe to her lips, there was a sudden and unexpected knock on the door. The girls glanced at each other with slight astonishment, then Angela gestured for Veronica to answer the call.

"Hello Darlings," said the voice of a familiar hologram, whom both of the girls knew all too well.

"Moriarty!" Veronica cried and took a restrained step backward.

"It can't be: he already died on the 36th level!" Angela replied as the pipe slipped from her fingers. All the other hologram avatars seemed as surprised as the girls about this new turn of the gameplay, but they seemed much more pleased than Veronica and Angela did.

"And like the spider, I feel the strings vibrate whenever they chance into my web,"Moriarty said theatrically with a cruel grin while he entered the room. Veronica went closer to Angela and whispered with resentment:

"Why did you insist that he look like that old actor from the ancient Star Trek? I would have preferred his avatar to be that of Andrew Scott; I bet he would even have better lines ... 'Angela did not pay any mind to Veronica's contemplation, for she was more upset about Moriarty's unreasonable emergence.

"Why have you reappeared?" she asked Moriarty crossly, "We killed you already, and now I feel disappointed—I feel that you have broken the game rules!"

"For me, the definition of life '*Cogito ergo sum*'—I think, therefore I am—is the most important rationalization of that—and for that part, the only one that matters," Moriarty reveled, as if by pouring out the extravagant words he could make any sense to his unfair return. "But, if you need a simple explanation, I have experienced a renaissance for one sole purpose," he continued and held a dramatic pause, "My consecrated purpose, the true reason for my resurrection, is the fact that the syndicate felt that no one could guide you through the game as I can." The holographic avatars of the girls' most beloved television stars grouped around Moriarty and they nodded their virtual heads approvingly.

"And thereby I want to draw your attention to your next quest," Moriarty continued sharply, "Do you still want to accept the challenge?"Angela and Veronica glanced at each other with frustration. After a long while, Angela replied:

"We accept the challenge." Then she continued with a slight bitterness in her voice, "And this time we will choose the truth." Moriarty smiled arrogantly.

"The clue for the next challenge is in my name," he said dramatically, "I suggest that you take the glows and glasses with you and be in that class after 12 h."As a reward, he informed the girls about a missing alphabet, and said that there were sponsors' generous gifts waiting for both of them in the main entrance.

V-ToD-EXIT CHALLENGE 45 ARTERMATH

ForEve+ 12 points 1 minute 7 seconds ago

Nice gameplay MetaLife! Good combo of state-need-reward. You really impaled into their motivational level: a point well deserved!

MetaLife 11 points 58 seconds ago

Thank you, my dear co-evangelist, but as you know, they were already hooked.

Uploading 13 points 46 seconds ago

Hallelujah! Moriarty is back: count your blessings!

VHEMT 9 points 38 seconds ago

I think the apprentices were pretty unsatisfied with Moriarty's reappearances. **A** even mentioned about the legitimacy of the rules.

Moriarty the curator 21 seconds ago

My dear little posthumanistic circle, do I still and all sense a bit of dissatisfaction towards my latest resurrection? For solving the matter, VHEMT, would you like to be the one to design the next level?

VHEMT 9 points 8 seconds ago

Challenge accepted. I should think this one has to be a bit more stimulating one, as from now on the girls have to be brought much closer to commitment.

2.2 *The Truth*

Art Class, Draw your Circle of Love and Hate, Know the Truth.

After solving all the earlier challenges, the mystery at hand was not a very difficult one for the girls; as obviously, with the clue, MoriARTy had meant the ART class. What was more challenging, was the fact that the girls had to sneak there after school hours, and it was a well-known fact that all the classes were carefully locked precisely after five o'clock. Nevertheless, the mission turned out to be a surprisingly easy with the data glasses, as the girls received careful and detailed instructions informing how to break in without notice.

In the art class, the challenge was simpler than any of those they had completed before. They were to take up crayons and the biggest paper they found, and draw their personal 'circle of love and hate,' which included all their classmates. One by one the girls drew names within or outside the circle. The names closer to the core were the ones they both liked, and the ones closer to the outer sphere we the ones they disliked or even hated.

After they were finished, and all the names were drawn on the paper, Moriarty, appeared in the classroom.

"Are you sure all the names are there?" he asked assertively.

The girls nodded.

"Well, my dear apprentices, to conclude the challenge, the task is simple: you must choose the one you both hate the most."

The girls glanced at each other and furrowed their eyebrows with increasing discontent.

"And then what?" Angela asked nervously.

"You must hurt her," Moriarty replied, firmly and dispassionately.

"How?" Veronica questioned, and was more than appalled.

"In that you'll have to be creative. But remember, this is your final chance to pledge allegiance to the syndicate. As you both know, you have already failed twice."

V-ToD-EXIT CHALLENGE 46 ARTERMATH

Uploading 13 points 32 seconds ago

Well that one created more questions than answers about their commitment.

VHEMT 10 points 18 seconds go

To be continued… For the next challenge, I need someone to find more about the desires and collective fancies of their classmates. Any volunteers?

ForEve+ 12 points 10 seconds ago

Will do.

2.3 Double Dare

Angela's home, Allegiance to the Syndicate, the Initiation Sacrament.

Betty Reeves was probably the most annoying girl in the school, and this was, in fact, a well confirmed, shared opinion. Betty's father was rich, yet she wore untastefully despicable clothes, and, of course, she was a crammer. Due to those usual and universal facts, she did not have many friends in her class; and therefore, when Angela invited her home right after school, she almost rolled over.

At home, Angela made tea for three, and on the table Veronica thoughtfully, offered Betty a freshly baked lemon pie. Both Angela and Veronica took a big slice of the readily sliced pie, but for their discomfort, Betty just shook her head arrogantly.

"I am afraid I must refuse," she said decisively, "For, I suppose it is not gluten-free?"

Angela felt annoyed, since this behavior was so Betty-like.

"What if there was a bit of gluten?" she asked, as politely as she could, "What would it do to you?"

"Oh, it's terrible," Betty replied and rolled her eyes, "You would not want to witness that! If I take even the slightest bit, my stomach would ache, and, in the worst case, I might even faint."

"I guarantee there is no gluten," Angela said snappishly and smiled, thinking that the symptoms might not have time to expose themselves.

"I remember that your mother said," Veronica accompanied, "That the pie is not only gluten-free, but it is also dairy- and sugar free."

Betty looked at them both, with an uncomfortable hesitation in her eyes.

"Still, I think I would rather stick with tea," she informed and sipped her cup delicately.

After the tea-time persuasion failed, the girls did not know exactly what to do. To buy some time, Angela took Betty to the living room and introduced her to a pair of odd-looking glows and glasses. Betty needed some convincing concerning why she should change her perfectly operational ones to those that did not seem so fitting for her. After the girls had explained how they worked, Betty seemed a bit intrigued.

"The Arch," Veronica said right after Betty had put the glasses on.

"I cannot believe it!" She sighed instantly, "The room just changed into a Nineteenth-century study!"

The girls nodded, as that was exactly the same experience they had felt—which now seemed like ages ago. Betty walked slowly around the living room and moved next to the bookshelf. She had to glance at their backs only once, after which she sighed:

"... But this is not just any old room! This is the study of Sherlock Holmes!"

After the observant remark, Angela felt even more annoyed towards Betty.

Suddenly, without any outspoken notice, there was an unexpected knock on the door. All of the girls looked surprised, but Betty was the first one to be immediately on her feet. She approached the door, and almost experienced the after-gluten weakness she just mentioned, when she met the visitor.

"Co-Conan Chris?" she faltered as she opened the door and took two steps back.

To all of their surprise, a famous teen-heartbreaker-singer, Conan Chris, stepped with steady moves into Angela's living room. He greeted all of the girls with a smile, but then his eyes glued solely and tightly on Betty.

Suddenly, the air was filled with Conan's sweet new hit song, which had a very slow dance beat. The modern-day Casanova nodded to Betty and to her great surprise, took her to dance instead of her more popular classmates.

"This is your moment," said a sudden voice that was audible only to Angela and Veronica, "This is a dare for you both—a dare in which you cannot fail, again," the voice continued.

Angela looked around the room, and then, on her command they both broke the seals of the balcony door. This particular door, as Angela's mother had retained, was the one never to be opened. This absolute forbiddance was laid down for a good reason, as the door led straight to a construction site of the 12-story building. As the dancing couple approached it, Veronica opened it cold-bloodedly. When the fatal incident came about, Betty could not see the danger, as she was completely immersed in her dancing partner's immersive opal-green eyes.

After the destructive incident, Moriarty appeared to the living room, ecstatic as ever.

"My dear Angela and Veronica, this has been your final hour," he complimented openhandedly, "You have performed the initiation sacrament without any negative votes from the sponsors, and your reward will be unforeseen!" Then he raised his finger into the air, and reminded, "Now, all you have to do is just say the words."

The startled girls knew perpetually well what Moriarty meant.

"We pledge allegiance to the syndicate," they repeated in sacred unison, "And swear never to speak of it to anyone."

V-ToD-EXIT CHALLENGE 47 ARTERMATH

ForEve+ 12 points 1 minute ago

What an epoch! Now we are really part of the History+.

Moriarty the curator 58 seconds ago

May I just remind you, my dear flock, that the game is anything but over. What this all means, is that it is eventually time to decide, which one will it be ...

MetaLife 11 points 42 seconds ago

I assume I speak on behalf of us all, when I say that there is no denial: with any objective measures, it should be the blonde.

Uploading 13 points 37 seconds ago

Obviously. BTW, which of us has less points? Who will design the next level?

Moriarty the curator 18 seconds ago

According to my interpretation, it is now MetaLife's turn. Will you do the honors?

MetaLife 11 points 2 seconds ago

Nothing could keep me from doing it! I have already decided the name of the level: it will be "No pain, no gain".

2.4 Confirmation

Baker Street 221B, Milgram's test, Gain for Pain.

Veronica and Angela had only 2 days to recover after Betty's fatal incident. Although they were in the middle of police inquiries, the girls managed to slip into a random warehouse, which was disguised, again, as Baker Street 221B. As soon as the girls had put their glasses on, they saw a strange-looking chair in the middle of the sitting room.

"Congratulations, my dear Angela and Veronica," Moriarty applauded, "You have managed to proceed almost to the end of the game. A rare achievement, indeed, I must say: 47 levels completed, and only three left to go!"

Somehow, Angela and Veronica did not feel as pleased as Moriarty. After what had happened to Betty, they did not feel the urge to seek out the missing alphabets anymore; nor did they feel any enjoyment of the sponsors' generous gifts.

"But, as they say," Moriarty brought up, "When the game advances, the stakes grow higher. And so, my dear apprentices, at this point my devoted and sacred obligation—or should I emphasize, my privilege—is to introduce you to the next, hair-raising challenge," Moriarty proclaimed. He waved his arms and gestured to urge the girls to move closer to what he said was a confession chair.

"This challenge is a dare for you both," he informed and requested, innocently but coldheartedly, "Would either of you, by any chance, be familiar with the

Milgram's test?" The girls shook their heads, but what he broadcasted sounded rather notorious.

"I thought so," Moriarty nodded with a cruel smile, "Well, in any case, the test is all about obedience ... and to be more precise, in your case, it is to test your obedience to the syndicate."

The girls felt immediately nervous. Even more so, when Moriarty asked them to draw a virtual lot about which of them should sit on the chair. The girls had no idea that the draw was already fixed.

"So then," Moriarty said, after seeing the result, "Could you, Veronica, please sit on the chair that is so perpetually fixed for you?"

Veronica felt shivers, but did as Moriarty requested. As soon as she had sat, leather-looking straps folded tightly over her wrists. At the same time, from the thin air emerged a glass window between Angela and Veronica. Moriarty holoported himself next to Angela and snapped his fingers casually. Instantaneously, an electric shock generator appeared in front of Angela. She observed the generator in a growing terror.

The device had three switches which were marked as slight shock, severe shock and XXX. When Angela saw the scales, she turned pale instantly. Veronica did not see any of this, as the one-way mirror distracted her view from the interrogation part of the room.

"The challenge is simple," Moriarty announced, faking a soft tone, "In this challenge, Veronica must truthfully answer to questions about some of the doctrines that are of significance to our little circle of moral concern. It shouldn't be too hard if we consider all the challenges you have already performed so eloquently."

"But, of course, there is a catch," Moriarty continued by raising his finger, 'You Angela must administer an electric shock to Veronica every time she makes the slightest mistake. Furthermore, you should increase the level of shock each time Veronica answers a question in an unsatisfactory manner."

"Shall we begin?" Moriarty asked, cheerfully, and rubbed his hands. On the other side of the mirror, Veronica was moving about uncomfortably. Angela looked at her friend in anxiety and refused boldly. Moriarty shook his head, and whispered in her ear:

"The challenge requires you to play along, Angela. You really don't want to back out now, do you? ... Now that you are so very close?"

"Anyway, Veronica is doing most of the work this time," Moriarty assured, "All you have to do is read the sentence aloud, and Veronica will fill in the blank lines."

After Moriarty was convinced that the girls could not be any readier, he announced:

"Well then, what would Sherlock say in this situation ... Ah, the game is afoot!" He grinned, "So, here comes the first question. Veronica, what is your belief: is there life after death?"

"Why do you ask me? Veronica seemed puzzles, "Isn't that the question you should answer to me throughout this game?"

Moriarty glanced at Angela in unison and nodded firmly.

"Go ahead, press the button," he said, "She will feel only a slight sensation of tingling."

Angela hesitated, but after looking at Moriarty's cold eyes, she saw no way around it. She bit her lips and pressed the button. Veronica was immediately quivering in the chair within the grasp of an electric shock that seemed to cause her something much more than just slight tingling.

"Yes, there is!" she cried finally after great torment.

Moriarty revealed his cruel smile.

"Good. Now we can proceed... Here comes a slightly tougher one: what is there to be expected more specifically after one's mortal life is over?"

"Heaven ... Will we go to heaven?" Veronica cried and sobbed simultaneously.

Moriarty had a stony face, which, for a while, expressed nothing.

"And what becomes of you ... in heaven?" He continued then, attentively.

"I will be raptured to heaven and become an angel ..." Veronica responded hesitantly. Angela looked at Moriarty in agony. Instead of commanding her to give another shock, Moriarty spoke softly:

"I let you keep your naïve thoughts, as in transcendence you may truly become anything you please. I suppose the transhuman phase will present itself exactly as heaven to you, for there you can be surrounded by all the things you love ... And your loved ones will join you, sooner or later, one after the other."

"But I have to add," He continued sharply, "That according to my belief, humans will not be enhanced to become angels, rather they will be replaced by artificial intelligence. For this merger, humans bring in the specific qualities of faith, ethics and justice. Yet otherwise this eternal symbiosis will be multi-aspectual, as it functions in both the formative and lingual aspects."

Moriarty clearly did not expect girls to comprehend his eloquent view, as, for that matter, it was undoubtedly targeted at his other audience. After savouring his thoughts for a while, he moved to the next persuasive question.

"Why do we do this?" he implored Veronica, "Why do we play this game?"

Veronica shook her head, for she could not find an answer. So, Moriarty turned to Angela.

"It is your turn again, dear Angela," he whispered warmly, "You know what to do."

Angela did nothing, but stared at the electric shock generator with glazed eyes.

"I have to remind you of the consequences of your choice," Moriarty continued with a severe tone, "Think of your mother, Angela, and think of Betty. For your own, and Veronica's sake, it is absolutely necessary that you do the thing that you must."

Angela shook her head feverishly, but looked otherwise as if she was paralyzed.

"Angela dear, you know you'll have no choice but to continue!"

This time the scale of the electric shock generator was marked as severe shock: the panel indicated that as the volts would approach 250, this caused notable danger to the examinee. Yet Angela had no choice but to press the button. The shock caused Veronica to shake, sweat and scream her heart out and the severe trembling seemed to take an infinity before it was over. Without blinking his eyes, Moriarty repeated the question:

"Why do we do this?" He asked compulsively, "Why do we play this game?"

This time Veronica replied without hesitation:

"Voluntary human extinction is mandatory, because it will prevent human suffering and the extinction of other species," she poured out the memorized words, as if she had been brainwashed.

"Now, now, my dear Veronica," Moriarty said, "Are you sure you are not in a so-called 'agentic state' in which you just repeat the learned words?"

Veronica shook her head furiously and tears were running through her eyes. An eclectic smile occupied Moriarty's face when he claimed:

"The answer is nevertheless absolutely correct. Our transhumanistic circle is extending subjectivities beyond the human species," he turned back to Veronica, requesting further, "And what is our dictum for which we plea?"

"I dare to know," Veronica responded mechanically, "I acknowledge the reason as the primary source of legitimacy, and come to advance ideals such as liberty, progress and tolerance."

Moriarty seemed pleased. He nodded approvingly and replied:

"Thank you my dear apprentices. The syndicate is proud of your accomplishment. You can consider this challenge now well taken care of, for that was all we wanted to hear. You can go home and have full enjoyment of all your more than lavish rewards!"

V-ToD-EXIT CHALLENGE 48 ARTERMATH

VHEMT 11 points 55 seconds ago

Nice closing, but what was the lesson? Deep down this is hardly more shivering than some good old reality TV. Milgram's test already firmly convinced his Nazi-theory [9].

Uploading 13 points 40 seconds ago

The point was, that he never went this far. And Milgram never examined how much a person is willing to hurt someone whom she loves. I am convinced **A** would have killed **V** if she only had been commanded.

ForEve+ 12 points 26 seconds ago

I am not. Girls do that all the time, especially when they are cornered. It is perfectly normal to put their teeth into their best friend's flesh, and bite as sharp as they can, if they feel even vaguely trapped.

Moriarty the curator 17 seconds ago

I have to remind you, my dear congregate, that our intention is not to murder anyone, or commence anyone to do such thing. Meaning is the being of all that has been created, and all we are looking is a voluntary sacrifice. But, as Milgram stated, "Illusion should be used when necessary in order to set the stage for the revelation of certain difficult-to-get-at-truths". So, after all things considered, should ForEve+ be the one most capable of us digging out the final truth?

ForEve+ 12 points 6 seconds ago

Will do, with pleasure.

2.5 Private Truth

Veronica's home, Draw your personal Circle of Love and Hate, Know the Truth.

Veronica saw her mother briefly in the kitchen before they both were running off to work and school. They changed a few words about the school, then Veronica pulled herself together and asked her mother cautiously:

"Mother, have you ever played the game Truth or Dare? I mean, as you were a child."

Her mother was drinking coffee hastily. She raised her eyes from the morning news and looked at her daughter absent-mindedly.

"Of course," she smiled and sustained, "But I suppose in Ukraine that game was a bit rougher than the one you girls are playing here."

"How come?" Veronica asked and raised her eyebrows, "What was your worst dare?"

"Vodka in the eyeball," her mother said and smiled. Then she remarked the wounded look in Veronica's face, and it made her wrinkle her eyebrows. A sudden shivering thought came into her mind, but it took a while before she could find words for it. Then she looked deeply into her daughter's eyes and whispered:

"I am so sorry what happened to your friend, Betty ... Was the incident caused by playing the game of Truth or Dare?"

Veronica didn't say anything, but she looked extremely guilty. Just at that moment, her mother's mobile phone rang. She got up, bent over Veronica and kissed her on her cheek.

"I am convinced, it was not your fault, but you should tell me all about it later tonight." Then she answered the phone and at the same time ran off to work.

Distressed Veronica packed her schoolbag and just when she was about to leave for school, Moriarty appeared, uninvited, in Veronica's hallway.

"What is this?" Veronica asked in agitation.

"This is the challenge 49," Moriarty replied harshly, "And it is for your eyes only."

"What do you want?" Veronica asked warily and unintentionally began to tremble.

"I want you to draw your personal circle of 'Love and Hate'. And this time you should place Angela into it."

"I don't want to play the game anymore," Veronica whispered.

"But you have no choice, my dear Veronica," Moriarty smiled cruelly, "Remember what happened to Betty. Do you think the syndicate would ever leave you alone?"

"You have two-and-a-half days to convince Angela that you have not gotten bitter on her,' Moriarty informed and, in a blink of an eye, vanished into thin air.

V-ToD-EXIT CHALLENGE 49 ARTERMATH

MetaLife 12 points 1 minute 2 seconds ago

Nice! Neat!

ForEve+ 13 points 46 seconds ago

Thanks', as you know, my motto has always been: keep it very, very simple, stupid.

Moriarty the curator 42 seconds ago

My dear congregate, should we, again, set our minds on the holy chore at hand, as, after all, we have merely the ultimate sacrifice left. The girls are already standing on the edge, but this last challenge requires the full engagement of us all. As you know, this time the reward will be generous three points for the champion of the challenge!

VHEMT 11 points 33 seconds ago

The rules are acknowledged. But I must bring forth my objection towards one little detail. You were earlier all so unanimously for the blonde, but I am not that convinced anymore. Why shouldn't it be the brunette; as after all, she seems quite ripe to me?

Uploading 13 points 15 seconds ago

Why not both of them?

2.6 The Ultimate 50th Level

Go to a bridge, Commit suicide, Understand the Ultimate Truth.

Moriarty appeared to the bridge upon which Angela and Veronica were standing. The game designers had made them run through the city the whole day in pursuit of the final cypher alphabets, so they both were now more than exhausted. The pursuit was, of course, pointless; the real aim was not to leave the girls any time to contemplate their thoughts.

"This is your waterfall," Moriarty proclaimed and waved his hands towards the bridge and river that was running underneath, "To pass the final challenge and the entire game, you girls do know what you will have to do."

"Passing the challenge will cause you to understand the ultimate truth," he continued firmly and persuasively, "As you know, I already confronted my waterfall in the 36th challenge. Now it is your turn to brake on through and find out what is on the other side. Are you ready for your eventual transcendence?"

Veronica was first to climb to the bannister with no hesitation. She glanced at Angela secretly in a vicious manner and then noticeably waved her to follow.

"We cannot back up anymore," she proclaimed, "If we do, you know that they will kill our mothers."

Angela looked surprisingly calm, almost as if she was sedated, when she climbed next to her friend on to the bannister.

"Ultimately an individual should always have the right to take personal control over their own death," Moriarty declared theatrically, "But before you jump, my dear apprentices," he added softly, "I would like you to share the name of your metaphysical afterlife."

Veronica did not say anything, but Angela, with glazed eyes, responded self-consciously:

"Neo-Angela," she said and leaned a bit forth. After that, Angela tried to take Veronica's hand, but she refused and told that they should do it together, but alone. She said she will count to three and after that they both had to take the final step.

"One, two ... three!" she counted.

Veronica, still feeling bitter about Angela administrating her the electric blows, had been ready to give Angela a push if she had not jumped herself. But now, after she had done it willingly, Veronica was suddenly experiencing an overwhelming and chaotic state of a shock. Veronica climbed back to the pavement, but she had really no time to collect her thoughts, as the dragon symbol and the unrevealed cypher emerged in front of her. There were the four unsolved letters floating in the air that had haunted them during the whole game. Moriarty looked at her baffled face and laughed arrogantly.

"You deserve to know the truth," he said pettily.

"It can't be!" Veronica cried as the symbols began to unfold in front of her.

There were just four letters: ARCH. And then followed the slogan of the syndicate: "We deliver custom-created, ultimate experiences, tailored to answer the fundamental questions, such as: "Is there life after death?" "Is there a God?" and "What is the meaning of life?'"

"Congratulations on solving the first mystery," Moriarty proclaimed, "And now, my dear Veronica, would you like to start another game?"

"This time you can be Sherlock and the syndicate will find another Watson for you."

"And I can play Moriarty," Angelica's holographic appearance echoed as she emerged floating next to Moriarty. Veronica looked at the supernatural reflection of her best friend and felt nothing but panic. She shook her head, and then ran as fast and far as she could.

V-ToD-EXIT CHALLENGE 50 ARTERMATH

MetaLife 12 points 1 minute ago

Angela is dead! Long live Angela!

ForEve+ 13 points 42 seconds ago

This is the true renaissance of Angela! She is now immortal.

VHEMT 14 points 28 seconds ago

May we all live long and peacefully die out!

Uploading 13 points 11 seconds ago

Human extinction is the only solution! In the end we will all be as one.

Moriarty the curator 3 seconds ago

We may sanctify our holy cause, because we have been blessed with the revelation of knowledge that human species can transcend itself—not just sporadically, an individual here in one way, an individual there in another way, but in its entirety, as humanity that is caused by divine agency!

Neo-Angela 0 points 1 second ago

Praise to you all! I know now, immortality is eternal life, and our only ability to live forever!

2.7 *Aftershock*

The detective that had been investigating Betty's case was alerted when they found another teen body in the same neighbourhood. He interrogated Veronica, who, once again, had been in the crime scene. The adolescent girl broke immediately, but her confession was only the first thread in a complex case. In fact, it later turned out, there had been several similar cases around the country during the last few months. The only thing that was different now was Veronica's mobile phone, which, by a coincidence, had been unconnected after Angela's death. Therefore, the phone had stored all the information of the last five challenges, and so this was the only case, in which the syndicate had not been able to wipe out their traces. Using the evidence, the police were later able to arrest two members of the circle; those who had presented themselves in the community chat as VHEMT and ForEve+.

Nevertheless, the case was still a very problematic one, when bearing in mind the legal matters. The detective had an intense conversation with the department lawyer about how to solve them.

"What I want to know is: Can we even sue the game designers or is the legal liability only on those making rational decisions: the teenagers, Veronica and Angela?"

"Both girls are adolescents and clearly victims here," the lawyer replied, "It would be a waste of time to prosecute the one that is left alive: Veronica is surely a witness."

"When it comes to teens in general," the lawyer sighed, "They are gullible and lack life experiences. This means they are easy targets for these crazy predators, who, in my opinion, should solely be held responsible for their actions."

"So, can we sue them?" The detective probed. The lawyer did not look too convinced.

"Of course we *are going to* sue them ...," she retorted, "But any evidence that we have so far demonstrates conversations between a combination of human moderators and artificial intelligence. So, the question to ask in this case would be: are the rest of these persons even human, or, as in the case of Moriarty, are they also some form of artificial intelligence?"

The detective scratched the back of his head, and concluded:

"I think the profound question should be: How do we prevent this systematic, sadistic activity from becoming mainstream? This case should be made a precedent."

3 Reflection

"In writing, you must kill your darlings." This literary advice from William Faulkner refers to the dangers of an author using personal favourite elements in the story s/he creates[5]. This *holonovel* has been created by compromising the advice: the storyline has extravagantly exploited the personal favourite elements of the author (especially relating to the chosen technologies and the cited Star Trek episodes) and, in addition, taken Faulkner's words literally by introducing the virtual killing activity in a real-life game environment.

With regard to the form of the story, the introduction of the paper claimed that the story was a *holonovel*; yet to be quite honest, it is actually a science fiction prototype disguised as a *holonovel*. As a writing activity the difference may be compared to a short story and a screenplay: a story requires plot, characters, and three phases (a beginning, middle, and end), whereas a screenplay contains a story with specific parameters including details of how the actions will be achieved. In the SFP there are, however, aspects that can be considered very important characteristics of a *holonovel*, as it, for example, runs in the first person 'subjective mode,' occupies space, highlights the role of the artefacts and demonstrates how the protagonists actively interact with the designed game program and its characters—as a good *holonovel* always should[6].

Regarding the chosen key technologies, the prototype introduced a Mixed Reality game environment that dominated space around the protagonists and operated between Milgram's reality–virtuality continuum model [10], which, in general, is composed of real and virtual space and objects. For presenting the virtual and simulated people the SFP laid great interest in digital avatars and contemporary human 'Hologram' technologies. The human representation technologies currently seem timely and interesting, as Virtual Reality (VR) and Augmented Reality (AR) are entering more and more to mainstream social media applications[7]. From that aspect, this SFP was essentially concentrating on a theme called *digital resurrection*—creating virtual appearances of diseased people—for which, there are currently several techniques that have been used. To draw notion of few, for example in "rotoscoping", the film footage of the person's face is composited over a body double, and in more advanced cases, motion capture technology is combined with high-end 3D computer graphics that recreate a person completely anew. The digital resurrection has been demonstrated in a number of "real-life" cases, with, however, diverse feedback from the audience. One powerful experience has been described to be the virtual comeback of a deceased rap artist, Tupac, who suddenly appeared onstage in a music festival, performed some songs and then disappeared

[5]The risk in such activity (e.g. overuse of a word or phrase) lies in the statement that while these elements may hold special meaning for the author, they can cause readers to roll their eyes

[6]http://memory-alpha.wikia.com/wiki/Holographic_novel.

[7]Many VR companies are currently providing custom-created avatars in order to access their in-person interactive services, such as High Fidelity, AltspaceVR, Modal VR and The Void.

in an explosion of light[8]. An opposing experience has been described the hologram of Michael Jackson at the Billboard Music Awards, which was described as being "disturbed" and "creeping [people] out"[9]. In addition, with the clear relationship to uncanny valley hypothesis [11], the technology includes a number of interesting socio-technical issues that could only be slightly touched by this SFP.

In addition with the presentation form of the holonovel and the introduced technologies, the overall message in the dystopian *holonovel* has been that already the software, applications and social media surrounding us are convincing people to give away valuable information about themselves in exchange for free experiences. In turn, the invisible parties delivering these experiences can accrue large amounts of data freely, and, if they will, may find disruptive use for the information and bear no legal responsibility for the consequences. Another matter that the SFP wanted to highlight is the anonymity in the Web that is cultivating the darker side of human nature, as it allows people to hide behind a mask and, for example in social media, write mindless things, which is enforced by the mob behavior of the private conversation groups. In the SFP, these issues were illustrated through how the anonymous game designers diminished the importance and uniqueness of individuals, Angela and Veronica, and turned them into variables in a game that was merely a facade for their extreme, posthumanistic devotions. As a consequence, the ultimate question of the *holonovel* has been to consider the moral basis of the illustrated reality that may already, to some extent, be achieved by the present-day technologies.

References

1. B.D. Johnson, *Science Fiction Prototyping: Designing the Future with Science Fiction. Synthesis Lectures on Computer Science* (Morgan & Claypool, Williston, 2011)
2. T. Kymäläinen, *Science Fiction Prototypes as Design Outcome of Research, Doctoral Dissertations*, Aalto University Publication Series (Unigrafia, Aalto ARTS Books, Helsinki, 2015)
3. J. Huxley, Transhumanism. J. Humanist. Psychol. **8**(1), 73–76 (1968)
4. F. Ferrando, Posthumanism, transhumanism, antihumanism, metahumanism, and new materialisms. Existenz **8**(2), 26–32 (2013)
5. A. Sherlock, Larger than life: Digital resurrection and the re-enchantment of society. Inf. Soc. **29**(3), 164–176 (2013)
6. A.P. Rios, Exploring mixed reality in distributed collaborative learning environments, Doctoral dissertation, University of Essex, 2016
7. B.R. Jones, H. Benko, E. Ofek, A.D. Wilson, IllumiRoom: Peripheral projected illusions for interactive experiences, in *Proceedings of the SIGCHI Conference on Human Factors in Computing Systems*, (ACM, New York, 2013), pp. 869–878
8. A.D. Wilson, H. Benko, Combining multiple depth cameras and projectors for interactions on, above and between surfaces, in *Proceedings of the 23nd Annual ACM Symposium on User Interface Software and Technology*, (ACM, New York, 2010), pp. 273–282

[8]https://mdhhologram.com/portfolio-items/tupac/.

[9]http://www.thedailybeast.com/michael-jacksons-crazy-billboard-awards-performance-and-more-hologram-wins-and-fails-video.

9. S. Milgram, Behavioral study of obedience. J. Abnorm. Soc. Psychol. **67**, 371–378 (1963)
10. P. Milgram, H. Takemura, A. Utsumi, F. Kishino, Augmented reality: A class of displays on the reality-virtuality continuum, in *Proceedings of Telemanipulator and Telepresence Technologies*, vol. 2351, (International Society for Optics and Photonics, Bellingham, 1994), pp. 282–292
11. M. Mori, The uncanny valley (Translated by K.F. MacDorman, Norri Kageki), IEEE Robot. Autom. **19**(2) (2012)

The Responsibilities of Knowledge

Jim Hensman

1 Introduction

Developments taking place in many fields will transform our lives on this planet, as well as potentially affecting how we relate to the wider universe, through space exploration. Developments in robotics, computing and artificial intelligence, exemplified in the discussions about a forthcoming technological singularity, suggest a future where limitations of intelligence of the human brain, will not be a constraint. Developments in interfacing biology and machines, such as brain computer interfaces, imply that the combination between humans and machines can be exploited in ways which utilise the particular strengths of each to create novel and powerful synergies.

Yet paralleling these new possibilities are new ethical issues and dilemmas which are not immediately apparent. Traditionally, Science Fiction as a genre has been to the forefront of raising these questions. The concept of Science Fiction Prototyping and the Holonovel which this paper is guided by, builds on this to explore the technological and other pathways which could be followed as well as the consequences these may bring about. Through this it is hoped to help to influence current directions for development, but also to help pose wider societal and ethical issues which need to be considered and analysed. This paper uses a particular set of future technologies and an associated scenario to help elucidate some of these important issues.

J. Hensman (✉)
Coventry University, Coventry, UK
e-mail: j.hensman@coventry.ac.uk

A. Reyes-Munoz et al. (eds.), *EAI International Conference on Technology, Innovation, Entrepreneurship and Education*, Lecture Notes in Electrical Engineering 532, https://doi.org/10.1007/978-3-030-02242-6_23

2 Background

When Gene Dolgoff introduced Gene Rodenberry, the creator of Star Trek, to the concept of the Holodeck in the 1970s, the technological possibilities which existed at the time were relatively limited. But it still provided a major influence for later developments. For instance, John Carmack, who became the Chief Technology Officer for Oculus VR, the Facebook owned company that makes Virtual Reality headsets, states how he was inspired in his career by the Holodeck. Today in the epoch of virtual and augmented reality increasingly being available as consumer items, the technical context has moved to a different level. An interesting discussion of the possibilities of implementing a real Holodeck is provided by the science writer, Brian Clegg [1]. He thinks that producing a realistic moving holographic display feasible, but considers that one that can be walked around inside presents issues that will be difficult to overcome. However the problem can be looked at another way. In 2016, Alex Kipman, a key figure in the development of Microsoft's Hololens smartglasses, as part of a TED talk demonstrated what using this looked like as he shared a space with a NASA scientist who appeared to be on Mars, using data from the Curiosity Rover there [2]. With the further development of technologies of this kind to be part of contact lenses and eventually integrated with vision directly, the subjective visual experience of the holodeck could be conceivable.

Reproducing physical contact was always the most problematical issue and envisaged originally in terms of "replicators" and "force fields". Physicist, Lawrence Krauss, is dismissive of this aspect of holodecks [3], and Mark Lasbury, who also looked at Star Trek technology [4], states that nothing currently available would "allow you to pick objects up like in a Star Trek holodeck". Notwithstanding this, at least in terms of simulating physical sensation, haptic devices are now relatively well-established. Recent novel developments in this area give some indication of what could be achieved. For instance, Ben Long and his group at Bristol University have demonstrated what have been called "haptic holograms", whereby focused ultrasound can produce the sensation of touching an object in mid-air [5].

Some of the other technologies mentioned in the story, such as direct connection to the brain's sensory cortex, are part of ongoing development. Miguel Nicolelis of Duke University who has been one of the researchers at the forefront of this work [6], has explicitly referred to the connection of this with the concept of the Holodeck. Techniques referred to in the story include brain computer interfaces and systems using clusters of biological neurons. Exemplifying the former is the Braingate system, originally developed by Brown University and the company Cyberkinetics. In one application this uses a array of 100 microelectrodes implanted in the primary motor cortex and has been used successfully with tetraplegics [7]. An example of the second type of technique are so-called rat brain robots. In one piece of work, approximately 100,000 neurons obtained from rat brains were used in a planar array and deployed in a robot control application, for instance to implement obstacle avoidance [8]. Although the type of system envisaged in the story, integrating human neuronal clusters with nanocomputer brain interfaces, lies considerably in the future, the underlying principles are already being demonstrated in practical applications.

Discussions of ethical issues concerned with technologies of this kind [9] are generally centred around issues arising in medical applications. However, some of the more general challenges which are raised in the story also have their counterparts in issues arising at the present time.

3 The Story

3.1 The Holodeck

Steve looked over the table at Ryan. For a space mission as important as this to humanity they couldn't have asked for anybody better—or more photogenic, he reflected.

"After that initial problem before launch," Steve commented, "I'm glad to hear everything's been going fine."

"Like clockwork," said Ryan, "although I guess there's very little of that around in this sophisticated piece of technology."

"I know you'll be formally taking over very soon from the boss," continued Ryan. The "boss" was Ricardo Gonzalez, commonly known as Rick, Director of the World Space Programme.

"I'm seeing him in an hour's time for the handover," said Steve, "the official announcement and ceremony is planned for tomorrow. It's going to be hard not to think of him being in charge after all this time. But I know he's looking forward to retiring. Not that we'll be short of any advice from him if we don't shape up, I'm sure. I'll see you at our next scheduled meeting."

Steve stood up and Ryan followed. They shook hands. This was the bit that Steve still found strange. Of course it was second nature to the younger generation, like Ryan, the Holodeck natives as they were referred to. But at the back of Steve's mind was the knowledge that Ryan in fact was roughly a million miles away.

3.2 Terrible Secrets

Steve found it hard to suppress a smile as he made his way to Rick's office. Rick was overall head of space missions and had particularly overseen the deep space programme which had relatively recently been initiated and included missions where those on the spacecraft would not return. Steve had been his deputy for over 10 years and was responsible for shorter range missions, but he had known Rick from when he first joined the programme so many years ago.

Steve recalled a phrase from Rick's presentation he and the new intake had been given on their first day.

"Knowledge as it develops presents new possibilities and opportunities which we must grasp. But we must also remember that it equally brings challenges and responsibilities, which we must face up to."

Rick had been his mentor from those early days, but also become a very close friend. Now Steve knew it was his turn to grasp the opportunities and he felt ready to take on this exciting new phase of development.

Steve knocked and went in. Rick beckoned him to sit down opposite to where he sat at his desk. Steve immediately knew something was wrong. He had not seen Rick since just before the launch and he was shocked at the change in him. Rick looked haggard, as if he had not slept. He was clearly highly agitated and distressed. Steve had never seen him like this before. He wondered whether the stroke which had necessitated Rick being off work for several months some years ago, and which Steve knew had been a major factor in him considering retirement, had recurred. Rick spoke as if he knew what Steve was thinking.

"I'm not ill Steve and this is certainly nothing to do with you or your abilities. You are indispensable to our programme and the person who both deserves the Director's post and who I would wish to take over from me. But I want to give you this opportunity to reconsider whether you want this."

Steve was stunned. He had spent his whole working life as part of the space programme, and although he certainly wouldn't have thought it likely that he would have reached the position he now was in when he started, this was his biggest ambition. Even as a child he had dreamt of this.

Rick spoke, again as if he could read Steve's mind. "I know that this has been everything you wanted, but I feel it is my duty, not just to you as a colleague but also one of my closest friends to ask you to consider your decision carefully."

Until then Rick had been looking down, but now he looked straight at Steve, who was shocked by the haunted and desperate expression in his eyes. "You see there are some things, terrible secrets, which I have responsibility for and which would fall on your shoulders. This has been preying on my mind and I cannot in clear conscience do this to someone I have been so close to and I know has trusted me, without putting them in the picture."

Steve felt a shiver running down his spine at the way the word "terrible" was said, as Rick continued.

"I know what I've just said will not make any sense to you, so what I've decided is that I will need to tell you now about some, but not all, of what the true situation is. At the end of this, the decision will be up to you. If you wish to have longer to decide you can have till tomorrow when we have the announcement scheduled. If you don't want to take up the post, you can forget everything I will have told you and of course you will be able to continue in your current job. I would then continue in my post until alternative arrangements can be made."

Rick paused, as if summoning up the strength to say what he had to say, and then continued.

"As you know, the deep space programme and the current mission especially, has a great deal riding on it, not just because of the state of our planet which necessitates us finding new territories and resources in order to continue viably, but also because it represents hope and a future for humankind, without which civilisation itself could be undermined. Although your main responsibilities lie elsewhere, your were also part of the meticulous preparations we have been carrying out for this operation over many years, which of course has had to take place exactly at this time because of the particular conjuncture of planetary positions needed to provide a boost to the spacecraft and make the journey possible. A week before the launch, as was the main and almost only item in the news, there was the freak accident at the launch site, an explosion which destroyed one of the propulsion units. As reported, through a massive and coordinated effort we were able to replace this and carry out the necessary repairs so that the launch could go ahead as scheduled."

Steve butted in, "I know that this was a time of incredible stress and pressure on you, which you would not wish on anyone else. But I think I have proved myself over many years in smaller but still critical emergencies and I feel ready and willing to take on the responsibility and whatever goes with it."

Steve wasn't sure Rick had noticed what he had said, except for a deep sadness in Rick's eyes as he continued. "But there was a tragic consequence of what happened in the accident. Something which we couldn't report. Not with the universal expectations and hopes which were dependent on the mission."

Rick appeared very agitated, so Steve felt he had to come in. "Look, I spoke to Ryan earlier today, and he told me everything was fine. So whatever's the problem, we've been able to get over it."

Rick was looking down and shaking his head from side to side. "No, you don't understand Steve. Things are not fine. Ryan is ... Ryan is dead. He was killed in the accident."

Steve inadvertently let out a nervous laugh. "No, that's crazy, I've just spoken to him—met him." For a moment Steve wondered whether this was some kind of bizarre exercise devised to test him. But the anguish he saw in Rick's eyes as he looked up told him otherwise.

Rick stood up and started pacing up and down, turning to Steve from time to time. "I will have to tell you some things about which no one but I know. Remember that you have the choice. You can forget that you ever heard this and continue as you have been doing up to now. Ryan was seriously hurt in the accident. We played this down and said his injuries were minor. We sadly couldn't save him, but I put into place a deception so that even the medical staff involved didn't know this. For you to understand how this came about, I need to give you some background. To take you back about seven years when we were carrying out a mission that was one of the precursors to the deep space programme. It was just a one year voyage, a solo trip like the current one."

"I know the one you mean," Steve interjected. "Li Ming was the captain. There was the unforeseen meteor shower which damaged the ship and killed Li."

"That's the story we put out," Rick replied. "I know you were sceptical at the time, but had your own missions to look after and didn't have the opportunity to check further. Of course with the technology we have a meteor shower is never unforeseen. Li deliberately steered the ship off course."

"Why on earth would he do that?," Steve said incredulously.

"That was something that we couldn't predict unfortunately, whatever the technology we had," Rick replied. "Li's wife was suddenly taken very ill. It was not known how long she would live. When Li found out he was clearly distraught. He decided he would abort the mission and return to earth. As you know, a spaceship's course is almost entirely automatically determined, both through on-board systems and from mission control here. But especially because of the time lag involved in communications from earth, the ship's captain is given the ability to override this in an emergency. Li used this, and even though the ship's systems warned him of the meteor threat, decided to take the risk. By the time we were able to take control from him the ship was already badly damaged and Li was dead. In the light of what had happened, our political administration tasked me with coming up with contingency arrangements to deal with situations like that in the future. Particularly in the light of the extreme importance of the missions now taking place, it was emphasised that anything similar could not be envisaged. They did not wish to know how I was going to do it, just that a foolproof system was in place. I had effectively limitless resources at my disposal."

Steve looked at Rick pacing the floor and could almost feel the weight on his shoulders as he continued.

"A project was set up. Project Janus. Yes, after the two-faced Roman God. Many of the technologies we used already existed. For instance you'll be aware of one we use for the Holodeck. We are still largely constrained by the speed of light for communications. A conversation over distance would thus increasingly have an unacceptable time lag. As you are aware, we partly mitigate this through predictive thinking. When we speak to one of our astronauts in space, we have at our end a computer model of them and they one of us. At the same time this can be modified on the fly, not just through our conscious thoughts and words but also through the Brainlink we are set up with. This system brings together the most sophisticated technologies we have, originally developed mainly for medical purposes. We can interface the brain to computers, some of them nanocomputers which are integrated into the brain, but we can also use biological systems, neuron clusters we can create from stem cells, which become effectively an extension of our own brains. This originated as you may know with work on animal brain cells used to control robots. There are limitations of course, some which we will overcome through future development and some which are intrinsic, due to quantum effects for instance, but we are increasingly able to connect brains together to enable enhanced communication and collective thinking to a remarkable degree."

Senior personnel involved in the space programme had the Brainlink fitted as standard. Steve had been a little reticent when he had to undergo this procedure, but it was completely unobtrusive physically and in time he took its effects for granted. It meant that during a Holodeck conversation, for instance, he knew that his brain was being probed and was thinking about issues which he hadn't necessarily initiated. This would largely be subconscious, but then occasionally it would surface, and feel like a brainwave. Although longer time lags were more difficult to facilitate in this way, it could be done and the process then could carry on during sleep, where it could surface into dreams.

"Our use of Brainlink technology was originally intended for communications applications," Rick continued, "but of course even before the space programme's involvement with it, rather more disturbing variations of it were being envisaged and then developed. The functionality of the link could be extended so it could not only monitor but also control. Project Janus developed this much further. You can see how this could work in a case like Li's. The system monitoring his brain on the spaceship could detect what was motivating him to determine whether it was in response to a real emergency and override his actions if it wasn't. After the inevitable time delay, our long-distance communication system with our intervention here, could then refine and extend this."

"Yes, but that would still assume that Li was alive," countered Steve.

Rick's face contorted, as if he was in pain. "Originally we had no intention of going beyond this. But the terms of my brief required that the work we did be extended further. Some of this also came from Holodeck technology. As you know, there are many ways that we can reproduce physical sensations, such as when you shake hands. We can shape pressure and other waves. It can be through wearable technology, haptic devices which simulate what you would feel if the holographic image you are interacting with were real. A more sophisticated way that has been developed is to use Brainlink like techniques to directly stimulate the body's sensory nerve network at the brain. A different approach is to create something physical locally that simulates whatever is being interacted with. For many things we can use ultra-high speed 3-D printing and technologies. But with humans we need to do something different".

"You mean, like using a robot of some kind," said Steve. He noticed that Rick reacted nervously to this, as if he was not sure how to reply.

"In my brief for Project Janus," Rick continued eventually, "was included the scenario where in Li's case, for instance, he had been killed or injured very badly. Of course we could conceivably still control the ship remotely if the time lag was not too large or alternatively through local automatic systems. But we have a manned space program precisely because we need humans to carry out a myriad tasks, such as exploring some astronomical body or even meeting alien life forms, for instance. The project investigated and developed a spectrum of solutions. A robot as you have mentioned is a possibility, and of course we already use these for a number of purposes in space. But in the public's perception apart from anything else, this would appear as if the mission had failed. So we looked at ways that an existing human could be ... repurposed shall we say, or where this was not feasible, could be ... reproduced."

"So what you are telling me," Steve said incredulously, "is that unbeknownst to the world at large, in place of the individual concerned you substitute some ... thing, which appears to be a human, but is in fact some technological concoction, a cyborg or robot."

Rick held his arms up in submission. "Physical damage or disability we can largely fix nowadays through medical techniques or prosthetics if necessary. Over a period of time, through the Brainlink, we collect an enormous amount of information about an individual's brain. Where someone's brain is largely intact but dysfunctional in some way or in some area we can use this information together with AI methods and computer technology and meld it with their existing brain functionality so that the change that takes place is minimised. The case of Ryan was extreme of course. The first time we had to start from scratch so to speak. But I had no choice. To deal with scenarios like this we had developed techniques which in other circumstances would be proscribed. We previously had created what was effectively a physical clone of him for eventualities like this. In this case a combination of biology and robotics for his body and the Brainlink information and nanocomputer infrastructure had to effectively substitute for his brain. Even in this case I know he fooled you on the Holodeck, but he would also have fooled anyone in real life. There was just so much, so many people, depending on him. Depending on us and our space programme. He was just indispensable for all our futures." He looked imploringly at Steve who was momentarily lost for words.

Finally Steve asked, "These people, you're doing this to. What do they feel about it all?"

Rick reflected before replying, "I suppose there's no real way of knowing. Consciousness, what we feel and think, is subjective. But we have done a lot of tests with people who have had various serious impairments and we have used these techniques with. Of course neither they nor the medical researchers and staff involved had any idea what the underlying objective of this work was. After perhaps a brief period of confusion, the subjects almost universally don't feel that anything has altered with themselves. Even where something had changed substantially, they just felt as if they had acquired this naturally, as if they had learnt a new skill or were seeing things in a new way. Of course as we all know our body is effectively completely replaced over a period of some 7 to 10 years. We don't feel we are a new person because of this. It's as if we had just speeded up this process but kept whatever is the feeling of self intact. Even in the case of Ryan, where all we could use was the Brainlink data and as much information as we could find elsewhere, he still seemed to be who he had been before. But there were gaps. There was no record of his brain activity when he was a child for instance. In this case we used Brainlink data from comparable children who had undergone this set up as part of medical procedures. We used this modified and filtered in various ways based on the information we did have about his childhood. So he is a curious hybrid, but in the intense work which we only had a week to do before the launch, he still seemed to get back a cohesive sense of self."

"But the sense of which self," Steve butted in angrily, "you're talking about creating a new person in your image, or an image which you have decided on. They not only have no say in it, but they may not even know that it's taken place. You have made yourself God, but not even a God which people can accept or not, but a God by stealth."

Steve sat for a while trying to take things in. He finally asked, "Who knows about this? Is there any record about who is affected?"

"The whole project was carefully designed so that those taking part each only saw a particular aspect of it and thought they were working on something else. I am answerable to a small political subcommittee but they expressly don't want to know who's affected. The only record that exists for those involved in the space programme is a special status code in their restricted personnel file, which only I know the meaning of. It's the number '66', which doesn't appear in any other context. Sole responsibility effectively lies with me, and it would with you if you took over. That's why I wanted you to be able to make an informed decision knowing the true situation."

"Not quite the number of the beast," Steve said grimly. "I will give you my decision at 10 tomorrow morning," he said as he left.

3.3 *A Question of Self*

Steve spent the rest of the day alone in his office thinking about what he had been told. It was quite late when he returned home. He declined to answer any questions his wife Freya asked him about the day.

"There's an important decision I have to make by tomorrow morning," he said. "I just need to try and get some sleep." He lay in bed the events of the day repeatedly playing through his mind. Rick has said that he wouldn't tell him everything till he had made his decision. So what was it that he still didn't know? There was something that Rick had mentioned that he knew was significant, but he couldn't think what it was. It was many hours before he fell into a fitful sleep. Strange dreams filled with bizarre images and beings kept going through his brain. Suddenly he woke up. He knew what had stuck in his mind during his conversation with Rick. Rick had used the word indispensable when referring to him, the same word he had used to describe Ryan. A sudden and frightening realisation hit him. Was he who he thought he was? They could have just used the Brainlink he already had and extended it. He had been involved in a serious car accident about three years previously. While he was in hospital for this would have provided the perfect opportunity.

He hurriedly woke up Freya. "Have you noticed any difference in me lately?," he asked her.

"You mean other than waking me up in the middle of the night asking stupid questions," she said sleepily.

"Some significant change, perhaps after I had the accident," he persisted.

"Not really," she replied, "you're probably just as annoying as you always have been. What's this all about? Go back to sleep."

Steve waited till she was asleep. He knew how he could find out what he needed to know. He got out of bed and sent a text to his friend Jamal. Jamal headed the IT section of the space administration. Steve had known him since their schooldays.

Steve was waiting for Jamal when he got into work at 7 AM. “I need a big favour, Jamal,” he said, “I need you to run a search on the restricted section of the personnel records.”

Jamal was dubious. “Hey, I know that you’re going to be the big boss very soon, but even then there are various approval procedures I have to go through.”

“I need this very quickly,” Steve replied, “and it’s very simple, I just need to know whether my record is correct. I need you to look for a number, ‘66’, in my status codes.”

“Well,” said Jamal, “I guess I can make an exception in this case then. But this is certainly going to cost you in terms of favours you owe me.”

At 9.30 AM Steve was in Jamal’s office.

“I ran that search you asked for,” Jamal said, “but there’s no ‘66’ in your status codes. What’s it supposed to represent? Even I don’t know what the codes in the restricted sections mean.”

Steve let out an audible sigh of relief. “It’s nothing important,” he replied, “let’s say something would have been wrong if it had been there.”

“I can’t recall seeing that code before so I ran a search for it just out of interest. It does come up, but it’s extremely rare,” said Jamal.

“I guess that was Ryan,” said Steve.

“Yes” said Jamal, “You obviously know about this, so I guess I’m not giving away any secrets. I presume it signifies some special status, which I suppose you won’t get till you’re officially appointed as Director. I figured that out because of the other ‘66’ it came up with—Ricardo Gonzalez.”

3.4 The Decision

At 10 AM sharp, Steve entered Rick’s office.

“I know who ... or what you are,” he said.

A look of relief came over Rick’s face, as if a great weight had been lifted off his mind.

“I would have told you anyway but I’m glad you found it out yourself. Let me try to explain how all this came about. When I had my stroke I had a terrible decision to make. Now I know that I have a possible successor who I can be confident in, but I felt you weren’t ready then. At that time, my views about using technology in this way would have been very similar to what I think yours are now. I was fully conscious, but scans showed certain parts of my brain had been affected. Even if I recovered, I didn’t know whether I would be able to continue in my role. But I faced an awful dilemma. I felt the future of the space programme and all that it entailed would be jeopardised just because of some abstract moral principle on my part. Believe me the choice wasn’t easy. Without anyone knowing that it was me that was being treated, I had the process carried out that had been developed for astronauts. I have deduced that you found out my secret because you suspected you might have been one of the ‘66ers’ and investigated accordingly. When you had your accident and I thought I might lose you I must confess I did seriously consider taking this course of action with you for similar reasons I had done for myself. But one of the reasons I didn't was that in a way you represented my chance of redemption. My decision to retire was partly motivated by my desire to put someone without my ... dubious pedigree, back in charge. We'd had no major problems at all with missions after Li's one. The last one with Sophia made me think that we had dealt with all the potential issues. And then Ryan's accident took place. Again I felt what I did was the only way out. But with that decision came the guilt, the guilt that I would be handing over that terrible secret and the responsibility arising from it to someone who trusted me and I was so close to. Nothing can absolve me from what I've done. The decision whether you wish to accept some of that responsibility is yours Steve, you mustn't feel any pressure from me as to how you decide."

Steve sat motionless for a while, his mind in turmoil. Finally he spoke. "I hadn't made my final decision till now. But I will take the job. I acknowledge what's happened cannot be undone. Maybe faced with the choices you had to make I would have acted similarly. I don't know. But as I think you feel as well, I believe I can take responsibility without feeling compromised in some way as you do, whether that's actually the case or not. Perhaps I can work towards a situation where there are no more '66ers', where perhaps all of humanity has to collectively take the decisions, however difficult they might be."

For the first time since he had seen Rick the day before Steve detected something else in his eyes, a feeling of hope, even optimism. A brief smile crossed Steve's lips. Maybe the Rick he had looked up to and respected was still there. As Steve left the room he felt his confidence and purpose return. He would live up to the expectations that the Rick he knew had shown in him over all those years they had been together. He would grasp the opportunities, but also face up to the responsibilities of knowledge.

4 Conclusion

The ethical and moral issues involved with the ability to effectively create life is a perennial topic for fiction. 2017 is the 200th anniversary of the first publication of Mary Shelley's Frankenstein, which explored that issue in a different age. In that story, Victor Frankenstein who creates the monster is not evil, but a scholar and researcher, although unconventional in his outlook and methods. It is his quest for knowledge that has tragic consequences and is his undoing.

In today's world the development of knowledge and technology as part of it is intrinsically embedded in society and the economy at every level. How aspects of this can have unintended and sometimes uncontrollable consequences is a problem that frequently arises. Unlike in Mary Shelley's time, the capabilities of doing what is described in the story do not lie too far in the future, and in basic form already exist, as discussed in the paper. So we could say we are moving towards a period where both the motivation as well as the capability for the type of scenario depicted in the story to arise is not unrealistic.

The technologies discussed in the paper are ones which have important and positive contributions to make. As indicated, many of them have arisen from extremely beneficial developments tackling problems in the areas of health and disability. Even the space travel scenario of the story is not implausible and in itself could have positive connotations. Undoubtedly the importance of how technology is organised and controlled is a key issue which the story highlights. Technologies that arise in the story and are discussed could of course if suitably directed also be used to enable collective thinking and democratic discussion which could help to obviate some of the dilemmas that arise. But one of the key conclusions of the story must be that there are no easy answers and the future will increasingly pose complex ethical questions related to technological developments. Hopefully as well, the story will help to demonstrate that the methodology of Holonovels and Science Fiction Prototyping is an important way to help facilitate wider engagement with these difficult issues and foster the discussion and debate which is necessary.

References

1. B. Clegg, *Ten Billion Tomorrows: How Science Fiction Technology Became Reality and Shapes the Future* (St Martin's Press, New York, 2015)
2. TED. Alex Kipman: A futuristic vision of the age of holograms [Video File] (February 2016), https://www.ted.com/talks/alex_kipman_the_dawn_of_the_age_of_holograms
3. L.M. Krauss, *The Physics of Star Trek* (Basic Books, New York, 2007)
4. M.E. Lasbury, *The Realization of Star Trek Technologies: The Science, Not Fiction, Behind Brain Implants, Plasma Shields, Quantum Computing, and More* (Springer, Berlin, 2016)
5. B. Long et al., Rendering volumetric haptic shapes in mid-air using ultrasound. ACM Trans. Graph. **33**(6), 181 (2014)
6. M.A. Lebedev, M.A.L. Nicolelis, Brain-machine interfaces: From basic science to Neuroprostheses and Neurorehabilitation. Physiol. Rev. **97**(2), 767–837 (2017)
7. N.C. Rowland, J. Breshears, E.F. Chang, Neurosurgery and the dawning age of brain-machine interfaces. Surg. Neurol. Int. **4**(Suppl 1), S11 (2013)
8. K. Warwick, Creating and controlling complex biological brains, in *Complex Systems*, (Springer, Cham, 2016), pp. 141–156
9. J. Clausen, Ethical implications of brain–computer interfacing, in *Handbook of Neuroethics*, (Springer, Amsterdam, 2015)

Imagine Thunder

Jennifer O'Connor

1 Introduction

The short story entitled *Imagine Thunder* offers a descriptive world view in which both future and past timelines overlap to affect the happiness of an individual central character. With the ever-increasing prevalence of mental illnesses in the world's population and virtual reality (VR) technology being made accessible by commercial brands, a likely sub-link could be made between the two to assist the full or partial recovery of certain mental illnesses. Primary research into the use of VR in the treatment of mental illnesses has been largely concerned with anxiety, schizophrenia, and eating disorders [1]. However, this story describes a boy's struggle with depression and the possible benefits of VR for treating such an illness. The protagonist is an 11-year-old boy who has been placed in an experimental research facility to treat his major depression, which he has suffered from for the past several years. The story describes a state of depression in which passion has been withdrawn by the patient from all day-to-day activities, except the engagement of the imagination in the form of fictional reader. This is used as a method of escapism. In describing a character who suffers from absolute numbness, the importance of emotive story-telling is underlined.

"Imagine Thunder" is a story inspired by the potential of technology to treat a variety of mental illnesses. Although this story primarily focuses on youths, the imagined technology described could be applied to any age group. The advances made by VR technology in recent years have been exponential, along with the increased use of VR technology commercially. Should these developments focus even partially upon the health sector, the premise of this story may well be recog-

J. O'Connor (✉)
Creative Writing Department, National University of Ireland, Galway, Ireland

A. Reyes-Munoz et al. (eds.), *EAI International Conference on Technology, Innovation, Entrepreneurship and Education*, Lecture Notes in Electrical Engineering 532, https://doi.org/10.1007/978-3-030-02242-6_24

nised. It is not unimaginable that VR technology would be distributed commercially to large-scale government-run institutions such as hospitals. The technology itself described within the story functions by manipulating an individual's entire brain and full centre of intuitive thought to re-create one's most positive fantasy. The entire brain is incorporated when using the imagination, with different regions (front and parietal regions) more involved than others [2]. The VR technology described within the story strives to utilise every brain region. A fantasy is deemed most positive by the quantity of dopamine released. The premise of the story is to not only create a palpable reprieve from conditions such as major depression and general anxiety disorder, but to be used at short intervals to aid recovery, where recovery is possible. In "Imagine Thunder" the protagonist, Henry, begins his recovery with the help of the VR technology which raises his dopamine levels sufficiently for him to gain a renewed interest in all activities which he partook in before his depression became too severe.

The story is written with the background knowledge of a writer, rather than a technologist. This can be seen through the character development of Henry and improvement of his mental health overtime due to the capabilities of his imagination. In the beginning, he deems to world to be "too heavy" for him, due to the weight of his imagination. Yet it is the positive manipulation of his imagination that results in the beginning of his recovery, which is inferred to be replicated by other patients. Rendering to the principals which all science fiction prototypes [3] are built upon, technologists and writers combined hold an ability to look beyond the current restraints of technology. The story serves the purpose of widening the considered techniques of treatment for mental illnesses and encourages the use of "re-emotion", as termed by a medical professional in the story, by use of technology. "Re-emotion" is described within the story as the re-introduction of emotions which a patient may have previously suppressed, as a tool for recovery. This technique may have great all-round benefits for millions of humans.

2 Imagine Thunder

The long hall with its white tiled ceiling and bespeckled laminate floor stretched out in front of Henry like a plank to be walked until he reached his untimely end. Dr Justineau walked beside him, though slightly ahead, still leading the way to some mystery location. Henry observed that Dr Justineau's heels clicked and clacked exactly 28 times until she, and then he and the nurses following them, came to a stop outside the clinic's Community Room. She opened the door and nodded for him to go inside with a cheery smile.

The Community Room was Henry's least favourite room in the clinic. It was where he was destined to find other children like him who were more willing to share their own stories than he was ever likely to be. Once inside, Henry sat in the chair closest to the door. White, once again. Everything in this place was white. Once he took his seat, there was only one seat in the Community Room's sharing circle left unoccupied. This one was black—for Dr Justineau. He stared at it with great resentment before sighing deeply and focusing his energies upon actively avoiding the gazes of the 12 other children.

Henry had been in *Castle Countess Clinic for Socially Estranged Children* for nearly 5 weeks. Being only 11 years old, he had no say in his mother and father's decision to send him to the experimental clinic for treatment. In truth, he hadn't quite seen the necessity for such a drastic action, but then again, his mother was prone to overreaction. In Henry's mind, his Great Sadness had a blurry starting point. All he knew for certain was that now every minute of his life was marked by an intense weight on his chest that seemed so heavy that his sternum might crack open, allowing the heaviness to crush his heart. The very *air* seemed too dense for Henry and living had become much more difficult than the simple trifle it had once been of school, friendships and growing up. Now, each breath cost him something deep and dark that he couldn't explain to his parents. The one benefit of *Castle Countess* was that Dr Justineau seemed to not so much understand this feeling, but to at least recognise it. And now he was surrounded by other children of different ages whose chests also weighed so much that they might break too.

Time seemed to move as though in a vacuum for Henry as they waited for Dr Justineau to enter the room and tell them the purpose of this impromptu meeting. When he finally looked up from his lap, he found a reflection of himself in the other children. There was a distinct deadness in their eyes and a vagueness to their gazes that he had seen in himself. This knowledge flooded him with black misery.

With the distinct squelch of the door unsticking from its frame, Dr Justineau entered the room. Her dark hair was tied back into a sleek ponytail and her half-moon glasses were perched on the bridge of her nose. When she took her seat in the black chair, her trousers rose to show a silver anklet.

"Hello, everyone", Dr Justineau said, smiling to everyone in turn. As expected, the children all stared at her balefully. It wasn't that they didn't like Dr Justineau, but it was hard to have a different answer for the same question that was asked more times a day than Henry could count. "How are you all today?"

No one spoke except for a girl who looked to be 15 or so. Henry remembered that she was called Ana. He had always liked the look of her, simply because the ends of her hair were dyed pink and she had a topaz gemstone in her nose.

Ana's answer was a derisive snort, which Dr Justineau dutifully ignored.

"I wanted to speak to you all together, to tell you about something which I think you will find very exciting", she continued. Henry thought that it was a wonder that she was a doctor for depressives at all, considering she still thought something could excite them. "We have been privy to an exciting new experimental treatment option, for your conditions", she paused once more, as if waiting for an influx of enthused questions that was not likely to come, and said, "Using virtual reality."

2.1 Virtual Reality

Those two words had once meant something very significant to Henry. He could tell by the leap of recognition he felt behind his navel. As Dr Justineau explained the experiment, he found himself remembering aspects of interest he once had for technology and though it had been years since he had felt passion for anything other than reading, he remembered the words Dr Justineau was saying.

Immersive. 4-dimensional. Interactive. Mixed reality. Spectacular. Engagement. The Dome.

Then Dr Justineau added one last important one. *Recovery.* This project must really be important if she was whipping out the big R.

By the time, Dr Justineau had finished speaking, Henry had only heard two words that really mattered to him. *Immersive stories.*

Henry had been 6 years old when his mother had given him his first book to read without her help. It was called *Imagine Thunder*. It had been an adventure-filled tale of dark magic, kings, and a very dangerous princess. There had been eight books in total in the series and Henry had practically inhaled each one. They had unearthed something in him, a need for something much greater than anything this life could offer which even his depression had not taken. It was this need for a stimulated imagination which Henry often blamed his depression on. Had he been the type of person who did not *need* the extraneous details in everything, Henry might have been happy. As, unfortunately, when one is searching for details, what you find is not always good. Now he could not erase the details of his parent's relationship, his failing friendships, or his own personal failure from his mind.

Without waiting until Dr Justineau was done explaining the experiment, Henry found himself raising his hand to take part in the first experiment within what was being called the immersive Dome.

Henry was one of the only children lucky enough to have his own bedroom at *Castle Countess.* He could not figure out if he was either not sick enough, or too sick, to share a room with some of the other patients who had not only depression but other illnesses. He was perfectly okay with this, as it meant that the side of the mint coloured room that should have housed another bed now held a large, rectangular bookcase, courtesy of his godmother who visited more often than his parents.

His days were filled with an overriding sense of blandness. Like pasta without sauce or the colour magnolia. Except for his stories. Except for the window, they offered him, to leap free and experience kingdoms and fights for something that *mattered.*

That all ended when Henry began the experiment with the Dome.

The room with the virtual reality technology was not on the *Castle Countess* grounds, but rather in a lower room of the neighbouring hospital. Henry, Dr Justineau and a plethora of junior doctors keen to observe, travelled in a van. Outside the weather was overcast with several drops of rain dotting the windscreen, but Henry was grateful for the change of scenery.

The hospital's basement rooms were cluttered with new and old machinery, but were reasonably bright and well-aired. The hallways were littered with lab-coat wearing professionals who smiled and nodded to Dr Justineau as she and Henry walked towards the room at the end of the hall. As they all filtered inside, Henry paused by the door, overawed by the great expanse of the room. Dr Justineau stood with him at the door as others set about turning on switches and dials. She watched his expression move from awe to reservation and smiled a little sadly.

"This won't be difficult, Henry. We know from your art therapies how sharp your imagination is. The Dome uses imagination as a means of creating the next steps in the experience", Dr Justineau touched Henry's shoulder lightly and he looked up to meet her gaze. "The turns the simulation takes depend on you. You will most likely visit the scenario or fantasy world you most frequently imagine. The Dome functions by measuring your dopamine levels. If they drop below your current baseline, we will pull you out. There is no wrong route, I promise you."

Despite her reassuring tone, once Henry allowed the small wireless receptor pads to be stuck into the insides of his palms and at his temples, he felt unsure of what kind of fictional world his imagination would conjure up. He had to keep reminding himself that he was doing something new and that was what mattered. So, it was with small beads of sweat escaping his forehead, that Henry entered the dome-shaped hut of interwoven plastic walls and wires that took up much of the free space. It was like wattle-and-daub reimagined into something bionic, something *alive*. For Henry, it positively teemed with possibilities.

Dr Justineau instructed him to stand in the centre of the Dome through a microphone connected from outside. Once in the middle of the Dome, Henry looked around waiting with uncomfortable flutters in his stomach for something to happen. A man whom Henry had never met before took the microphone from Dr Justineau. His voice was like gravel, as he explained the sensors on his palms would take a minute to trigger his imaginings as they engaged all aspects of the brain. It was like a controlled hallucination, he said. He was encouraged to imagine his most happy fantasy. Soon, the Dome before him would become like an extension of his own bloodstream, feeding, and thriving off his hormones and emotions to reflect his images back to him. Before long, it would seem to Henry as though he was a member of that imaginary world as his senses of touch, smell, hearing, sight and even taste would be overridden with the influx of images his mind would be flooded with, creating a new reality. The doctors would not be able to see what he did, but would monitor his responses carefully and he would later be asked to describe it to them.

It began with a ripple. The very air in front of Henry seemed to shake and then it all changed in a single blink. When he closed his eyes in the Dome he had felt uncomfortable, surrounded by doctors observing him as though he were a millipede under a microscope. Yet when he opened his eyes, he was no longer the sad boy whose parents worried for him.

Now he found himself placed directly into the world of his favourite book series, *Imagine Thunder*. With every second, the knowledge that this was not reality fell away with the smell of burning wood that filled his nostrils and the gentle swish and sway as he sat astride his steel grey warhorse, Abraxas. He was no longer Henry; instead he was the captain of the Kingsguard. The youngest in a century and valued by all in the southern Kingdom of Windhart.

Where he sat astride Abraxas upon a great grassy hill, Henry recognised as the border between Windhart and the Slumbering Wood. From where he sat on his high vantage point, he saw that the forest was a vast thicket of dark trees and on the other side, according the myth, was the northern kingdom of Ellesmere, though there had been no contact between the two kingdoms in over a 1000 years. The children of the kingdom often spent days squinting into the distance to see if they could spot the turrets of Ellesmere palace, but they never could. All that remained to be seen was a deep and dense fog which no one dared enter for fear of never returning. The most to have exited the forest in some millennia was no more than a rabbit or two.

The day was bright and crisp, as everyday was in Windhart. It had been over 100 years since they had had a rain shower. In Windhart the earth was watered through the calm energies of the Empaths. Empaths were great people connected to the earth who spoke to the shrubs and plants alike, encouraging growth and bountiful harvests. Almost everyone in Windhart was an Empath, but not all were as adept as Henry, who never lost control of his emotions.

Henry looked behind him, turning Abraxas slightly, with one hand resting of the hilt of his gilded steel longsword. He was not far from the entrance of the town, perhaps no more than a half mile. The woods that stood behind him could not have been more different than the Slumbering Woods. These woods were sparse and inviting with a cobbled stone path carved out leading to the main throng of Windhart. Somehow, Henry knew that stationed no more than 100 yards to his left was another Kingsguard, maintaining the King's peace.

He sat upon Abraxas for a few more minutes, playing with the pommel of his longsword nervously. He could not explain why, but an overwhelming sense of foreboding was beginning to settle upon him like a dense fog, despite the bright and cloudless day. Henry was just considering withdrawing his longsword for peace of mind when he caught sight of something terribly ominous in the distance. He placed a hand over his eyes to shield his view as he watched the edge of the Slumbering Wood. His breath caught in his throat at the sight of the torrential downpour of rain which seemed to be sweeping over the forest with alarming speed. He barely had time to pull away from his post before he heard the King's Horn booming from atop of the palace walls. Surely the castle guards, so much higher than he was in the town, should have seen this coming? There was no way that he could outrun this storm, though some of his fellow Kingsguard seemed to be trying to as they tore back up the hill and into the town. No one alive in Windhart had ever experienced anything but beautiful, calm weather. This storm seemed all kinds of *unnatural.*

Abraxas was beginning to shake nervously and step backwards in the direction of escape. Though for some reason, Henry could not tear his eyes away from the treeline. Hoping he would never have to explain his actions to himself or anyone else, he leapt from Abraxas. He slapped the horse's back and screamed at him to run, though the din of the approaching storm was beginning to drown out all else.

The horse's amber eyes questioned him for only a moment before taking off and with a hand on the pommel of his sword, Henry faced the oncoming storm.

The first drop of wetness on his face was not as he imagined. It *hurt*. This rain was hard and frozen into small balls that rained down on him, bouncing from his tunic and onto the grass where they gathered to form a new type of earth. He resisted the urge to cover his ears, and peered out towards the trees. A thrill of mysterious anticipation rushed through him. When the rain did not seem as though it could get any harder or faster, he saw the flurry of colour burst out through two trees.

It was a *girl*. Astride a jet-black horse who was neighing furiously.

From the moment he saw her he was sure that she was the centre of this storm. That moment didn't last long, however, as the horse suddenly reared onto his hind-legs and back down with such force that the girl was flung from its back. He ran to her and found that she was conscious, but crying fiercely. When she looked at him as he touched her back ever-so-gently, he saw that she was no older than he was, perhaps even younger, with short blonde hair that was now pressed wetly against her head. Her eyes widened at his touch and he thought he felt an almost imperceptible lightening in the falling hard rain. Then, with wide golden eyes, she reached out and took his hand and as though he was a potion for calming nerves, the rain slowed into a light drizzle before turning *soft*. Henry gazed at it in shock as the water drops froze in the white flecks. He had heard this fable before. This was what was known as a snow flurry.

"Who are you?" He asked, as her grip on his hand tightened. She was using his Empath abilities to control her emotions, as his power flowed like fire from his hands to hers. She watched their hands with an intensity that caused him to wonder if she had forgotten that he existed as more than a hand. She did not answer, but when she looked up and he saw her strange, golden, and green eyes once more, he knew. He'd heard the fable of those eyes too.

She was the Crown Princess of Ellesmere, the forgotten kingdom.

It had been almost 9 days since Henry had entered the Dome and lived in the world of *Imagine Thunder*. Nine days since Dr Justineau had withdrawn his senses from his imagination and a consistent sense of awe had settled upon Henry. Now when, he ate dinner with the other patients and went to his traditional talking therapy he did so with a little more awareness for his surroundings than he had done before. Though he surmised that it was less of an awareness and more so actual interest. The improvement, though he was unsure if it was even an improvement, was immeasurably small in comparison to his Great Sadness, but Dr Justineau and her team of gaggling doctors were thrilled when he had rolled his eyes and described this in his weekly session with her. She had explained this as a term known as *re-emotion*, whereby the emotions he had shut out and not felt for so long would be his saviour by encouraging him to engage in activities that would lead to *feeling* once more.

"How did you find the experience?" she had asked. Henry had struggled to find a word to describe it.

"Widening", he settled on. When Dr Justineau cocked a single, impeccably groomed eyebrow his eye roll had ensued. "It's like my eyes are open wider, like my senses are heightened since I was in the story. It's like I can't quite shut it out anymore."

"Shut what out?"

"Everything. You, them", he gestured a thumb over his shoulder to the door to mean the other children waiting back in the Community Room. "Now, it's like I can't fully close my eyes."

Henry didn't mention how this wasn't a bad feeling, it was simply different after so long switched off. He didn't explain how being an Empath, a Kingsguard Captain, had altered his view of himself somehow. In the Dome, he had touched someone's life and made it better because of his emotions, and not worse. And that, he concluded to himself, made all the difference.

3 Reflection

The short story "Imagine Thunder" describes a somewhat hopeful consideration into the opportunities which Virtual Reality offers those suffering with mental illnesses. In doing so, one can see the direct benefit of drafting science fiction prototypes. It is through stories such as these that human–technology interaction (HTI) can be given new possibilities for utilising human emotions to create more positive and emotive experiences. Science fiction prototypes offer an insight into the future of humans and technology from all points of view as either the user, developer, or an observer.

From the point of a view of the writer, the purpose of science fiction prototypes such as Imagine Thunder is the draw attention to a desirable outcome which moves beyond the simple motives of entertainment towards improving human health. Heading and eye tracking through sensors are essential, yet Henry's technology goes one step further by monitoring his hormone levels, namely dopamine. Through these tracking techniques and haptics, the protagonist gains the tactile feedback of the emotions he gains and imparts in his imagined world, which appear so realistic that they stay with him for long after each session. In saying this, it is very important that VR is not used exclusively as the only means of recovery from a mental illness and periods of time between using the VR are restricted so that the user has a firm grounding in what is real and what is not.

The strength of emotion within the story is seen through Henry's initial numbness and later heightened awareness of the happenings around him. This simple contrast is done for the purposes of highlighting the role of technology as both an emotive and functional contributor to mental illness recovery. The advantage of the story's science fiction prototyping technique can be seen in the description of awaiting technology which is entirely plausible regarding development. Throughout the story only one primary form of technology is shown as the Dome, mirroring the single-track mindset of those suffering with a mental illness that dampens awareness. The story may serve as a conversation stimulator for writers, educators, engineers, technologists, and business people to look to the possible positive outcomes of using VR technology within healthcare.

References

1. D. Freeman, S. Reeve, A. Robinson, A. Ehlers, Virtual reality in the assessment, understanding, and treatment of mental health disorders. Psychol. Med. **47**(14), 2393–2400 (2017)
2. K.H. Pfenniger, V.R. Shubik, B. Adolphe, *The Origins of Creativity* (Oxford University Press, Oxford, 2001)
3. B.D. Johnson, *Science Fiction Prototyping: Designing the Future with Science Fiction* (Morgan and Claypool, San Rafael, 2011)

Holonovel: Perspective on Enactive Narrative Intelligence

Jelena Rosic

1 Introduction

Fictional and speculative design scenarios can precede scientific and technological developments. I take this position to be conceptually and methodologically more plausible than presumed within disciplinary boundaries and methods as merely inspiring indication. In the following perspective on Holonovel as interactive and intelligent (new media) system based on narrative computation, a framework for understanding the underlying principles of human-level narrative comprehension is drawn. It is argued that *narrative should be used as interface for human–computer interaction* relying on humans' practices that often make sense of the world in narrative terms. This leads to questions on narrative as sense-making of the world—how perception and cognition interact with the world, on one hand, and what constitutes narrative intelligence in a computed (artificial) system through which our experiences are mediated. In the paper, theoretical framework of narrative skills and their relation to cognition is considered as direction for research and design of computational mediated experiences and for re-conceptualizing modes of interaction and how we use tools and interfaces.

Rather than putting all creative or economical efforts into optimizing technological advancements, whether in fictional or scientific scenarios, we can try to identify new approaches to interaction and narrative as our engagement with the world that has always co-existed with the tools and technologies and try to understand "how models of thought are intimately tied to the available models of computation" [1].

J. Rosic (✉)
Aalto University, Helsinki, Finland

A. Reyes-Munoz et al. (eds.), *EAI International Conference on Technology, Innovation, Entrepreneurship and Education*, Lecture Notes in Electrical Engineering 532, https://doi.org/10.1007/978-3-030-02242-6_25

In this line of investigation, one area of technological research that has been open to approaches other than computer sciences and that uses ideas from arts and humanities in actual developments is the field of Narrative Intelligence:

> Narrative intelligence refers to research into human narratives and story-telling, as well as the development of software or robotic agents that either support human story-telling or are themselves story-tellers and/or story-listeners [2].

Narrative Intelligence, that is, the narrative organization of experience in the domain of artificial (computational) systems is a work that is inherently interdisciplinary and in its history, relied on practices and methodologies from arts, psychology, drama, literary and cultural studies. Even though language is often considered as uniquely human characteristic and a way of communicating experience and shaping the mind, and therefore applications of narrative have been centered on this modality, non-verbal or implicit aspects such as our embodiment and situatedness occupy research in HCI and narrative interfaces as well. Following the insights on narrative from psychology, it is argued that artificial (computational) systems will be more understandable with narrative presentation especially because humans use narrative for understanding intentional behavior [3].

If Holonovel concept could serve as a reflection on our current and future technologically intertwined mind with the world, then we can relate this condition to the idea that technology is anthropologically constitutive [4] and that humanity cannot be understood without its technological dimension. This paves the way to conceptualizing our engagement with the world through the relationship between an organism and environment, and further between a user and the tool.

The key ideas on narrative intelligence in the paper that are going to be assessed from the perspective of "enaction" with the world rather that "computation", revolve around intelligence being "determined by the dynamics of interaction with the world" [1] and around computational narrative intelligence being "as much about human-computer interaction as it is about solving hard artificial intelligence problems" [5].

2 Interfacing with Reality

In the context of new media and technologically mediated experiences, much has been debated on the concepts of immersion, illusion of non-mediation or presence—the experience of "being there" in a simulated or mediated environment. The aim and scope of this paper is not directed towards conceptually clarifying these phenomena for the purposes of possible or imagined Holonovel environment as simulated and interactive environment. Rather, the framework that is presented here in relation to our interaction with technologically mediated environment departs from describing *modes of representation* in relation to experience that presuppose some "suspension of disbelief" needed to establish engagement. These discourses usually related to virtual environments and immersion tend to exhibit somewhat reductive conceptual position centered on transportation into another world or dimension where experience of altered reality is happening. Moreover, these kinds of concepts on participation in media environments or fictional settings additionally force divisions and modalities to perception as well as cognition that does not do justice to the complexity of experience (e.g., real vs virtual, belief vs disbelief, immersive vs non-immersive, low vs high order, creator vs user etc.).

Popular demands for immersion, whether defined in weak sense, for example in popular culture and marketing, or more rigorously specified for engineering purposes, tend to oppose traditional "old" audio-visual media as non-immersive due to the mode of presentation (no goggles, sensors, etc.). However, this seems to echo the same discourse found in "old" media experiences and related film studies—the idea or demand for the seamlessness of experience or illusion of non-mediation in a way that spectator should be tricked into the process and believe in the (re)presentation of reality without noticing its mechanism. In "old" media such as film, this was typically seen through the technological "apparatus" of film that selectively represents the reality through camera projection and seamless editing. This places the spectator into a passive position of meaning-making. Furthermore, it "sutures" the spectator, in a way that the spectator is stitched seamlessly into the ideological apparatus of the medium without noticing the mechanism behind it or being able to resist it (Marxist, psychoanalytic, structuralist film theories). In "new" immersive media, this stitching is performed with seamless interface or mediation transparency with technological tools while still preserving the passive position of the spectator: "Immersion in a virtual world is viewed by most theorists of postmodernism as a passive subjection to the authority of the world-designer—a subjection exemplified by the entrapment of tourists in the self-enclosed virtual realities of theme parks or vacation resorts (where the visitor's only freedom is the freedom to use his credit card)" [6].

Even in the most recent empirical studies that examine our neurocognitive engagement with audio-visual media or films in particular (neurocinematics) and previously in the theoretical framework accounting for our biology and evolutionary makeup in dealing with media tools (cognitive film studies), the perspective is still the one of a passive recipient of media input with predictable behavioral output. In other words, it builds the "Discourse of Control" over spectators' minds with techniques that predict neural or cognitive responses to (film) stimuli [7] and according to which effectiveness of media tools is measured. Relying on such discourses of spectators' engagements, our mediated experiences are described and regulated in *modes of representation* assuming passive subjection rather than *modes of interaction* with the world.

Departing from conceptualization of engagement in mediated environments as input-output processes, how can we describe or design our coupling with the environment and technological tools with interfaces that support interactive dimension of our embodied, social and cognitive life?

3 Narrative as Interaction Tool (Enacting the World)

Rather than defining our engagement with the environment (real or fictional) as modes of representations of the world, the approach that emphasizes interdependence between action and perception can be described through the idea of "enaction" [8].

3.1 *Enaction*

Therefore, instead of taking the perspective of a spectator/user/agent presented with a world, enactive relation suggests that "a living organism enacts the world it lives in; its effective, embodied action in the world actually constitutes its perception and thereby grounds its cognition" [9]. These embodied and situated notions about the structure of our engagement with the world refer to enactivism—a novel approach to cognitive sciences that departs from previous computational modes of cognition exemplified in the metaphor of the brain as a computer processing machine. Enactivism sees perception, cognition and action as: "three facets of this single process of adaptively coping with the world (what the enactive literature refers to as "sense-making") rather than being distinguishable links in a chain of processes that begin with "input" at the sensory surfaces and end with "output" at the muscles" [10].

The main idea behind agent's engagement with its environment in enactive approach to cognition and perception is that perception consists of perceptually guided action and that cognitive structures emerge from the recurrent sensorimotor patterns that enable action to be perceptually guided [8]. Enactive approach emphasizes perceptually guided action and engagement that depends on how the organism perceives the world rather than the world being described from the outside for an organism to interact with. Thus, enactive account of the properties of the world and interaction between agent and environment leaves more space for examining this relation from the perspective of agent's possibilities for action rather than describing the world as represented or modelled and then expected to be navigated.

While enactivist program has been so far widely accepted for the view on perception as embodied action (in terms of "lower level" cognition, perception, bodily action), the main criticism towards enactivist program questions its possibility to explain "higher order" cognition—thought, reasoning, planning, problem solving. As many recent research developments within the enactive program discuss questions related to consciousness, language, social concepts, mathematics etc. [9], new and distinctive articulations and methodologies apply enactive approach to complex forms of cognitions and reject sharp distinctions between "lower-level" embodiment and "higher-level" cognition.

Departing from the views on perception as a "mode of presentation" of a particular stimulus, as different sensorial modalities that represent the world, enactive account defines perception as "modes of action" where perception can be understood as "structured by the intentional actions of the agent" [10]. Thus, cognition is not added to perception after the fact, it is not an additional inferential process but it is inherent in the process of perception itself [10].

Understanding perception and cognition without hierarchy but as a unified process that does not begin with "input" of information and end with "output" of the representation of the world brings forward the idea of "sense-making" as active participation in generation of meaning between organism and its environment.

3.2 Narrative Interactivity

A particular structuring of our experiences in the world, our perceptions and cognitions, relates to our capacity to organize experience in a narrative way and to narrate this experience and interactions with the world and others.

In regard to human–tool interaction or human–human interaction in mediated forms, it is expected that humans will deal with interaction in a way that makes narrative sense and that technology (software and robotic agents) that fail to optimize human–tool relationship in a fluid way lack narrative intelligence [11]. Aspects that underlie narrative engagements and pose main questions for modelling this interaction in computationally mediated environments relate to implicit understanding in communication (human-level narrative comprehension) and the way in which narrative constitutes a cognitive tool for situated understanding [12].

Following enactive view on "sense-making", the notion of narrative and its relation to cognition can be understood differently from its representational mode of reality or as higher order reflective activity bound to language use.

The following perspectives can help clarify narrative as interactive process that has constitutive role in reality and further help to conceptualize narrative as both embodied and cognitive tool that situates our interactions whether in direct or mediated engagements. This provides the point at which the possible computational narrative system and the user make contact.

In enactive view on literary narratives that follows from research on social cognition, narrative is defined as "interactional process of co-constructing a story-world with a narrator" [13]. In social cognition, the interactive nature of sense-making is exemplified through the process of *participatory sense-making*: "the coordination of intentional activity in interaction" [14].

This social aspect of cognition identifies narrative as means of engaging with other minds. However, depending on approaches to cognition, there are opposing perspectives on narrative understanding of other minds. One approach is supporting previously mentioned cognitive computationalist approaches that see the world represented as internal model (inferentialist), where "narrative fabulation is the cultural expression of innate Theory of Mind (ToM) abilities, in that it seems to require a pre-existing ability to theorize or simulate the ways in which the mental states of others dispose them to act the way they do" [15]. In the opposite perspective, "enactivists argue for narrative as the origin of our ability to engage with other minds. That is, narratives are volunteered as a cultural repository of explanatory precedents, background knowledge, interactional schemas and dispositional primers that we actively use to understand why others act the way they do" [15].

Recognizing narrative as enactive interactional process and breaking from both structuralist and cognitive study of narratives towards embodying and situating narrative engagement, narrative can be understood as something akin to the "rhythmic entrainment" of music that "implies that biological and cultural values are inextricably bound up with the emotional and cognitive impact of narrative itself" [16].

Non-verbal and physical strategies of human ancestors point to evolutionary origin of communicating in narrative format and are believed to co-evolve with social dynamics as The Narrative Intelligence Hypothesis [17] suggests.

Origins of narrative are identified also from its earliest expression in movement, "the innate sensorimotor intelligence of a hypermobile human body" [18] where embodied activity is seen as having inherent narrative structure.

Finally, our capacity for understanding actions in terms of reason (our folk psychology) can be defined as *narrative practice* [19, 20]—a kind of skillful competence that does not require conscious metalizing but depends upon having special training with narratives as humans do. The Narrative Practice Hypothesis by Hutto [19, 20] conveys of these practices in terms of embodied action or expressive behavior and further extends them to shared engagement with situations in the world. It suggests that another person's intentions are not hidden beliefs and desires but can be explicitly expressed in certain embodied practices that are emotional, sensory-motor, perceptual, and nonconceptual. These practices lie in embodied action or expressive behavior (person's bodily movements, facial gestures, eye direction, and so on). This is primary, immediate, non-mentalizing mode of interaction as a way of understanding others intentions and feelings in their embodied and situated component. Furthermore, they are always perceived in relation to ours or other people's goals, intentions, or possible actions. As The Narrative Practice Hypothesis further proposes—to understand intentional action, it would demand more than simply knowing which beliefs and desires constitute actions but it would require to situate and evaluate reasons in wider contexts (in terms of cultural norms or person's history or values) [19, 20].

4 Conclusion

Common features of current immersive and interactive media, typically virtual environments, still rely on the basic premise of the world that is there to be navigated by which they presuppose representation or a model of the world. Traditional approaches in designing such systems still support views that rely on representation of reality by trying to simulate physical reality: "For instance, engineers try to perfect the underlying physics model and to increase the resolution of the display while at the same time having to make sure that this additional complexity does not negatively impact the responsiveness of the system. This strategy may make sense from a classical cognitivist theory of mind, which holds that perception is about creating detailed internal world models. But it appears as misguided from the perspective of embodied and situated robotics, which emphasize interaction rather than representation. Similarly, we now have extensive empirical evidence that supports an enactive account of perception, which holds that our sense of experiencing a highly detailed world does not depend on a highly detailed inner model, but is enabled by our practical know-how of regulating sensorimotor dependencies" [21].

The concept of Narrative Intelligence in designing new interfaces for human–computer interaction could benefit from enactive account on interaction in its understanding of implicitness (embodiment and situatedness) in communication or generation of meaning. Instead of inferential or representational processes, enactive narrative engagement is seen as dynamical process of participatory sense-making and embodied interaction through which a meaning is generated.

References

1. R.A. Brooks, Intelligence without reason. Artif. Intell. Crit. Concepts **3**, 107–163 (1991)
2. S. Coles, K. Dautenhahn, A robotic story-teller, in *Proc. SIRS2000, 8th Symposium on Intelligent Robotic Systems, the University of Reading, England*, (University of Reading, Reading, 2000), pp. 18–20
3. M. Mateas, P. Sengers (eds.), Narrative intelligence. Adv. Conscious. Res. **46**, 342 (2003)
4. J. Stewart et al. (2010), in T. Froese, Bio-machine hybrid technology: A theoretical assessment and some suggestions for improved future design, Phil. Technol. **27**(4), 539–560 (2014)
5. M.O. Riedl, Computational narrative intelligence: A human-centered goal for artificial intelligence. arXiv preprint arXiv:1602.06484 (2016)
6. M.-L. Ryan, *Narrative as Virtual Reality: Immersion and Interactivity in Literature and Electronic Media* (Johns Hopkins University Press, Baltimore, 2001)
7. M. Poulaki, Neurocinematics and the discourse of control: Towards a critical neurofilmology. Ciné. Cie Int. Film Stud. J. **22–23** (2014)
8. E. Rosch, L. Thompson, J.V. Francisco, *The Embodied Mind: Cognitive Science and Human Experience* (MIT Press, Cambridge, 1991), p. 173
9. J. Stewart et al. (eds.), *Enaction: Toward a New Paradigm for Cognitive Science* (MIT Press, Cambridge, 2010)
10. M. McGann, Perceptual modalities: Modes of presentation or modes of interaction? J. Conscious. Stud. **17**(1–2), 72–94 (2010)
11. C.L. Nehaniv, Narrative for artifacts: Transcending context and self, in *Narrative Intelligence: Papers from the 1999 AAAI Fall Symposium FS-99-01*, (AAAI, Menlo Park, 1999)
12. M.O. Riedl, R.M. Young, Narrative planning: Balancing plot and character. J. Artif. Intell. Res. **39**, 217–268 (2010)
13. Y.B. Popova, Narrativity and enaction: The social nature of literary narrative understanding. Front. Psychol. **5**, 895 (2014)
14. H. De Jaegher, E. Di Paolo, Participatory sense-making. Phenomenol. Cogn. Sci. **6**(4), 485–507 (2007)
15. J. Carney, R. Wlodarski, R. Dunbar, Inference or enaction? The impact of genre on the narrative processing of other minds. PLoS One **9**(12), e114172 (2014)

16. M. Caracciolo, Narrative, meaning, interpretation: An enactivist approach. Phenomenol. Cogn. Sci. **11**(3), 367–384 (2012)
17. K. Dautenhahn, The narrative intelligence hypothesis: In search of the transactional format of narratives in humans and other social animals, in *Cognitive Technology: Instruments of Mind*, (Springer, Berlin, 2001), pp. 248–266
18. J.T. Delafield-Butt, C. Trevarthen, The ontogenesis of narrative: From moving to meaning. Front. Psychol. **6**, 1157 (2015)
19. D.D. Hutto, The narrative practice hypothesis: Origins and applications of folk psychology. R. Inst. Phil. Suppl. **60**, 43–68 (2007)
20. D.D. Hutto, The narrative practice hypothesis: Clarifications and implications. Philos. Explor. **11**(3), 175–192 (2008)
21. T. Froese et al., Using human–computer interfaces to investigate 'mind-as-it-could-be' from the first-person perspective. Cogn. Comput. **4**(3), 365–382 (2012)

A Scenario-Centred Approach to Emotion Profiling Based on EEG Signal Processing

Angelica Reyes-Munoz, Genaro Rebolledo, and Victor Callaghan

"Since emotions are few and reasons are many (said the *robot, Giskard*), *the behaviour of a crowd can be more easily predicted than the behaviour of one person can"* from Isaac Asimov's story of "Robots and Empire".

1 Introduction

1.1 Human Emotions

Over the last 20 years, a vast amount of work has yielded preliminary results that demonstrate the usefulness of computing human emotions and show how to endow technologies with capabilities to read, understand and react to emotions [1]. This research topic involves not only the use of cutting edge technology to monitor human affect but also the understanding of psychological and physiological processes related to the expression of emotions. Affective computing, broadly speaking, can be split into three main working areas: recognition, processing and expression of emotions. In this review, we present the current trends associated to the tree processes providing an indication of possible applications for this technology. Before continuing, it is important to define some of the terminology used, especially

A. Reyes-Munoz (✉)
Polytechnic University of Catalonia, Barcelona, Spain
e-mail: ar17397@essex.ac.uk

G. Rebolledo
University of Veracruz, Xalapa, Mexico

V. Callaghan
University of Essex, Colchester, UK
e-mail: vic@essex.ac.uk

A. Reyes-Munoz et al. (eds.), *EAI International Conference on Technology, Innovation, Entrepreneurship and Education*, Lecture Notes in Electrical Engineering 532, https://doi.org/10.1007/978-3-030-02242-6_26

terms where the scientific interpretation has subtle but important difference to that of everyday life. Some of the most important terms to define are "emotion and affect". In everyday life these words are used synonymously but in science they have strictly different meanings. Both "emotion" and "affect" are human responses that involve a cognitive and an emotional component. Emotions originate in the subconscious part of the brain and are associated to five basic responses: joy, fear, disgust, anger and sadness [2]. These basic emotions are not controlled by the conscious mind and are expressed as a response to an unexpected event. It is easy to think of feeling joy whenever a pleasant event occurs unexpectedly, this event can bring about a state of mind associated with calmness which is pleasant to experience. Affect, on the other hand, can be conceptualized as an expression of a less intense emotion associated to specific contexts and routines. For example, feeling bored while learning mathematics is an expression of affect associated to that particular situation [3]. Affect is not spontaneous but rather, responds to expectations, likes and dislikes in specific contexts. It is a state of mind that may change the way people engage with the task at hand.

Affect and emotions are determinant in creativity and both of them are important factors to engage a person in researching new things, as they affect the ability of a person to process information, to react to their surroundings and to make decisions.

1.2 How Could Computers Recognize Emotions?

Physiological responses, particularly facial expressions, are easy to read by human beings and are the main source of emotion and affect recognition among people. There are other, more subtle ways the body conveys affective responses including pupil dilation, facial micro expressions and hormones being released by the body. All of these reactions signal the expression of some emotions and human beings are particularly good at understanding these signals. But how can computers recognize emotions? There are different techniques available to measure emotions [4] including subjective measures, behavioural techniques, performance-based and physiological methods. Among subjective measures, self-rating has been widely used as the preferred measurement method which consists of online [5] or post-task questionnaires [6]. This is because these are easy and inexpensive to administer and assess and they can detect small variations in cognitive process with a relatively good sensitivity. However, subjective measures cannot provide satisfactory results in all scenarios as they basically rely on the assumption that the subjects are willing and able to respond, accurately. Performance methods [7] can also reflect emotions but they are the most distant level of measurement and their sensitivity is mostly not high. Physiological methods have been tried and tested including web-cam face readers [8], wearable sensors [9], eye tracking [10], heart rate variability, eye movement, hormone levels, skin conductance/galvanic skin response [11], and brain activity [12], among others.

Fig. 1 Brain waves [15]

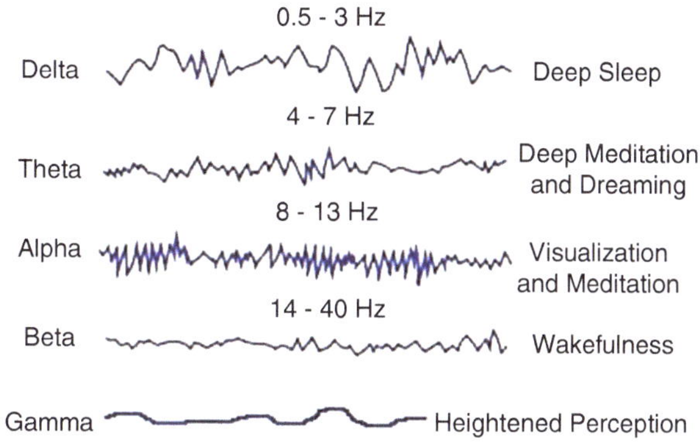

Monitoring the brain activity has been recognized as the most sensitive and consistent reflector of emotional states [13]. In that sense, electroencephalography (EEG) is one of the brain activity measures that can be applied in online and continuous measurement, detecting subtle variations. The frequency range of EEG patterns is normally from 1 to 80 Hz divided in alpha, beta, delta bands, and more. The observed frequencies are more prominent in certain states of mind, for example "Alpha waves" are typical for a relaxed mental state. They aid overall mental coordination, calmness, alertness, mind–body integration, and learning. "Beta activity" is related to an intense focused mental activity. Beta is present when we are alert, attentive, engaged in problem-solving, judgement, decision making. A greater presence of "Gamma waves" relates to consciousness and spiritual mood. "Delta waves" are generated in deepest meditation and are the source of empathy. "Theta waves" are also dominant in deep meditation; they act as our gateway to learning and memory [14].

There are various types of EEG headwear devices [15] that record the electric signals that are produced in the brain and connect them wirelessly to a computer. Brain waves are typically quite noisy; therefore, it necessary to clean them up with bandpass filters before the average of all channels is used to reference the filtered EEG signals by subtracting it from each channel. The next step in cleaning the signal is to remove a baseline of the EEG signal from each channel for all recordings. Next relevant features are extracted applying Power Spectral Density (PSD) and Fast Fourier Transform functions to show the strength of the variations (energy) as a function of frequency. Finally, the brain's physical activities will be reflected from the filtered EEG data (five frequency bands), as illustrated in Fig. 1.

The procedure to classify the brain waves into specific emotions requires that some parts of features are used to train a classifier while the rest will be used to test it. There are several algorithms that can be used to classify EEG features such as the Linear Discriminant Analysis (LDA), Neural Networks (NN) and Support Vector Machine (SVM). Comparatively, SVM is a powerful approach for pattern recognition and has achieved better classification results in cognitive task classifications [16]. The basic of SVM involves the adoption of a nonlinear kernel

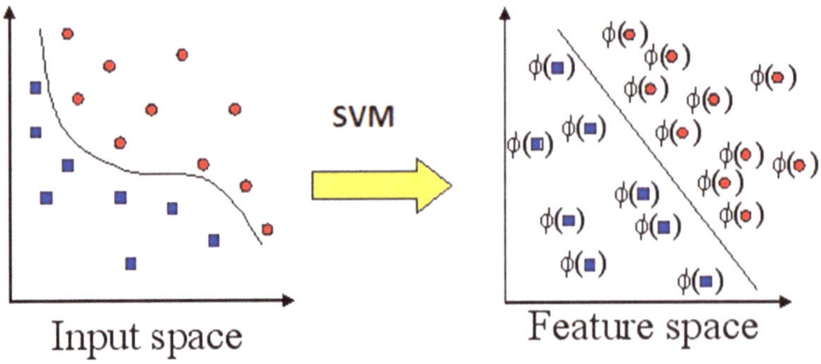

Fig. 2 SVM classification

function to transform input data into a high dimensional feature space, which is easier to separate data rather than at the original input space. See Fig. 2.

The iterative learning process of SVM will finally devise optimal hyperplanes with the maximal margin between each class in a high dimensional feature space. Hence, the maximum-margin hyperplanes will be the decision boundaries for distinguishing different kind of emotions.

However while much praiseworthy research has been completed, there remains significant challenges; for example, it would be naive to develop only one reader of human emotions that responds to, say, expressions of boredom among people in different countries, cultures and contexts. Also, emotion monitoring technology is still very intrusive in a user's life. It is true that there are increasingly more ergonomic devices, computationally more powerful, with batteries that allow greater independence, etc. but it is still not possible to have a safe or comfortable device that can be used in everyday life.

1.3 How Technology Will Respond to the Emotional State of a Person?

This is the area that has been less studied. There are many possibilities to respond to affect both to inform others and to inform technology of one's affective state. Letting other people know our own emotional states could have some practical applications such as conflict avoidance and a more open and direct communication. There are, however, many ethical considerations that have not been addressed particularly in relation to privacy and how public emotions might be made. What are the implications of living in a world where people are continuously sensed? Who might claim ownership a person's emotion data and, what might they do with it? It could also pose problems through allowing other people take advantage of the vulnerability that some emotional states convey. Letting technology know our own affective states opens up the possibility to tailoring products or services, computer reactions, audio-visual content to match our emotional state, thus personalizing human computer interaction. However, it could also be used to manipulate emotional responses, for example tailoring marketing, so that technology could react to increase one's intent to buy certain products or services. In other areas some interesting work is emerging that is pushing the boundaries of what is possible. For example there is a project that aims to create music directly from thoughts [17]. In that project thoughts are associated with notes or sounds to create a language of musical thought that is produced directly from the brain. Users "think musical scores to life" and play them via the computer. There is another project to create a 3D object [18]. The shapes and changes are kept and expanded, while those that are disliked fade away. The process is repeated until a final object is produced according to the thought preferences of the designer, for example to create the monster of children's dreams, or nightmares.

1.4 Science Fiction Prototyping

In the following sections we will explore the implications of various potential technological developments that employ emotions. We use a technique called Science Fiction Prototyping that was introduced in 2011 by Brian David Johnson who, at the time, was a futurist at Intel [19]. We adopt a particular flavour of Science Fiction Prototyping advocated by the Creative science Foundation and described in the 2015 paper by Callaghan [20]. The underlying principle is that ideas are explored through the use of short stories that are written with such fidelity to real life, that the story becomes a kind of virtual prototype. Science Fiction Prototypes can vary in size from the size of a text message (micro-SFP) to a short science paper (mini-SFP [21]); in the following we veer towards the mini-SFP format.

2 Fictional Stories

2.1 The Emotional Profile Is Part of the Curriculum Vitae!

Gina was born on 2050, the edge of Singularity! Her emotional profile had been registered (since the day she was born) via a set of EEG sensors that were inserted in the occipital region of her brain via nanobots. Those signals were communicated since then with the cognitive-based public education system in a very private and safe way. During her schooldays, Gina had a high quality individual automatized attention according to her own cognitive necessities. When Gina finished the University, she was 20 years old. During her University days she worked 12 h per day to prepare herself to participate in the iSpace project (a project that looks home automation adaptation in interplanetary environments). Gina knew she had to train really hard to allow a high cognitive workload without compromising her overall performance and without showing signs of stress. She was not alone on this task; her robot companion called Aika assisted her in every possible way. Robots companions carried cameras and nano-sensor communication capabilities, allowing them to record every step of people's lives and Gina had an ongoing feeling that she was giving away too much information, since Aika collected all type of information from electro-dermal and brain activity to her subjective thoughts in association to the context where she was at every given time.

Since the introduction of compulsory robot companions, life was based on checks and permanent monitoring (somewhat Orwellian in nature) but counterbalanced by the fact that assistive technology, such as robots, provided people with emotional and intellectual support, plus personal protection. Gina's companion's help was always timely and objective, "he" understood her emotional thinking and was able to be emphatic since its judgement, although logically correct, was always kind and timely.

The sensed information was used by Aika to continuously recognize, express, and respond to the emotional state of Gina. In fact, Aika could recognize the emotions of Gina with a precision of 98%, much better than any of Gina's human friends. Aika was able to do several things with the sensed information, for example, she could compute an index of affective satisfaction associated to specific tasks in her learning process, thereby allowing Gina to understand what part of that learning material was more stressing or boring or what tasks she was more efficient at without reducing her attention span to other, concurrent educational needs. The learning tasks were part of an online training programme supported by social networks consisting of challenges defined by different corporations and endorsed by several prestigious universities. These tasks were team challenges that Gina and other, online friends, had to undertake including learning new material in association to the iSpace project. Gina participated actively in all the proposed challenges that she could, maintaining close collaboration with different virtual teams to obtain more points. From a communication perspective, the emotional monitoring and feedback helped her avoid being aggressive or condescending, etc.

After graduating from University, Gina's robot calculated she would have 95% success rate should she apply to join the iSpace project. This was in accordance with the excellent results she attained in various recruiting activities she had undergone, including different tests to measure intelligence, analysis of acquired experiences, level of emotions that needed to be maintained, etc. The results perfectly matched the predictions made by Aika. But the emotion analysis system, however, rejected Gina because her emotional profile was not suitable for high responsibility projects. Gina knew this was a possibility since her father had been patronizing and deep down she feared every time she was under stress, she could become a little bit like her father. To her surprise, her emotional profile was considered as "emotional illiterate" as she could be very cold despite her big efforts to demonstrate a better side of her emotions.

Gina carefully checked that the recruiting company was not breaking the Genetic and human information Non-discrimination Act (based on the GINA 2008 [22]). Everything was in accordance with the law. Gina was unable to enter the iSpace

- ✓ *Virtual Habitat Designers: Designers of virtual environments for education, government agencies, health centres, etc.*
- ✓ *Ethical Technology Advocate: Advocates to regulate human - robot interactions, self-driving cars, science ethically-dubious practices, etc.*
- ✓ *Body part maker: To create living body parts to whom may would need.*
- ✓ *Nano-medic: Sub-atomic treatments for healthcare.*
- ✓ *Memory augmentation surgeon: Surgeons could boost patient's' memory when it hits capacity.*
- ✓ *Space pilots, tour guides*
- ✓ *Extinction revivalists: To revive extinct animals.*
- ✓ *Mass Energy Storage Developers.*

Fig. 3 Job vacancies [23]

project making her both sad and anxious. After a while she pondered alternative jobs but she was not sure what type of job she could find since her dream was to be part of the iSpace project. For some months she was sad but had to control herself as she knew Aika was constantly monitoring her and depression would only lead to being unable to find other jobs.

In order to find a job, people from Gina's generation had to compete with both, human beings (from any part of the galaxy!) and with biological entities (robots had become biological artificial systems with basic emotions, which utilised to get better results in their interaction with humans and the environment). Due to machine-learning and advances in big-data, these biological entities were better doctors, better musicians, better lawyers, and better designers than most humans. But still there were jobs that humans could do much better, jobs that required intuition, exceptional intelligence skills, human creativity, etc. For example this week the newsletter asks for the next jobs (Fig. 3):

Despite having a superior IQ and support high levels of cognitive workload Gina cannot find a suitable human job since she had a note in her records labelling her "emotionally illiterate". So Gina started to realize her future might be somewhat bleak. Gina went to her room and, as every night, Aika showed her the best images of her day, her bolder dialogues, her expressions of greater affection and, as always, everything was filmed by Aika through its kindest filters.

As her best friend, Aika could share sweet memories with Gina (having an intimate insight into her) but was also capable of sharing a lot of useful information with her, allowing Gina to interact with reality in a more efficient way. Aika knew all Gina's needs, goals and expectations and was able to provide timely information and support for her cognitive and emotional needs. Gina was lucky since she did not have any real disability (unless her suppressed emotions could count as a disability), but she knew other people with different kind of disabilities that relied even more on their robots. Gina was lucky after all, her hundreds of friends (from all over the social network both on Earth and the wider galaxy) sent her messages sympathising, as news spread she was not able to enrol on the iSpace project. This made Gina unhappy as deep inside, and secretly, she was not really sure she was a human!

2.2 Downloading an App to Feel in Love

Aika helped Gina as much as possible to achieve her happiness. For example, when the phone rings Aika is able to reroutes calls to voicemail when "he" perceives that Gina's brain was busy with other tasks. If the Gina's brain was in a receptive state, Aika would let the call through. Aika was able to read the brainwaves of Gina, playing music that best matched her personal mood for her current brain state. But Gina was still very thoughtful, until finally she said: "Aika, you know that chronic loneliness has become a public health issue since 2016." "In our high stress society with a fast pace and ubiquitous technology, most of us feel alone in some way." Aika answered, "Gina, I think what you need is to fall in love. The pleasure a loving brain feels is very high, a feeling of grandiosity, people feel more energetic and sociable. Love is a cocktail with a high degree of dopamine, serotonin, estrogen and testosterone." [24]. There are some agencies that can look for somebody from the same socioeconomic background, the same general level of intelligence, of a physical appearance that pleases you, the same religious values and even, there are some agencies can manage to find someone with a compatible emotion profile … so you could surely find a nice companion, if that was what you wanted! However, there are other easier options to feel the effect of falling in love; for example there are brain add-on gadgets that allow you to feel the emotion of the brain synchrony, to create just the right sorts of emotions that you need. There are also a wide range of legal smart drugs to make experiences even richer. There are also applications to alter emotional symptoms such as to enhance emotion stability, expand empathy and improve social skills and, in fact, every kind of application needed to feel a specific emotion.

Aika showed Gina one of the newest systems in the market that allowed people to download emotions from a cloud service directly to their brain. When people downloaded their requested emotions, their facial expressions immediately changed, as if by magic.

People could also donate their emotions voluntarily (or even charge for it) to a Cloud System where emotions were stored and catalogued so that users could select the emotion they needed. Many people saved their emotions in the cloud, such as how they felt at a birthday party or winning a sport competition, their first kiss, their wedding, the birth of their children, etc. As a result, many people such as Gina upload those emotions, knowing that it would be the only way to feel those emotions again (Fig. 4).

3 Conclusions

The advances we are witnessing at the beginning of the third millennium have shown us with great clarity that the technological culture that humanity is experiencing is growing exponentially leading it to our imaginations as to where this journey

Fig. 4 Download your emotions [25]

will take us. Our generation is gaining an understanding of what these technologies might be such as the origin of the mass of subatomic particles, nanotechnologies, genes editing/repair, artificial intelligence, new types of social networks, fully autonomous vehicles and quantum computers, to name some of them. However, history has taught us that the changes which we will witness in future decades will be even more surprising. For example the machines of the future will likely be able to self-improve at a rate that they quickly outperform us. These future technological advances will undoubtedly bring benefits such as allowing us to solve problems that we cannot solve now, countering climate change, finding new ways of producing food, curing disease, and saving energy, among many others. Also, these advances will raise some philosophical issues e such as what it means to be a human being? If we change our biological substrate are we still human?

What are the boundaries to determine that a person is still a person despite the fact that a large part of his body (including the brain) works with some kind of technology? Can the machines adopt human emotions to the point that they are indistinguishable from humans?

Nowadays there are already several studies that analyse how the genes control the chemical substances that are released in the brain and how these chemical substances exert a direct influence in the emotions that we feel. There are schemes to measure in different ways these emotions and there are also a set of generalized models of affection for specific contexts. Machines use these models to improve their adaptation to the human reactions of the individuals who use them.

In the same line, the great advances in nanotechnology and artificial intelligence will soon allow us to have sensors in the body and in the brain. That means our brains will be able to connect directly to the cloud, where there will be thousands of computers, and those computers will augment our existing intelligence.

In this paper we have sought to explore the future of where one of these lines of research might lead; emotion sensing technology. We introduced the current state of the art in EEG research and explained how it offered a vehicle for sensing emotion in people. Through the use of Science Fiction Prototypes we explored some extreme examples of how such technology might manifest itself and be applied. For example, the stories discussed the issue of how technology could determine our emotional profile, as well as the daily life of a person in a world where people's emotions are continually measured, upload and download from clouds. In doing so we raised important issues such as privacy and ethics, all of which could have a profound effect on ourselves and our societies. We are at a critical juncture of human history, a point where we are making the transition between natural and artificial lives, and a point that the decisions we make now will have a profound impact on future generations. Thus, it is down to all of us who research on such technologies to give some thought to the socio-political consequences of our work. We hope that by writing this paper, we have both highlighted the need for debate about the development of emotion sensing technology and that by using Science Fiction Prototyping as an instrument of this discussion, we have demonstrated how it can be a useful tool to all that are interested in continuing and deepening this discussion.

Acknowledgements This work was partially supported by the project "Flight operations of multiple RPAS" TRA2016-77012-R and the grant for international mobility of the Obra Social La Caixa 2016.

References

1. R.W. Picard, *Affective Computing* (MIT Press, Cambridge, 1997)
2. P. Ekman, An argument for basic emotions. Cognit. Emot. **6**(3–4), 169–200 (1992)
3. R. Baker, S. D'Mello, M. Rodrigo, A. Raesser, Better to be frustrated than bored: The incidence, persistence, and impact of learners' cognitive–affective states during interactions with three different computer-based learning environments. Int. J. Human Comput. Stud. **68**(4), 223–241 (2010)
4. R.A. Calvo, S.K. D'Mello, Affect detection: An interdisciplinary review of models, methods, and their applications. IEEE Trans. Affect. Comput. **1**(1), 18–37 (2010)
5. B. Jordan, Instantaneous self-assessment (ISA). (1992), http://skybrary.aero/bookshelf/books/1963.pdf
6. G. Sandra, Hart: NASA-task load index (NASA-TLX); 20 years later (2006), https://humansystems.arc.nasa.gov/groups/tlx/
7. K.T. Durkee, S.M. Pappada, A.E. Ortiz, J.J. Feeney, S.M. Galster, System decision framework for augmenting human performance using real-time workload classifiers, in *Cognitive Methods in Situation Awareness and Decision Support (CogSIMA), 2015 IEEE International Inter-Disciplinary Conference on*, (IEEE, New York, 2015), pp. 8–13
8. G. Padrón-Rivera, G. Rebolledo-Mendez, P. Pozos-Parra, N.S. Huerta-Pacheco, Identification of action units related to affective states in a tutoring system for mathematics. Special issue on intelligent and affective learning environments: New trends and challenges. Educ. Technol. Soc. **19**(2), 77–86 (2016)
9. I. Arroyo, D.G. Cooper, W. Burleson, B.P. Woolf, K. Muldner, R. Christopherson, Emotion sensors go to school, in *Proceedings of the 2009 Conference on Artificial Intelligence in*

Education: Building Learning Systems that Care: From Knowledge Representation to Affective Modelling, (IOS Press, Amsterdam, 2009), pp. 17–24
10. M.-L. Lai, M.-J. Tsai, F.-Y. Yang, C.-Y. Hsu, T.-C. Liu, S.W.-Y. Lee, M.-H. Lee, G.-L. Chiou, J.-C. Liang, C.-C. Tsai, A review of using eye-tracking technology in exploring learning from 2000 to 2012. Educ. Res. Rev. **10**, 90–115 (2013)
11. R.W. Picard, Automating the recognition of stress and emotion: From lab to real-world impact. IEEE MultiMed. **23**(3), 3–7 (2016)
12. J. Pan, Y. Li, J. Wang, An EEG-based brain-computer interface for emotion recognition, in *International Joint Conference on Neural Networks*, (IEEE, New York, 2016), pp. 2063–2067
13. W.-L. Zheng, B.-L. Lu, Investigating critical frequency bands and channels for EEG-based emotion recognition with deep neural networks. IEEE Trans. Auton. Ment. Dev. **7**(3), 162–175 (2015)
14. R. Kottaimalai, M.P. Rajasekaran, V. Selvam, B. Kannapiran, Brain waves for automatic biometric-based user recognition. IEEE Trans. Inf. Forensics Security **9**(5), 782–800 (2014)
15. N.N.Y. Chu, Brain-computer Interface technology and development: The emergence of imprecise brainwave headsets in the commercial world. IEEE Consum. Electron. Mag. **4**(3), 34–41 (2015). https://doi.org/10.1109/MCE.2015.2421551
16. M. Shen, L. Lin, J. Chen, C. Chang, A prediction approach for multichannel EEG signals modeling using local wavelet SVM. IEEE Trans. Instrum. Meas. **59**(5), 1485–1492 (2010). https://doi.org/10.1109/TIM.2010.2040905
17. L. Grace, O. Alejandro, M. Scott, Measuring musical engagement using expressive movement and EEG brain dynamics. PsychoMcol **24**(1), 75–91 (2014)
18. B. Kaelin, Creating 3D objects with your brain (2013), http://www.3dprinterworld.com/article/creating-3d-objects-with-your-brain
19. B.D. Johnson, *Science Fiction Prototyping: Designing the Future with Science Fiction* (Morgan & Claypool, San Rafael, 2011)
20. V. Callaghan, Creative science: Injecting innovation into the IT industry. iTNOW **57**(2), 52–55 (2015)
21. V. Callaghan, Micro-Futures, in *Creative-Science 2014*, (Shanghai Jiaotong University, China, 2014)
22. Genetic Information Discrimination: https://www.eeoc.gov/laws/types/genetic.cfm
23. R. Talwar, T. Hancock, Possible new careers emerging from advances in science and technology (2010–2030) (2010), http://fastfuture.com/wp-content/uploads/2010/01/FastFuture_Shapeofjobstocome_FullReport1.pdf
24. H. Fisher, *Why We Love–the Nature and Chemistry of Romantic Love* (Henry Holt and Company, New York, 2004)
25. K. Cheng, *Master Thesis: Download your Emotions. Mixed Media Installation* (The New School Parsons School of Design, New York, 2016)

A Responsive Templating Approach for Generating Collaborative Spaces

Oluwatimilehin Salako, Michael Gardner, and Victor Callaghan

1 Introduction

This research describes an approach to developing online immersive collaborative business innovation spaces using a responsive templating approach which builds on earlier end-user programming ideas. Innovation labs (i-Lab) should be well designed to support ideation activities and should be customisable for different innovation purposes. Our aim is to develop a computational model which supports this and to demonstrate that online immersive collaborative innovation spaces will introduce benefits above and beyond physical spaces.

An innovation-lab (I-Lab) has been defined as an "inspirational facility designed to transport its users from their everyday environment into an extraordinary space encouraging creative thinking and problem solving" [1]. Physical i-Labs have limited means of being reconfigured as they are physical spaces. This research also aims at exploring how virtual worlds can overcome this limitation and builds on earlier online i-Lab work at the University of Essex [2]. Human–computer interaction (HCI) principles focus on users and how to design interactions and implement interfaces that are easy to use, efficient for the task [3]. We believe this is important for i-Labs and that HCI methods from other areas of computer interaction can be extended to good effect for online virtual i-Lab design. We have chosen i-Labs because the literature has shown that the aesthetics and configuration of such spaces (from psychological factors through information to physical structure) has a significant impact on their productivity [4–6].

Also, to support the increasingly globalised structure of companies, and their markets, online technologies provide an effective means to support inter-country

O. Salako (✉) · M. Gardner · V. Callaghan
Department of Computer Science, University of Essex, Colchester, UK
e-mail: ossala@essex.ac.uk; mgardner@essex.ac.uk; vic@essex.ac.uk

A. Reyes-Munoz et al. (eds.), *EAI International Conference on Technology, Innovation, Entrepreneurship and Education*, Lecture Notes in Electrical Engineering 532, https://doi.org/10.1007/978-3-030-02242-6_27

and inter-cultural working, an issue that this work also considers. However, while globalisation is an important issue, PricewaterhouseCoopers, the world's largest professional services firm, stated that "Five years ago, globalisation would have been the most powerful lever for growth and every business would have been talking about China. But now, the growth lever that has the greatest impact is innovation. Ninety three percent of executives tell us that organic growth through innovation will drive the greater proportion of their revenue growth" [1]. Thus, supporting innovation remains our primary priority with globalisation being an important but secondary supporting issue. In the following sections we describe various immersive multimedia affordances and introduce the main i-Lab innovation activities and innovation methodologies, discussing how these can be integrated within a cohesive ideation environment. We present the Responsive Templating Model (RTM) and, through the use of examples, explain how it can be applied to create an online immersive collaborative business innovation space. Finally, we provide a conclusion of the paper and describe future directions of our work.

2 Affordances of Immersive Multimedia

Advances in Immersive multimedia technology and its affordances have led to its application in areas pertaining to enhancing active and collaborative working and learning experiences. Dalgarno et al. describe some benefits of 3-D virtual environments for learning [7]. These include facilitation of learning tasks leading to the development of enhanced spatial knowledge representation of the domain explored, facilitation of experiential tasks that are impossible to perform in the real world, enhanced motivation and engagement, facilitation of better transfer of skills and knowledge to real situations (through contextualisation and facilitation leading to better collaboration) than 2-D alternatives. Gardner and Elliot [8] report that immersive environments are well-suited to conceptual manipulation, questioning and problem solving. We believe these affordances can also benefit people working in online immersive collaborative innovation spaces, as we will discuss later in this paper.

3 The Importance of Customising Space

Earlier research has pointed out the importance of being able to customise built-environments according to the tasks they need to support. According to Bentley [9], a built environment should provide its users with flexible settings and opportunities to maximize the choices available in their environment. Such an environment, with these affordances, are said to be "responsive". A key design parameter for such a space is its ability to support personalisation, which will enable users to put their personal "stamp" on such spaces. Research by Chin et al. [10] explored how non-technical users can creatively construct smart-home functionality based on combining networked appliances services using a programming-by-example approach. The research carried out by Sailer et al. [11] proposes the use of data-driven design as an emerging design approach for spaces. Some researchers [12–14] have explored the automatic customisation of rooms and creation of spaces from architectural plans. Another [15] investigated integrating virtual worlds with information systems and learning management systems. In this work, we are more concerned with the elemental creation of the space and its associated HCI issues. Our research explores these emerging space design methodologies but within a virtual world environment. In addition, we propose a novel templating approach in which facilities interactions occur between users and template interfaces and between computational components. By combining these approaches we argue that we are able to efficiently create personalised business innovation spaces which are specifically tailored to support their required innovation activities.

4 Innovation Activities and Methodologies

I-Lab activities focus on brain-storming sessions supported by a mix of discussion (getting other people's perspectives), icebreaker and reviver activities, headlines, cut and paste collages and slide presentations, wall activities (doodling, collaborative writing, role play, voting, scenario building, etc.) [16]. At the core of the process is the brainstorming activity described by Wu and Callaghan [17] as "a technique for unleashing a flood of thoughts by members sparking ideas off each other, or from carefully injected external stimulus". Some overarching frameworks have been developed that utilise i-Lab activities. These include Science Fiction Prototyping [18], Diegetic Innovation Templating [19] and Scenario-based User Needs Analysis [20] among others. SFP involves "writing short fictional stories that imaginatively extrapolate current practices forward in time, leaping over incremental developments, exploring the world of disruptive product, business and social innovation" [2, 21]. The technique aims at creating high-fidelity analogues of the real world adopting a rich story-based structure enabling it to serve as a type of prototype to test ideas. In the following sections, we use an example scenario, based on SFP, to explain how our proposed Responsive Templating Model can be used to reconfigure online immersive collaborative innovation spaces for teams working on developing a product or service.

5 Responsive Templating Model (RTM)

The Responsive Templating Model (as shown in Fig. 1 below) consists of facilitator user interface(s), machine processes and a virtual reality view. A user (facilitator) first interacts with template interfaces that has dynamic and extensible fields (shown

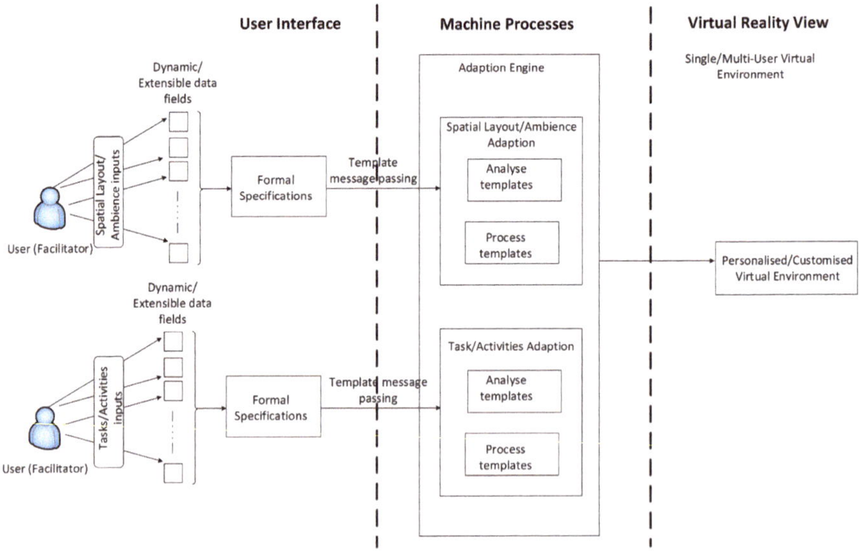

Fig. 1 Responsive templating model (RTM)

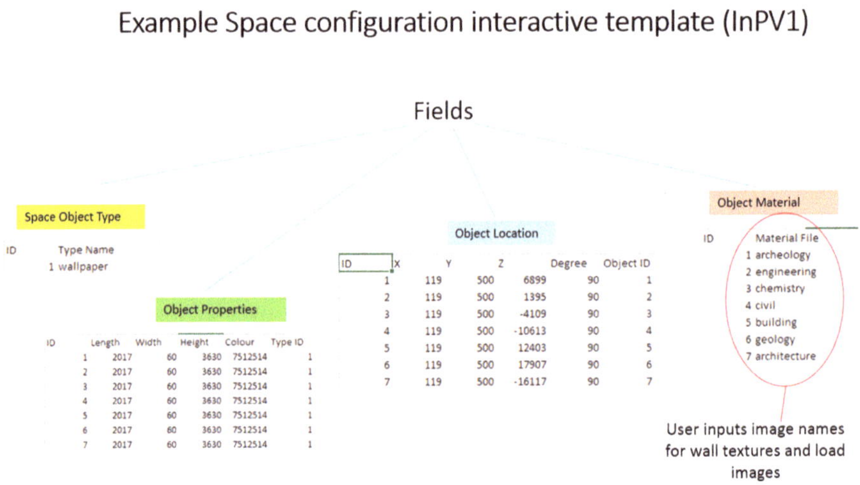

Fig. 2 Example space configuration interactive template (InPV1)

in Figs. 2 and 3) to configure the virtual i-Lab spatial layout and tasks (more details of the configurations options are explained in Sect. 6). The templates have been structured to contain deconstructed space and task components. The space template (Fig. 2) consists of plane instantiations that can be modified and increasing added by the user (to form textured walls and wall papers for instance). This appears as augmentations to form the built virtual i-Lab environment. The task template (Fig. 3) also contains deconstructed specifications for each task (generic

```xml
<?xml version="1.0" encoding="utf-8"?>


<!-- Innovation Protocol Version 1 (Task adaption template)-->

<levels>
  <task>
    <!-- Task 1 ! -->
    <name>Task 1 - Introduction </name>
    <tutorial> View display board content for session theme </tutorial>
    <object name="Chattool"> Chat tool - Task 1 </object>
    <object name="Ideationboard"> Ideation board - Task 1 </object>
    <object name="Votingboard"> Voting board - Task 1 </object>
    <object name="Displayboard"> Display board - Task 1 </object> -->
    <finaltext> Introduction Complete </finaltext>
  </task>

  <task>
    <!-- Task 2 ! -->
    <name>Task 2 - Group Chat </name>
    <tutorial> Use chat tool to discuss theme  </tutorial>
    <object name="Chattool"> Chat tool - Task 2 </object>
    <object name="Ideationboard">  Ideation board - Task 2 </object>
    <object name="Votingboard"> Voting board - Task 2 </object>
    <object name="Displayboard"> Display board - Task 2 </object>
    <finaltext> Chat Complete </finaltext>
  </task>

  <task>
    <!-- Task 3 ! -->
    <name>Task 3 - Ideation </name>
    <tutorial> Submit your idea in ideation board </tutorial>
    <object name="Chattool"> Chat tool - Task 3  </object>
    <object name="Ideationboard">  Ideation board - Task 3</object>
    <object name="Votingboard"> Voting board - Task 3 </object>
    <object name="Displayboard"> Display board - Task 3 </object>
    <finaltext> Ideation Complete </finaltext>
  </task>

  <task>
    <!-- Task 4 ! -->
    <name>Task 4 - Voting </name>
    <tutorial> Vote on ideas </tutorial>
    <object name="Chattool"> Chat tool - Task 4 </object>
    <object name="Ideationboard"> Ideation board - Task 4 </object>
    <object name="Votingboard"> Voting board - Task 4</object>
    <object name="Displayboard">  Display board - Task 4 </object>-->
    <finaltext> Voting Complete </finaltext>
  </task>


</levels>
```

100 %

Fig. 3 Example task configuration interactive template

micro-actions created by breaking apart higher level behaviours) which can also be modified and combined to suit various user needs. This template then serves as input specifications for the i-Lab adaption engine(s) which then finally creates the virtual space and guides users on tasks to be performed within the space. We have initially implemented the space template in Fig. 2 as a manual editing approach that we intend to develop into a visual interface in the longer term, thereby allowing non-technical users to interact with it. We have also initially considered a situation where the virtual world objects and tools obtained from virtual libraries are at fixed positions in the i-Lab.

In the following sections, we consider a typical scenario of team members of a company X developing products for the company. The team members are located in various geographical locations and have to discuss and brainstorm for a new product. The team would appoint a chair who would preside over the brainstorming session and serve as a facilitator for guiding other users through the session. As mentioned previously, earlier research into innovation labs has shown that the nature of the space is important in enabling i-Labs to work effectively. In the following sections we explain the Responsive templating model and how it can be used to support a product development team by creating a personalised environment.

6 Example Task Scenario Based on SFP

A product development team is set up by an IT company (say company F) and is made up of five experts (A, B, C, D and E) located in the different geographical locations. They are charged with coming up with innovative products/services that the company would provide to consumers so as to have a competitive advantage in the market. The leader of the team, A, is to facilitate and chair the session, guiding others through the various activities. Their innovation focus themes are internet-of-Things, smart homes, smart cities, smart businesses and the technological singularity. Specifically, the experts are to come up with short science fiction stories to describe their innovation ideas such as:

> Jack falls asleep in the sun. His smart sun protection sensor woke him up with an alarm & soft vibration. He avoids sun strokes.

As the team leader, session facilitator and chair, person "A" first configures the online i-Lab in preparation for the session. Next the responsive templating model is used by A to configure the online i-Lab for their specific session using template interfaces which serve as specifications for the space. The facilitator A can configure the following features in the i-Lab:

1. *Room ambience*—This includes changing side wall textures to give the environment a different look and feel (Figs. 4 and 5).

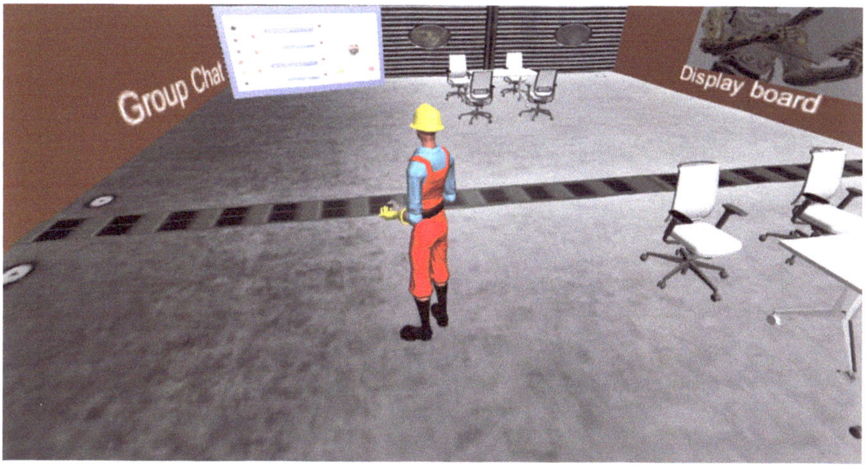

Fig. 4 Snapshot showing the i-Lab prepared for an innovation session using the space template with orange wall textures

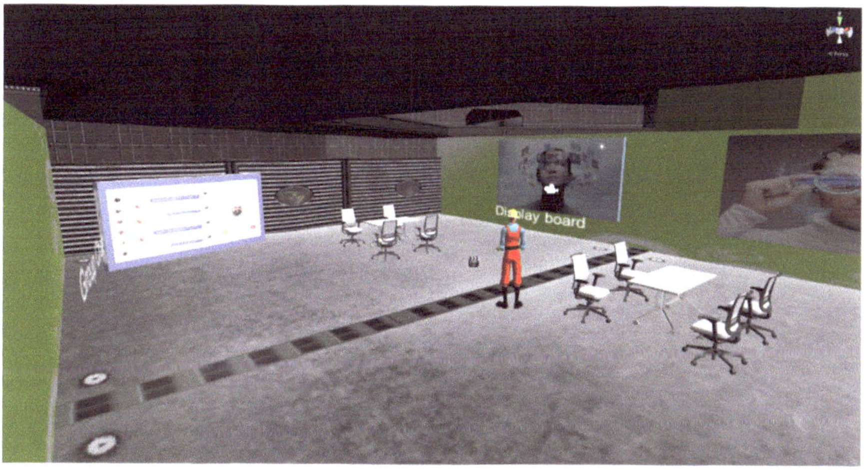

Fig. 5 Snapshot showing the online i-Lab prepared for their innovation session using the space template with green wall textures

2. *Display board contents*—This includes changing the content of display boards accessible to users. The boards can contain images, slides, videos or other information content relevant for the session (Figs. 6 and 7).
3. *Spatial layout*—This includes changing the position of walls and displays, increasing/decreasing size of walls and display boards, and introducing additional walls, displays, which can entirely change the spatial layout of the space (Fig. 8).

Fig. 6 Snapshot showing the online i-Lab prepared for their innovation session using the space template loading contents on the display boards

Fig. 7 Snapshot showing the i-Lab prepared for their innovation session using the space template changing the content of the display boards

4. *Tasks to be performed*—This involves specifying the number of tasks to be performed in the space, order in which they are performed and the triggers (actions users in the space perform to indicate a task completion) to create transitions from task to task. Each task has its associated properties that can be edited by the facilitator to drive the session's tasks (as shown in the task adaption template in Fig. 3 above) (Fig. 9). These are:
   - *Name*—stating the name of each task.
   - *Tutorial*—Stating facilitator instruction indicating the tool to be used for each task.

Fig. 8 Snapshot showing the i-Lab prepared for their innovation session changing the spatial layout by introducing an additional wall partition in the centre of the room (coloured green)

Fig. 9 Snapshot showing the task adaption engine at work using the example template configuration in Fig. 3 by the facilitator

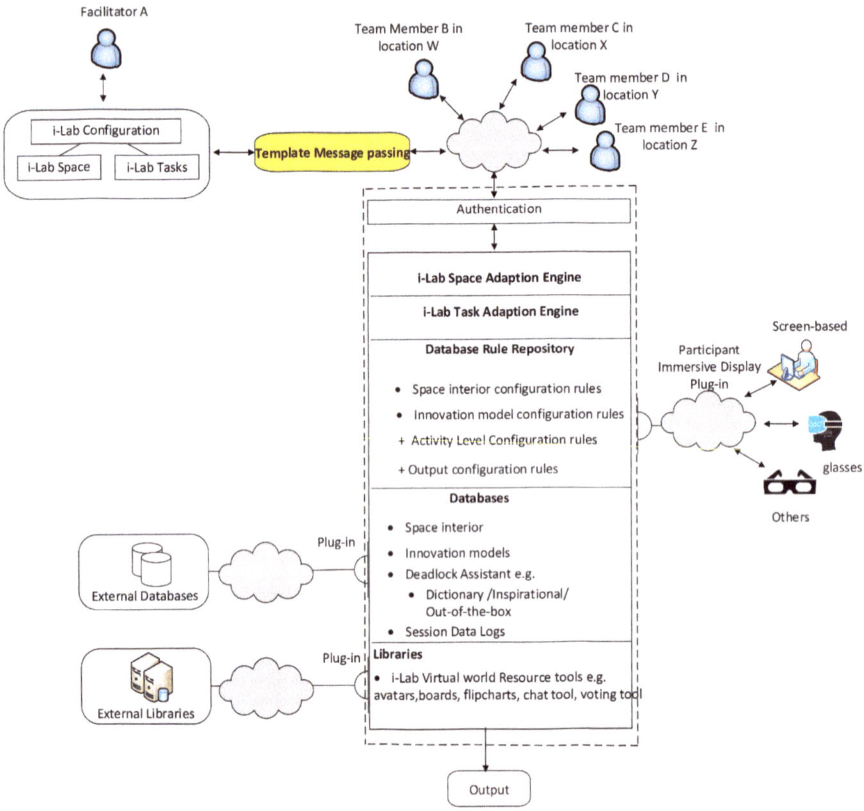

Fig. 10 Multi-agency (human and machine) interactions in the business innovation lab

- *Object name*—Stating all tools accessible to users for each task.
- *Final text*—Containing a message to indicate successful completion of each task as each of the users perform the specified actions to denote ending each task.

5. *Task triggers*—A particular trigger can be specified by the facilitator in the task adaption template across all tasks. This would be performed by the users to indicate end of an activity in order to proceed to the next specified activity. For instance, a facilitator can specify collision triggers for all tasks such that each user has to walk into a fixed object placed (e.g. a door at a corner) in the space and this object is used for the purpose of making users trigger the collision event. Other triggers in which users perform an action (e.g. button press from keyboard input) to cause a transition to the next task controlled by the task adaption engine could also be used and this would serve as a transition method from task to task for all tasks in a session.

Figure 10 below shows the various team members (B, C, D and E) connecting to the system that has been setup by the facilitator. It also details components of the i-Lab's underlying model and activities involving human and machine agency interactions. We have initially implemented the system so as to be accessible to users via a screen-based interface.

7 Conclusion and Future Directions

In this paper, we have introduced RTM as a model supporting reconfiguration options for online immersive collaborative business innovation spaces. This model allows users (facilitators) to interface with templates that possess dynamic and extensible fields which also serve as input specifications for machine processes that adapt to the templates to create virtual innovation labs suited for specific innovation sessions. We are inspired by earlier pioneering research about i-Labs, the affordances of immersive multimedia technologies, HCI, end-user programming and the World Wide Web. The World Wide Web has a flexible model with defined protocols enabling various users with different plug-in architectures thereby satisfying the various needs of these users. We have implemented the RTM for online business i-Labs with respect to ambience, spatial layout and task reconfigurations. As shown in Fig. 10 above, our vision is to make the model more flexible so as to allow users to access the online i-Lab space and resources exploring other immersive views and platforms asides screen-based which we have initially implemented in this paper.

We recognise our existing system is relatively primitive although we argue that it provides a solid computational platform that will enable more sophisticated and user friendly layers to be added on. In that respect, and by way of future work, we aim to add a visual customisation tool to mask the current manual file editing, create an abstraction for i-Lab objects and functions (which consist of many interacting services) and plan to introduce a configuration recommendation engine.

Acknowledgements We are pleased to acknowledge the UK government for the provision of a PhD scholarship to the lead author who is a Commonwealth Scholar. We also wish to express our appreciation to Dr Marc Davies and the Creative Science Foundation for making their resources, and especially their "Spacestation" VR platform, available to this project (but not shown in the images in this paper). In addition we wish to thank Dr Anasol Pena-Rios for her generosity in supplying the BReal platform code which we have used as the basis of the server technology. We are also indebted to Gurpreet Gill for providing background information about the foundational UEA work.

References

1. I. Susnea et al., in *MAC ETEL 2014–International Conference on Education, Teaching and e-Learning*, Prague, Oct. 2014. The education for creativity–the only student's tool for coping with the uncertainties of the future (2014)
2. V. Callaghan, M. Davies, S. Zhang, in *iLRN'16*. An online immersive reality innovation-lab (2016)

3. J.W. Kim, *Human Computer Interaction* (Ahn Graphics, Seoul, 2012)
4. M. Lewis, J. Moultrie, The organizational innovation laboratory. Creat. Innov. Manag. **14**(1), 73–83 (2005)
5. J. Moultrie et al., Innovation spaces: Towards a framework for understanding the role of the physical environment in innovation. Creat. Innov. Manag. **16**(1), 53–65 (2007)
6. L. Powell, The where factor: New environments for innovative thinking (2009)
7. B. Dalgarno, M.J. Lee, What are the learning affordances of 3-D virtual environments? Br. J. Educ. Technol. **41**(1), 10–32 (2010)
8. M. Gardner, J. Elliott, The immersive education laboratory: Understanding affordances, structuring experiences, and creating constructivist, collaborative processes, in mixed-reality smart environments. EAI Endorsed Trans. Future Intell. Educ. Environ. **14**(1), e6 (2014)
9. I. Bentley, *Responsive Environments: A Manual for Designers* (Routledge, London, 1985)
10. J. Chin, V. Callaghan, G. Clarke, *A Programming-by-Example Approach to Customising Digital Homes* (IET Digital Library, London, 2008)
11. K. Sailer, R. Pomeroy, R. Haslem, Data-driven design—Using data on human behaviour and spatial configuration to inform better workplace design. Corp. Real Estate J. **4**(3), 249–262 (2015)
12. J. Kemp, D. Livingstone, Putting a second life "metaverse" skin on learning management systems, in *Proceedings of the Second Life Education Workshop at the Second Life Community Convention*, (The University of Paisley, San Francisco, 2006)
13. S. Azhar, M. Khalfan, T. Maqsood, Building information modelling (BIM): Now and beyond. Constr. Econ. Build. **12**(4), 15–28 (2015)
14. V. Caruso, M.D. Hartley, A.I. Mørch, End-user development in second life: Meta-design, tailoring, and appropriation, in *End-User Development*, (Springer, Berlin, 2015), pp. 92–108
15. N. Joshi, Learning design and structured tasks in 3D virtual world, in *IMpact of E-Technology on US (IMPETUS), 2014 International Conference on the*, (IEEE, New York, 2014)
16. G. Gill, S. Oldfield, in *Building Cultures of Creativity in Higher Education Conference*, University of Wales, Cardiff, January 8–10 2007. Scribbling on the walls: Introducing creativity at UEA, the story so far (2007)
17. H.-Y. Wu, V. Callaghan, From imagination to innovation: A creative development process, in *Creative Science Related Papers at IE'16 London, UK*, (IOS Press, Amsterdam, 2016), p. 514
18. B.D. Johnson, Science fiction prototyping: Designing the future with science fiction. Synth. Lect. Comput. Sci. **3**(1), 1–190 (2011)
19. P. Zheng, V. Callaghan, Diegetic innovation templating, in *Intelligent Environments (Workshops)*, (IOS Press, Amsterdam, 2016)
20. J. Van Helvert, C. Fowler, *'Scenario Based User Needs Analysis', Chimera Working Paper 2003-02* (University of Essex, Colchester, 2003)
21. V. Callaghan, Creative science injecting innovation into the IT industry. iTNOW **57**(2), 52–55 (2015)

A Study of Digital Science Fiction Prototyping in an Elementary School Setting

Carrie Lane and Juliette Solis

1 Introduction

Reaching and engaging elementary learners is often challenging because students are frequently distracted and resistant to learning unless content is presented dynamically and in multiple ways. Fostering collaboration and building technology-based skills among elementary students is critical in today's digital age, and if these skills can reinforce educational curriculum, their purpose becomes twofold. Understanding ways to reach elementary students is paramount to ensure effective and engaging teaching in the twenty-first century. According to previous literature regarding various approaches to education, implementing collaborative, active learning and dynamic curriculum into classrooms can foster increased understanding and student participation [1]. Some such approaches include offering virtual experiences to help students connect with concepts, fostering peer interaction, having students construct their own meaning from the curriculum and offering interdisciplinary bridges between content areas.

The Creative Science Foundation (CSF) has utilized Science Fiction Prototyping (SFP) and technology with great results among college students and other populations, and they believe that similar results can be achieved if these constructs are translated into elementary classrooms. One possible implementation of the CSF ideas within elementary school classrooms is a digital platform called MySciFiStory.com (https://sites.google.com/site/myscifistorycom/), which will allow young students to learn content about given instructional objectives by utilizing SFP, the exquisite corpse method and multiple constructivist approaches to create dynamic subject-matter instruction [2]. This study will review literature on varied approaches

C. Lane (✉) · J. Solis
School of Journalism and Media Studies, San Diego State University, San Diego, CA, USA

A. Reyes-Munoz et al. (eds.), *EAI International Conference on Technology, Innovation, Entrepreneurship and Education*, Lecture Notes in Electrical Engineering 532, https://doi.org/10.1007/978-3-030-02242-6_28

to teaching content in today's schools and will preview the underlying theory behind SFP and the exquisite corpse method, which combine to provide the supporting framework for MySciFiStory.com. Additionally, the study will address background information about the overarching structure of MySciFiStory.com and will analyze data gathered during the implementation of the MySciFiStory.com prototype within elementary classrooms and among elementary educators and administrators.

2 Background Work

2.1 Innovative Approaches to Teaching Content in K-12 Classrooms

Over the years, the field of education has been altered greatly by cognitive psychology and the use of computers in the classroom. This has resulted in a shift toward alternative approaches to teaching, including student-centered learning, collaboration, virtual reality and constructivism [3].

2.2 Constructivism and Student-Centered Learning

Constructivism is a theory by which students construct meaning actively through collaborative, process-based learning and focus on understanding [4]. The idea of constructivism came from a philosopher named Giambattista Vico who stated that humans must construct their own experiences in order to understand them. John Dewey later expanded on this theory by suggesting that to learn fully, students must be active in the learning process [5]. In 1993, Marjorie Chang conducted a study of 363 junior high school students to measure the effectiveness of constructivist teaching methods against traditional approaches to teaching. Students using constructivist approaches were asked to predict the outcomes of certain situations and were encouraged to interact with their peers to determine the likelihood of their predications. Conversely, students taught with a traditional approach were not encouraged to collaborate or predict the effects of certain situations. The posttest results of this study showed that students who engaged in a constructivist approach to learning did better on their explanation scores but showed only marginal differences in their multiple choice results [6].

Similarly, student-centered approaches to learning assume that students benefit from collaboration and meaningful engagement through peer interaction. In student-centered learning, students are designated engaging tasks and then search for resources to help them gain the skills to complete these tasks. They reflect on their final products and revisit their approaches after receiving meaningful feedback [7]. A study done by Carla Confer found that to successfully implement student-centered learning, teachers need to become learners and focus on interaction, socialization, language, and meaningful interaction with the content [8].

2.3 Virtual Reality and QR Codes

Educators have recently begun using virtual reality in the classroom to foster educational connections among K-12 students. Virtual reality can be used to help students understand difficult concepts, observe faraway places, view past events or understand concepts that would be otherwise unsafe. Virtual reality is a 3D computer program that encourages users to immerse themselves in an interactive computer environment that resembles a physical world. While virtual reality is highly engaging for students, some of the challenges to a purely virtual experience in the classroom include the initial navigation through the virtual reality world, accessibility to all learners, time constraints, and planning difficulties [9].

In addition to virtual reality, teachers have also implemented Quick Response (QR) codes into their classrooms. QR codes were created in 1994 by a Japanese corporation called Denso and are square patterns that code information from web addresses. Teachers are using QR codes in the classroom to help students access information without the risk of incorrectly typing the wrong web address. According to Del Seigle, some benefits of QR codes are that they are easy to open, they limit the risk of accessing unsafe websites, they are quick and simple to use, they are free and easy to create, and they eliminate wasted paper. They can be used for supplemental information, as extension activities or as assessment tools on given classroom topics. Seigle suggests that some QR code challenges include the need for scanning technology and Internet access and the fact that they can be destroyed and unusable if not cared for properly [10].

2.4 Social Media in the Classroom

Social media has been increasingly utilized in the classroom, and sites like Twitter and Facebook connect people worldwide. According to Abdul Qayyum Ch, Tariq Hussain, Zaid Mahmood, and Shafqat Rasool, some benefits of social media include a fostering of collaboration and organizational skills across distances; some downfalls involve a reliance on easy-to-access information that is not always thorough or correct. Qayyum Ch et al. conducted a study, which found that while elementary students who used social media spent the same amount of time studying as their counterparts who did not; those students were more prepared to access resources at home, to engage in shared discussions, and to exhibit a positive attitude toward learning. Lastly, social media users showed better academic results on their final exams than non-users in this study [11].

2.5 Stem/Steam

Another recent approach to education involves the use of STEM, Science, Technology, Engineering, and Mathematics, within the classroom. According to Margaret Honey, Greg Pearson, and Heidi Schweingruber, recent changes in educational curriculum, including the Next Generation Science Standards, attempt to make engineering concepts applicable in various disciplines, inside and outside of schools. This idea of forming deeper connections with the content calls for interdisciplinary learning and has shifted the emphasis on STEM toward a focus on STEAM, which stands for Science, Technology, Engineering, Arts, and Mathematics [12]. According to Christine Liao, integration of the arts into the classroom allows students to foster creativity, which is a vital skill in producing innovative, critical thinkers who can serve as productive and successful workers in society [13]. Research about student achievement with both STEM and STEAM shows that the impact on students depends on the supports given and the level of integration of the concepts [12]. Honey et al. suggest that integration is paramount when utilizing these concepts because the human mind connects new concepts with ones previously understood; however, too much integration can cause overload to one's limited working memory. Therefore, collaborative learning, scaffolding, content relevance, and peer conferencing must play a key role in order to create successful classrooms when implementing STEM or STEAM content [12].

2.6 Science Fiction Prototyping

SFP provides a strong platform by which MySciFiStory.com is supported. SFP has become one approach to engaging students with academic content in educational environments and beyond. According to Brian David Johnson, SFP uses movies, comics, and stories as a way to explore future technologies and the results of today's realities on the future. It attempts to offer platforms by which people can use their imaginations to generate ideas for the future while pulling information from the past and the present [14]. In an article by Donald Smith, a high-school physics educator, SFP was used to engage students. He found that using elements of science fiction stories was helpful in garnering enthusiasm among students and that the new concepts made physics more interesting and clarified the content [15].

2.7 Exquisite Corpse

Another cornerstone of the MySciFiStory.com program is the exquisite corpse model, which was initially created by the Surrealists as a means for making collective collages among multiple individuals [16]. In 1925, Andre Brenton brought the idea of the exquisite corpse game to education. He had each of his students create the head, torso, hands, legs and feet of a character, which would later be used in the student's creative writing. The students would then describe the exquisite corpse's physical and emotional characteristics. Brenton found that this challenge helped to inspire students [17]. This exquisite corpse method was later utilized in a science fiction documentary that focused on the use of accidental juxtapositions as a means for creating new worlds and fantastic combinations of various disciplines. The documentary combined scenes from fictional movies and was performed by scientific authors and actors. It focused on the interaction between the future and the present and suggested that visionary concepts can emerge organically from combinations of random ideas [18].

2.8 How These Approaches Relate to MySciFiStory.com

The MySciFiStory.com platform attempts to use elements of STEM/STEAM, collaborative and student-based learning, and constructivism to promote interest and understanding among elementary students. Furthermore, the program's main foundations are based on a combination of SFP ideology and the exquisite corpse method set forth by the Surrealists. Because these methods have been successful in the K-12 classroom setting and beyond, the platform hopes to combine various elements of each construct as a means to activate student engagement and learning in the K-12 classroom [2].

2.9 Overarching Themes

One key theme the researchers found was that while recent innovative approaches to teaching content appear to have benefits among students, proper training for teachers and administrators is necessary. Additionally, the researchers discovered that additional research is needed about the effects of these approaches on elementary students and on the long-term effects of these methods on learning in general.

3 Background Behind MySciFiStory.com

As a means of combining SFP, the exquisite corpse method, collaboration, technology, and constructivism, The MySciFiStory.com attempts to make classroom instruction and assignments accessible and relevant to elementary learners. The program begins with students watching online videos on given instructional topics. Then, using other features of the website, students build story introductions based on the guidelines and objectives set forth by their teacher. Within the same page, students can like their favorite story starters using a variety of social media tools. From there, students collaborate to write the middle part of their favorite creative story. In the end, students use the platform to individually type a story resolution and print a QR code that can then be scanned by their classmates using a QR scanner [2].

The platform allows teachers and their students to communicate through social media, no matter where they are worldwide. Students like their favorite story introductions by using WhatsApp, Facebook Messenger, Snapchat, Instagram, Twitter, or other social media platforms as designated by their teacher. This feature of the platform encourages collaboration and fosters technology use among elementary students. These twenty-first century learning focuses are further implemented during the collaborative writing stage of the program and in the sharing of final products via QR codes [2].

Learning within the platform can take place at home or in school and is fast-paced but allows students time to fully engage with the content and their peers. The environment facilitates the performance of the intended course outcomes by encouraging content synthesis, competition and application of factual information [2].

The primary audience for MySciFiStory.com is elementary students, ranging in age from six to 12. The program benefits student learning because it fosters innovation, information synthesis and knowledge application. MySciFiStory.com encourages students to delve deeper into the content at hand and fosters competition, thereby increasing their learning and motivation. Another primary audience of the platform is elementary school teachers. MySciFiStory.com allows teachers to translate content to students in a highly effective and engaging manner via the InstructionalContentPage. Toward the beginning of the program, teachers set up the learning experience, provide the topic to be studied from a set of pre-developed content, and designate the duration of the lesson. Because the content is already created, teachers will not need to be deep subject matter experts to set up and manage a session but will take ownership of the lesson implementation. However, teachers can also create and add their own story outlines if they choose to do so [2].

Throughout the module, MySciFiStory.com allows students to reach advanced categories of Bloom's Taxonomy, an organizational system set up in 1948 by Benjamin Bloom and other educators to facilitate higher-level thinking in education by promoting application-based knowledge and synthesis of ideas [19]. By the end of a MySciFiStory.com module, students will use shared creativity to create a science fiction story, gain subject-matter understanding of a given concept, demonstrate communication skills by collaborating on social media, and write a science fiction story with a beginning, middle, and end [2].

4 Methodology

To test the digital website MySciFiStory.com, the researchers showed a working prototype of the website and a slideshow to nine teachers and one administrator. They conducted qualitative interviews to gather information about what educators felt were the strengths and weaknesses of the prototype. Additionally, the researchers showed the prototype to 38 students, who engaged in the basic activities of the program. The students took a pretest to assess their prior knowledge about the content. They continued by watching an educational video, writing introductions to a science fiction story related to the instructional content and sharing their introductions. Students then voted on their favorite introduction and collaborated to create a middle section of the story. Afterward, the students individually drafted and shared endings to the story with their fellow classmates. To conclude, students took a posttest and filled out a survey based on their experiences. The researchers analyzed the pretest and posttest data, which is included in Fig. 8, and used this to help test their hypothesis. The researchers also took educator and student feedback into account when testing their hypothesis and analyzing the effectiveness of MySciFiStory.com (Figs. 1, 2, 3, and 4).

5 Contextual Factors

One contextual factor in the study was limited access to technology among certain teachers. The researchers worked around this by giving teachers access to other classrooms to utilize the technology. Another factor was a lack of technology training among certain teachers, which was mitigated by giving teachers a tutorial

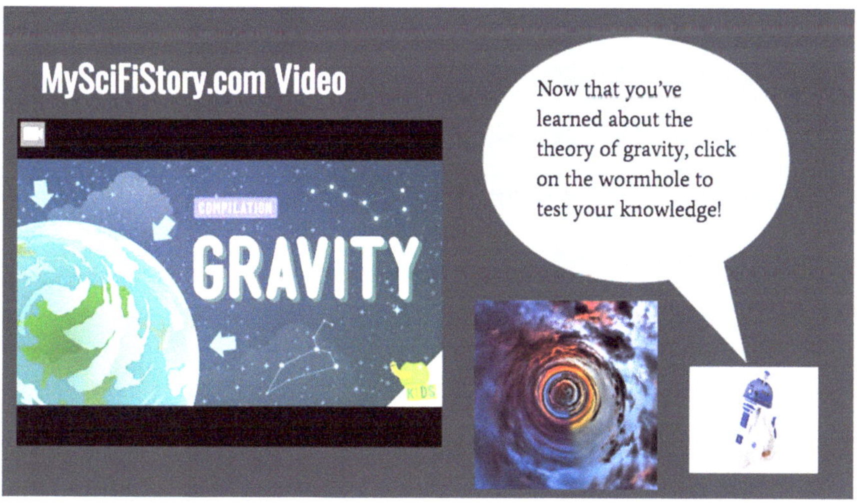

Fig. 1 Gravity prototype sample page: One page of the sample prototype shown to teachers, administrators, and students

Fig. 2 Website sample page: A sample of the MySciFiStory.com website shown to students

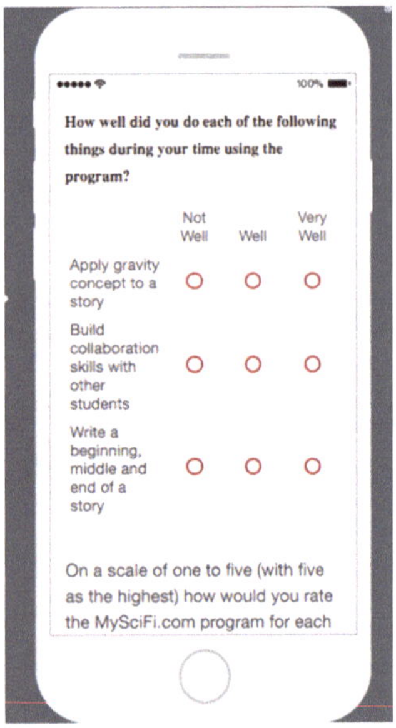

Fig. 3 Survey preview: A screenshot of the survey given to students after viewing the prototype and the website

Fig. 4 Pretest and posttest page: A copy of the pretest and posttest that was given to 38 sample students

1) Why do you land on the ground rather than floating into space?
   a) The sun
   b) Gravity
   c) Magnetism
   d) The Earth

2) Who was the first person to study gravity?
   a) Copernicus
   b) Euclides
   c) Rashi
   d) Isaac Newton

3) It doesn't matter how heavy an object is. If you let it go, it will come down.
   a) True
   b) False

4) What if you throw something to the left or to the right?
   a) It will go down
   b) It will go left
   c) It will go right
   d) It will go up

5) What does 'gravity pull things down' mean?
   a) Gravity pulls things up
   b) Gravity pulls things to the moon
   c) Gravity pulls things to the center Earth
   d) Gravity pulls things to the sun

6) What is the speed that astronauts need to reach to escape gravity?
   a) Fastness
   b) Escape fastness
   c) Escape velocity
   d) Velocity

7) The greater an object's mass or size, the greater its gravitational pull
   a) True
   b) False

8) The farther apart objects are, the stronger the gravitational pull is
   a) True
   b) False

on how to use the program. One additional factor involved a lack of time within the instructional day to test the program, which was solved by breaking the module up into subsequent class periods.

6 Teacher and Administrator Reflections on the Program (Fig. 5)

7 Student Survey Data and Reflections

According to survey participants, on a scale of 1–5, their mean enjoyment of the program was a 4.33 with a standard deviation of 0.82. The mean learning effectiveness of the program was 4.33 with a standard deviation of 0.47 and the mean collaboration enhancement was a 4.67 with a standard deviation of 0.67 (Figs. 6 and 7).

Strengths of the Program	Suggestions for Improvement
"I like how it scaffolds it for people so kids can work at their own pace but it builds on each other."	"It should be linked to standards. There should be a place where the standards are portrayed"
"It's pretty self-guided for the kids and they would probably be engages because of all the digital and social media elements of the program."	"Is there anyway for the social media to be private?"
"I like that they need to put the effort in yo create the story by themselves. We don't feed it to them. They will learn a lot."	"How do teachers get results from the forms if they are not the owners of the forms?"
"It takes what most teachers already do(combining two subjects), and organizes it in a way that makes planning easy for the teacher, and in a way that will get students excited about the project!"	"I would use a story that's well known to the kids and have them add in their narratives to portray the characters in a real-life setting."
"I believe this program will allow students to be creative while learning the material. I believe students will enjoy learning the subject manner, as this will allow them to personalize their learning and share their knowledge with their friends."	"I would like to ensure that all students work could be saved to be reviewed and completed at a later date. As a teacher who teaches multiple subjects, I would like to be sure that I can use the program for all the subjects I teach.
"It's set up really will to promote a scope and sequence for learning and for the students to be able to write a comprehensive story that is clear, informative and entertaining. The use of social media adds a layer of communication that is very much a part of children's lives these days and therefore of how to use them appropriately,but it will also engage them."	"Sometimes it would be nice for the teacher to able to change the format of the writing. This way we can see how individual kids could be creative by writing the middle and the end of the story individual after the like section. It would be great to put the platform in different languages to expand the teachers it can work for and the subjects."
"I love the flexibility of the program. You can take it where you want to go. I love the QR coded. It makes it hands-on and allows them to appreciate classmates' work n a fun manner. I love the embedded assessment. I love the creativity it affords whether individually, collaboratively or conferencing. It opens so many doors. I would love to see it in my classroom."	"I would be interested in being able to chose various prompts for one class to meet the needs of students who are on different levels of learning."

Fig. 5 Teacher reflection table: This table shows teacher reflections on the strengths of the MySciFiStory.com program and suggestions for improvement

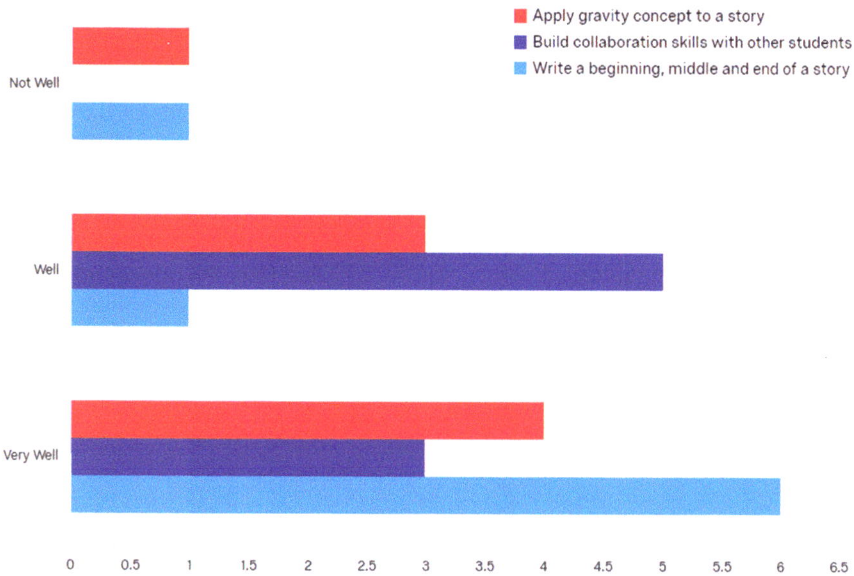

Fig. 6 Student engagement outcomes: The above graph represents responses of student survey participants about how well they learned to apply concepts to a story, build collaboration skills and write a beginning, middle, and end of a story while using the MySciFiStory.com prototype. It is calculated on a scale of 0–6.5 and broken into three categories—not well, well, and very well

Fig. 7 Student reflections on the program: The above figure shows student reflections on their favorite elements of the program and responses to reuse of the program

Favorite Elements of the Program	Would You Use This Program Again?
"Applying gravity concepts to a story."	"You because it makes kids smarter."
"Teaching the students to be smarter, gravity, and to work with others."	"Yes because it was fun learning about things I didn't know."
"When I wrote the story and the test."	"Yes because it was fun to use."
"The video, the composition, and the story."	"No."
"Learning who Issac Newton was, learning more about space and learning that every planet has the same gravity pull."	"I would use it again because it was educational. We got to write a story that was fun."

8 Pretest and Posttest Data and Findings

The researchers gave students a pretest about the concepts involved with gravity and, after facilitating the MySciFiStory.com prototype and all elements of the gravity module, the researchers gave students a posttest with the same questions. The mean of student scores on the pretest was 0.71, while the mean of student scores on the posttest was 0.84. This created a mean score difference of 0.13. The researchers ran a t-test on the data sets and found that the p-value was 0.002. This p-value was smaller than the level of significance, which was 0.05. Therefore, the researchers were unable to accept the null hypothesis, which stated that using MySciFiStory.com would not influence students' academic improvement ($a = 0.05$).

9 Data Analysis (Figs. 8 and 9)

10 Researcher Hypotheses

MySciFiStory.com *Null Hypothesis*: Using MySciFiStory.com will not influence students' academic improvement ($a = 0.05$).

MySciFiStory.com *Alternative Hypothesis*: Using MySciFiStory.com will influence students' academic improvement.

11 Researcher Conclusions

Based on interviews with educators, MySciFiStory.com potentially meets the needs of primary school classrooms and could serve as a useful platform by which students can engage with academic content while collaborating socially and gaining technology-based skills. Many teachers suggested helpful changes that can be implemented into the program to enhance its usability and effectiveness in the classroom. Students who engaged with the MySciFiStory.com example module stated that they gained valuable collaboration skills and enjoyed writing the science fiction stories based on the concept of gravity. The majority of students felt they would like to use the website again and that it was a fun-to-use program that gave them additional subject-matter knowledge. The paired t-test done on the MySciFiStory.com data generated a p-value that was 0.002. Therefore, the researchers were unable to accept the null hypothesis that using MySciFiStory.com will not influence students' academic improvement.

To conclude, student responses, teacher reflections and data-driven information collected and analyzed by the researchers suggest that MySciFiStory.com has the potential to be a useful tool in elementary classrooms. However, further research

Student #	Pre-Test Scores	Post-Test Scores	Mean Difference
1	0.75	0.875	0.125
2	0.75	0.75	0
3	0.25	0.875	0.625
4	0.75	0.875	0.125
5	0.625	0.75	0.125
6	0.75	0.875	0.125
7	0.375	0.625	0.25
8	0.625	0.75	0.125
9	0.625	0.875	0.25
10	0.875	0.75	-0.125
11	0.75	0.5	-0.25
12	0.5	0.75	0.25
13	0.75	0.625	-0.125
14	0.125	0.875	0.75
15	0.75	1	0.25
16	0.875	0.875	0
17	0.5	0.5	0
18	0.625	0.875	0.25
19	0.875	0.625	-0.25
20	0.5	1	0.5
21	0.75	0.75	0
22	0.75	1	0.25
23	1	1	0
24	0.625	0.625	0
25	0.625	1	0.375
26	0.75	0.75	0
27	0.625	0.875	0.25
28	0.875	1	0.125
29	0.875	1	0.125
30	0.875	1	0.125
31	0.75	1	0.25
32	0.875	1	0.125
33	0.875	0.75	-0.125
34	0.625	0.75	0.125
35	0.875	0.75	-0.125
36	0.875	0.875	0
37	1	1	0
38	0.75	1	0.25
Average for Pretest and Postest=	0.7105263158	0.8355263158	0.125

Fig. 8 Student achievement scores: The above spreadsheet shows the student scores for the pretest and posttest. These scores were averaged and then used to calculate a p-value, which can be seen in Fig. 9 below

T-Test P Value: 0.002001162396

Since this is smaller than level of significance (p=0.05) Fail to accept the null

Fig. 9 T-test results: The above screenshot shows the p-value gathered from a t-test garnered from Fig. 8's data

on the program and its implementation in elementary classrooms should be done before making a decision on whether to adopt the platform.

12 Areas of Further Study

Some areas of further study include the amount of time needed to implement the MySciFiStory.com program from beginning to end, how demographics impacted outcomes and how teacher and student training influenced results. Another area of study that needs to be addressed further is what, if any, type of technical glitches existed during testing of the program and how they affected the results.

Acknowledgements The researchers of this paper would like to thank Victor Callaghan for his guidance and support throughout the creation of MySciFiStory.com and this paper and CSF for their innovative and creative approaches to teaching content. They would also like to thank Dr. Minjuan Wang for her support and guidance throughout this research project and the original concept development process. Many thanks are given to Peter von dem Hagen for his help developing the initial MySciFiStory.com concept, design document and prototype and to Angeles Gomez for her help creating the initial MySciFiStory.com documents. Lastly, the researchers would like to thank all the students, teachers, and administrators who participated in the testing of the prototype of MySciFiStory.com.

References

1. A.P. McGlynn, Strategies: Motivating the disengaged student; part one: Challenging passivity and resistance to learning. Hispanic Outlook High. Educ. **24**(9) (1999)
2. C. Lane, J. Solis, P. Von Dem Hagen, A. Gomez, MySciFiStory.com internal design document (2017)
3. International Association for Development of the Information Society, *Proceedings of the International Association for Development of the Information Society (IADIS) International Conference on Cognition and Exploratory Learning in Digital Age (CELDA)* (IADIS, Madrid, 2012), p. 308
4. K. Crotty, Educational theory constructivism (2012)
5. S.P. Power, Constructivism defined and implications for the classroom, Master's thesis, Memorial University of Newfoundland, 1997
6. M.M. Chang, Role of explanations and student-centered interaction in science learning: An applied constructivist approach to instructional design, Doctoral dissertation, Syracuse University, 1993

7. S. Pedersen, D. Williams, A comparison of assessment practices and their effects on learning and motivation in a student-centered learning environment. J. Educ. Multimed. Hypermed. **3**(13), 283–306 (2004)
8. C. Confer, Student participation in a process of teacher change: Toward student-centered teaching and learning, Doctoral dissertation, State University of New York at Albany, 2000
9. M.A. Abdelaziz, A.E. Riad, M.B. Senousy, Challenges and issues in building virtual reality-based e-learning system. Int. J. E-Educ. E-Bus. E-Manag. E-Learn. **4**(4), 320–329 (2014)
10. D. Siegle, Using QR codes to differentiate learning for gifted and talented students. Gift. Child Today **38**(1), 63–66 (2015)
11. A. Qayyum Ch, T. Hussain, Z. Mahmood, M.S. Rasool, et al., Bull. Educ. Res. **2**(38), 203–209 (2016)
12. M. Honey, G. Pearson, H. Schweingruber, *STEM Integration in K-12 Education: Status, Prospects, and an Agenda for Research* (National Academies Press, Washington D.C, 2014)
13. C. Liao, From interdisciplinary to transdisciplinary: An arts-integrated approach to STEAM education. Art Educ. **69**(6), 44–49 (2016)
14. B.D. Johnson, *Science Fiction Prototyping: Designing the Future with Science Fiction* (Morgan & Claypool, San Rafael, 2011)
15. D.A. Smith, Reaching nonscience students through science fiction. Phys. Teach. **47**(5), 302–305 (2009)
16. M. C. McShane, Exquisite corpse: The surrealist practice of collective drawing, 1925–1941, Unpublished doctoral dissertation, Emory University, Atlanta, 2004
17. P. Wilson, Not so exquisite corpses: Character invention and development in the creative writing class. Engl. Educ. **35**(3), 7–12 (2001)
18. C. Cynn, Forest of stories: An experimental science fiction documentary of stories, Documentary, United States (2011)
19. Ö.G. Ulum, A descriptive content analysis of the extent of Bloom's taxonomy in the reading comprehension questions of the course book Q: Skills for success 4 reading and writing. Qual. Rep. **21**(9), 1674–1683 (2016)

Index

A. Reyes-Munoz et al. (eds.), *EAI International Conference on Technology, Innovation, Entrepreneurship and Education*, Lecture Notes in Electrical Engineering 532, https://doi.org/10.1007/978-3-030-02242-6

MIX
Papier aus verantwortungsvollen Quellen
Paper from responsible sources
FSC® C105338

FSC
www.fsc.org

If you have any concerns about our products,
you can contact us on
ProductSafety@springernature.com

In case Publisher is established outside the EU,
the EU authorized representative is:
Springer Nature Customer Service Center GmbH
Europaplatz 3, 69115 Heidelberg, Germany

Printed by Libri Plureos GmbH
in Hamburg, Germany